SO-BZF-553

99 TEST EQUIPMENT PROJECTS YOU CAN BUILD

99 TEST EQUIPMENT PROJECTS YOU CAN BUILD

BY THE EDITORS OF 73 MAGAZINE

TAB TAB BOOKS Inc.

BLUE RIDGE SUMMIT, PA. 17214

FIRST EDITION

THIRD PRINTING

Printed in the United States of America

Reproduction or publication of the content in any manner, without express permission of the publisher, is prohibited. No liability is assumed with respect to the use of the information herein.

Copyright © 1979 by TAB BOOKS Inc.

Library of Congress Cataloging in Publication Data

99 test equipment projects you can build.

    Includes index.
      1. Electric instrument—Design and construction—
Handbooks, manuals, etc. I. 73 magazine for radio amateurs.
TK7878.4.N56   621.3815'48   79-18223
ISBN 0-8306-9749-7
ISBN 0-8306-9749-9 pbk.

Preface

At this very moment, the junk box in your workshop is hoarding valuable items just waiting to be put to work.

All of the 99 test equipment projects explained in this book can be made from various items in your junk box. Even if you would have to purchase a few necessary parts, most of the gear can be built for less than $10. How's that for beating inflation?

The first chapter starts you off on the right foot by explaining how to build and use test gear. After reading the chapter, you'll become a troubleshooting wizard!

Your home-brew equipment will quickly gain the look of professionalism.

All these projects are cheap and easy. Why not get started right away building such equipment as soldering iron controllers frequency counters, oscilloscopes, voltmeters, ammeters, capacitance meters, calibrators and all types of bench and lab power supplies.

Whether you are an experimenter, hobbyist, ham radio oerator or just like to tinker with electronics enjoy all these projects. And they are, indeed, junk box projects you can be proud of.

Contents

Chapter 1
How To Build
and Use Test Gear

Have you ever built a project and then felt disappointed with its appearance after it was done? Do you admire the beautiful projects that grace the pages of most electronics books? Would you be interested enough to spend a little time learning how to make your projects look better, be easier to build and service and perhaps work better? One of the problems facing electronics people who like to build their own gear is that there is more emphasis on *circuitry* than on nuts-and-bolts construction. You see this whenever you read a construction book. All too often you get a lot of "how it works" theory, a few chapters of "connect the green wire to point C" construction, a "how to use it" section and schematics. That leaves a lot of open avenues for construction—great for experienced constructors, but a stumbling block for less knowledgeable people. We are going to get you started with the basics in electronic construction, and well down the road to successful project building.

A test bench can be designed using homemade equipment. Everything can be built by using ordinary tools and techniques. Granted, large and costly projects such as an oscilloscope and frequency synthesizer are beyond the abilities of most people, but some can be done at home. Why not build your next project like a pro using the following methods?

Good tools are the most important part of electronic project building. They save you time (your time is valuable!) and temper by making the work easier. Here is a *minimum* list of tools you should have:

- Needlenose pliers
- ½" blade diagonal cutters
- Adjustable wire strippers
- 25 watt soldering iron for ICs
- 100 watt soldering gun or iron for wires
- 1 lb 60/40 rosin core solder
- Screwdriver set
- Nutdriver set (especially ¼" unit)
- Heavy duty jackknife (for deburring holes)
- ¼" hand drill or drill press
- Set of drill bits
- 12" square/ruler

Check over your tools and be sure that all cutting edges are sharp. If you have to add tools, get good quality ones. The extra cost of good tools pays off in the long run. They stay sharper and don't break so easily. You will probably have to add tools to your present ones to handle the demands of different projects (e.g., chassis punches, etc.), but this list represents the bare minimum.

Okay, we're ready to start. We'll take each stage of the construction process step by step. To highlight the process, we are going to assemble a 5 volt, 1 amp power supply along the way. A 5 volt power supply is a great addition to any lab that works with digital ICs!

Appraise the Project. The most logical way to start an electronic project is to appraise it for the best way to build it. When you find something you would like to build, you should start by looking over the circuitry and any method of construction that may be shown. If you are a newcomer to electronics, you may want to build a kit the first time out and then start building projects out of books or from schematics. This makes electronics a lot easier if you start with a "paint by numbers" kit and gradually work up to more challenging projects.

Start by reading over the project or by checking out the schematic. When you are reasonably familiar with it, ask yourself the following questions:

- How am I going to assemble the electronics?
- Am I going to build the completed project in a cabinet?
- Are there any critical areas in the electronics that require special care, e.g., high gain amps?
- Are there any special requirements in the mechanical construction, e.g., shielding?
- Can I get all of the parts?

If you are building the project from a project in a book, you can answer the first four questions simply by copying the author's finished unit. The fifth question must be answered by you. If you can't get all the parts, or if they cost more than you can afford, don't build it. Instead, set the project aside, and tackle it in the future if you really have your heart set on building the item. If you are building your project from a schematic or custom building a book project, you'll have to answer these questions yourself and provide the solutions. Experience is the best teacher here. The schematic should give you some clues. Table 1-1 lists some pitfalls to watch out for as well as the corresponding solutions.

Needless to say, the list could go on, but Table 1-1 is a sample. Make allowances for these things. Leave room for metal shielding, bypass capacitors and heat sinks. Lay out parts carefully to keep input and output separate on high gain amplifiers (that includes i-f amplifiers) and leave space for heat sinks if necessary. Watch lead dress in logic circuits and VHF-UHF circuits, too. Good grounds are also very important. Keep these things in mind, along with anything else you can dig up on the project. All of the items mentioned here should influence how you build your project.

Table 1-1 Project Pitfalls and Solutions.

Pitfall	Solution
High gain amplifier or tuned amplifiers	Extra shielding may be required
High gain amplifier or tuned amplifiers	Inputs and outputs well separated.
High gain amplifier or tuned amplifiers	Power supplies remoted or shielded.
AF or RF oscillators	Good shielding and bypassing of power leads.
AF or RF oscillators	Sturdy mounting of coils and capacitors.
Digital logic ICs	All power supply leads must be kept short and well bypassed with capacitors.
VHF-UHF circuits	All leads must be kept short.
Power supplies or power handling circuits	Good heat sinks for all power devices.
Power supplies or power handling circuits	Heavy wire where necessary to minimize losses in power.

Collect the Parts. Now that you are reasonably familiar with what you are going to build, you can get the parts. There are many sources of electronic parts, of course, but you should start with your junk box. Don't have one? Start collecting old radios, TVs and other cast-off electronic devices and strip them for parts. You'll need hardware such as nuts and bolts, so save all that you can get. Junk boxes are good for the basic stuff you need for a project. If yours is well equipped, you might be able to build an entire project, such as a power supply, but this is rare. For any ICs or other semiconductors and parts, you may have to turn to your local dealer, so get to know him well if you don't already. Another parts route open to you is the surplus mail order dealer listed in the back pages of most electronic publications. If you haven't tried these dealers, you are missing out on some great bargains. But beware of reject or retested components. They can cause more problems than you would beleive! Pros use quality, name brand parts. This one move often saves hours of troubleshooting later! All you have to do at this time is collect the electronic components. Leave the cabinet selection until later if you are "rolling your own" project, or buy the one called out if you are duplicating someone else's device.

Once you have all the parts, you can test them if you desire. Test any used parts that show signs of being hot; otherwise, this step is optional. Many people check all their components to save troubleshooting later, and that pays off with parts of poor quality. But this shouldn't be necessary if you use good quality parts to begin with.

Select the Cabinet. Now that you have all the electronic components together, the time has come to select a cabinet to house your project, and perhaps a chassis as well. The secret of success in selecting the right housing for your equipment is *advance planning*. The idea is to get a cabinet that is large enough to house all of the parts of your project, plus allow room for easy servicing and future modification. You do not want the cabinet to be too large. This is an unnecessary expense and oversize cabinets mean excessive bulk. Here's how to select the box or chassis that is right for you with a minimum of fuss.

The first step is to visit your dealer and find out what cabinets and chassis are available to you. You might also want to write the manufacturers listed in the Appendix of this book for catalogs. Next, lay out the parts that normally mount inside the cabinet on a table. This normally includes circuit boards, large caps and transformers. Lay out the parts like you were assembling the unit into a cabinet. Allow at least 1" clearance around the circuit board (if used) and any adjacent parts. Separate heat producing parts such as transformers, power resistors and power transistors at least 2" from any other parts. The back cover is a good place for resistors and transistors, while the transformer may be mounted toward the rear. This is just a first fitting, so you don't have to place the parts exactly. Measure the height, width and depth of the layout and you have the *minimum* case dimensions. Consider what components you have to add to the front and rear panels. If they would interfere with the parts layout you made, add more space to the *minimum* dimensions. Meters and speakers are great space hogs in this respect! Continue to add parts to the front and rear panels, making corrections to your minimum dimensions as necessary. Be conservative in your estimates. A little extra room in the layout makes construction and servicing much easier. You may also suddenly discover the space for shielding, as in the case of a radio receiver project. Consider using chassis in your more complex projects. You can mount your PC boards on top, over a suitable sized cutout, and this will make construction and servicing a snap. Chassis are also used as shielding boxes. This may be necessary in a project where many sensitive circuits must be placed in the same case. This frequency synthesizer is a project in point. It uses six fully enclosed chassis boxes to isolate the many VHF frequencies the circuits generate from each other.

Finish up your case or chassis selection by taking your dimensions and selecting a box to fit. Since you will probably have some oddball dimensions, you may have to look for a larger box or an odd sized one. With practice, selecting a case can be done quickly, with just a few measurements and the catalogs.

Build the Electronics. This is the largest step you'll probably have to make, but we are going to simplify things a bit by taking a look at the highlights of electronic construction.

To start with, there are three basic ways to assemble an electronic circuit. You can use a PC board, a perfboard or mount all the components on the project's cabinet. Often two of these techniques are combined, as it is common to mount small components such as caps, resistors or transistors on a PC/perfboard and then mount large components such as transformers and speakers on the cabinet. Which construction method is best for you? The choice is fairly easy—if a PC board is available, or if you feel you can make one, use it. PC boards are also recommended for complex digital circuits (about 10 chips or more) and critical circuitry such as high gain amplifiers. Perfboard construction is a handy method of construction for simpler, less critical circuits. Construction may be a little harder than a PC board because you have to figure out the placement of each part as you wire it up. On a PC board, the designer has determined layout for you and assembling a PC board (nicknamed *stuffing*) is often like building a kit. It's easy!

Another method of construction is assembling all the parts on the project's cabinet and wiring them up. This method works fine when there are few parts and most of them are made for chassis mounting. Small parts are often mounted on terminal strips to keep them from touching the chassis. You will often see this method used in power supplies such as this one and other simple projects.

Laying out your circuitry isn't necessarily difficult if you don't try to rush construction. Just take your time and use some intelligent planning and the results should be good. If you are using a commercial PC board, you can skip this part; just stick the parts in as per the pictorials and solder them! But if you are working with perfboard, or laying out a PC board, follow these hints to ease your job.

The schematic, and any other circuit information available, has the most powerful influence on how a circuit should be built. For example, if pictures or drawings are available, you might get by building your circuit from these. Or, at the very least, these illustrations can give you ideas on how to lay out the circuit to suit your own needs. So it goes without saying that you should read over any texts and illustrations available on your project before starting! There just might be enough information available to skip this section! Also you want to look for *problem areas* —parts of a circuit that are sensitive to component layout. Examples of these problem areas include the following:

- grounds in HX to VHF circuitry
- power supply bypassing around digital or linear ICs
- component lead lengths

A good author will point these things out and probably more, so when a suggestion is made to handle these problems, take heed! *Caution*: If you see many problem areas in a project and you aren't sure you can handle them all, get a PC board if available, or drop the project. This can save you grief!

Suppose you are building a project from just a schematic. Now you have a challenge! There are basic ways to go about building the circuitry. First, select a board large enough to hold all of the parts, then refer to the schematic. You can often lay out the parts on the board just like the schematic—you might consider this. This makes complicated circuits easier to trace, but a well-drawn schematic is required. Otherwise, try sticking the major parts (e.g., transistors, ICs, etc.) on the board. Then stick in the smaller components around the pins of the IC or transistors they would connect to. Try to position the parts for shortest lead length. And remember how you appraised the circuit to begin with? Try to account for any pitfalls you found at that time. You may do this technique for just one stage at a time, as in complicated circuits, or do an entire circuit at once! The following are some tips to use in your layout and construction. Leave plenty of room for all parts, avoid layered construction, or the placement of resistors on top of capacitors and always use sockets on ICs. When you have a layout that satisfies you, wire up the parts. Use #18 to #24 bare tinned copper wire for all grounds and power supply leads if possible. When you are done, check your work and plug in any ICs. You might be able to check out the board to see if it works, too.

Tackling the Cabinet. Now you get to lay out, machine and label the cabinet. And in the bargain you'll get some exercise! Start by laying out the cabinet. You should have a general idea of what goes where in the cabinet from the section on its selection. Now improve on that by collecting the parts that would mount on the front panel. Oh yes, don't forget the box you selected! Play chess with the parts by placing them on the outside of the box and moving them around until you get an arrangement that looks aesthetically pleasing. Lay out controls in a symmetrical manner—in a straight line (if you have many controls, stagger them). Balance them so they are neatly centered between the ends of the box. If possible, group the controls by function. Mark all hole locations. Masking tape works well here. Then follow the same procedure for the bottom and back of the box. Be sure that there is still room for all of the parts. If you need ideas for your cabinet layout, why not check out some commercial gear? This can be very helpful if you are stuck.

Once you have the locations marked, you can start drilling them. A center punch is recommended to punch all hole locations for greater accuracy, but this is an option. When you have all cabinet holes drilled, use the knife to deburr them. Then tackle special holes, such as square ones for displays or round ones for meters. You can cut these either by drilling holes around the inside edge of the cutout, punching out the slug and filing to size or by using a chassis punch. A sabre saw could also be used, but it would mark up a painted panel. The choice is up to you. After you are done, deburr any leftover holes and wash the box. Use detergent and water if the box is painted, or a scouring pad and detergent if it is bare aluminum. Finish up by drying the box thoroughly.

You might want to paint the box. For best results, warm up the box to about 30° to 40° C. Then use your favorite color of aerosol spray to do the job. Follow the instructions on the can and you should get good results. Incidentally, it is often cheaper for you to buy an unpainted box and paint it yourself. Let the box dry in a warm dust-free place overnight. Then, take it and apply a light coat of clear acrylic spray. This will make application of the labels easier. Let it dry for several hours.

Now you can apply decal labels to the cabinet. If you don't have any, you should be able to get them from the larger electronics distributors or perhaps by mail order. You can also get alphabet sets from most drafting suppliers for low cost. Typical names for decals are Letraset® or Technilabels®. They come in sets for experimenter, ham, etc. Common words are already spelled out for you and they are very easy to use.

The drafting alphabet sets have names such as Paratipe® and Zapatone® and have only letters—you must make up the words yourself. Applying these labels to the cabinet is easy—they just rub on with a blunt pencil. If you get a word or letter in the wrong place, it easily comes off by placing a piece of cello tape over it, rubbing it and pulling it off. Be sure that you allow plenty of clearance for the knobs when you apply the labels. After

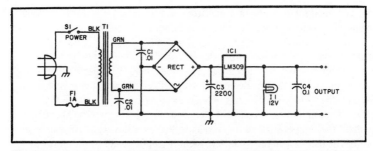

Fig. 1-1. A simple 5 volt, 1 amp power supply. C1, C2-0.01 uF disc capacitor; C3-2200 uF 25 volt electrolytic capacitor; C4-0.1 uF capacitor; F1-1amp 3AG fuse and holder; I1-12 volt, 50 mA lamp and holder (Radio Shack 272-322 OK); IC1-LM 309K voltage regulator; RECT-6 amp, 50 piv bridge rectifier; S1'SPST toggle switch; T1-12.6 volt, 1 amp filament transformer; Misc-cabinet (LMB442 used in example), 3 wire cord and plug, binding posts, wire, etc.

you label the front you may wish to label the rear, too. This will complete the professional appearance of your equipment! Spray the outside of the cabinet with clear acrylic spray when it is labeled to your satisfaction and let the box dry.

Final Assembly. You are probably getting pretty excited by now because you are in the homestretch! Now you get to connect up a lot of loose parts and see if the project really works.

Complete the final assembly by mounting all the parts on the front panel. Install the knobs when you finish. If you did everything well so far, the project should go together like a kit—fast and easy! This is something to shoot for. Continue with the rear panel, too. Then mount the circuit board in the bottom of the box (if a board was used) with ½" or longer metal spacers. Use at least four. Mount any other parts as necessary and the mechanical construction is done.

Complete the wiring and you are home free. Refer to the schematic for details (Fig. 1-1). It may be necessary to loosen or disassemble parts to wire them, so you might want to keep this in mind. Despite the best plans of mice and men, this problem crops up from time to time, so keep a screwdriver handy! Note the power supply.

Checkout. The moment of truth has come! You plug it in and there is a very good chance your project will show signs of life. If so, congratulations on a good job! If not, you'll find a carefully built project much easier to troubleshoot than a haywire rig with a maze of carelessly laid out wires and components. You really gain with a properly built project; it is more likely to work, it is easier to service and it looks a heck of a lot better to boot!

Handy Dandy Soldering Iron Cooler Offer

If you enjoy finding solutions to the nagging little everyday problems found in your favorite hobby, then you'll love this gadget. It is a solution to a little problem that has bugged hobbyists for years. Surely many of you have been faced with this same irksome situation: what to do with the mini-iron that you need for working on transistors and ICs when it is up to working temperature and you need the hand to do something else.

In other words, where can you put it down? You've probably tried ashtrays, saucers, boards with nails, etc. Well, finally something has been done about it—and it's not a factory-made stand. It's perfect for two reasons. First, it's cheap! (Besides, it's more fun figuring it out for yourself.) Second, the factory-built stands only give you a place to put the pesky iron down. If you are going to build a holder, you might just as well build one that controls the temperature of the iron as well. Never buy what you can build and never build a gadget to do one job if it can be made to do more.

This is a project built 100 percent from genuine junk box parts, and not the kind you have to buy. The base is an old 3″ × 5″ × 2″ steel black crackle finished (at least it used to be) box that had been used for at least five or six other projects before finally finding respectability as the centerpiece of the workbench. Large green and red jeweled pilot lights of the 7½ 115v type are used in the top to show that the device is on (green), and that the soldering iron is on high heat (red) when it is withdrawn from the cradle. The lights aren't really necessary, but if you have a lot of them, they fill up two holes in the top of the box that need filling.

The cradle for the iron is made from scrap aluminum, and the heat shield is a piece of scrap perforated aluminum sheet rolled into a tube and fastened to the cradle. This assembly is fastened to a piece of ¼″ square steel rod. Two holes are drilled through the cradle and the square rod and hit with two pop rivets. The rod is then drilled at a point just aft of its balance center so that when the iron is removed it will tilt forward of its own weight. A little three-sided post out of scrap aluminum can be made and popriveted to the top of the box. This provides the pivot point for the cradle and also a place to fasten the microswitches. The microswitches are turned off by the weight of the iron on the cradle. As you will note in Fig. 1-2, this opens the circuit across a diode, thereby cutting down the heat on the iron when it is at rest and shorting it out when the iron is removed. Removing the iron also

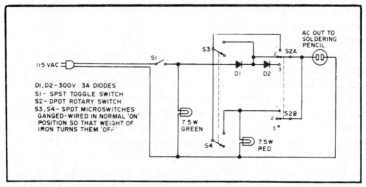

Fig. 1-2. D1, D2-300 V 3A diodes; S1-SPST toggle switch; S2-SPDT rotary switch; S3, S4-microswitches gang-wired in normal ON position so that weight of iron turns them off.

lights the red signal light showing that higher heat is being applied. It also informs you when you don't put the iron back in the cradle properly.

Of the two switches on the front, the toggle switch controls line voltage, while the double pole triple throw rotary switch selects one of three modes. Number one position is direct line voltage or high heat. In this position the iron is connected at all times directly to the 115v AC line. In number two position the iron is in series with a 300v 3a diode while in the cradle, and directly to the line voltage when it is lifted to be used. The third and last position puts the iron in series with a diode at all times, cutting the heat still further. The third position is best for soldering tiny leads close to the bases of delicate solid state devices such as HF transistors. The number two position keeps the iron reasonably hot, but not at a high temperature. Within five or six seconds after lifting it from the cradle it is up to full working heat.

The number one position could have been eliminated, but the switch position was there so it was wired in. It is rarely used because it causes tips to oxidize very quickly. The wiring diagram is, hopefully, self-explanatory. This simple tool will probably be the most used addition to your bench.

Cordless Iron Tips

No, not the kind of tips that heat up, the kind of tips that keep the latter heated up! In a recent advertisement, the Wahl soldering iron people set forth the claim that many users have said that their $20 cordless soldering iron is worth its weight in gold. For numerous tasks around the hobbyist's workshop it is almost indispensable, especially once you get used to having one of them. For example, they are great for antenna jobs 150′ from the nearest AC outlet, small quickie soldering jobs when you don't have time for the landlocked iron to get perking and for mobile or vacation jobs to fix that intermittent when it happens. The extra fine tip that's available for these irons is great for making accurate joints on those little bitty circuit board pads that seem to be in the vogue now (you know, the ones that turn to goo with an American Beauty). One model (Wahl 7500) comes with a charger stand. All you have to do is drop it in the stand to revitalize those two little 1200 mAh nicads inside.

Did you ever wonder whether or not the two contacts on the iron were making good connection with the matching contacts in the stand? After all, the iron element draws about 5 or 6 amps when the cells are fully charged, which represents about 10 to 15 minutes of actual soldering, so it's important to keep it up to maximum in order to have it when you need it. You can jiggle the iron around in its stand just to make sure. The charging stand as it comes from the factory consists merely of a 120v Ac to 2.5v Ac transformer and the two mating contacts (Fig. 1-3). On units that have the non-rapid charge cells, the maximum safe charging rate is 1/10 of 1200 mA, or 120 mA. Notice the clever placement of the diode in the iron itself so that reverse charging from the stand is impossible. Also, if the charging terminals on the iron accidentally short to something metal, the nicads won't discharge because the diode won't conduct that way.

Back to the point. We installed an LED (light emitting diode), polarized, in series with the red charger lead inside the charger itself. We chose to limit the charging current to 90 mA (nice and safe) with 30 mA through the LED (bright enough and also safe)—hence the 12 ohm, ½ watt resistor paralleling the LED. Now when the iron is dropped into its stand, it's taking a charge by the fact that the LED is releasing photons.

The actual modification is quite easily accomplished in one evening. Use a diffused red LED that requires a 3/16" mounting hole deftly drilled in the front of the stand. After checking the fit and deburring the hole, apply a couple of drops of epoxy from the inside to hold the LED nicely. Clip the LED leads to a convenient length and make small loops in them to act as tie points for the resistor and charger leads. Incidentally, there is a rather wide variance in light output, current drain and chip centering in surplus LEDs, so check the device before committing yourself with epoxy. You may find that you will have to choose from several devices to get the desired results. All that's required to custom tailor this idea to your hardware is a meter that will read at least 100 mA, a few test leads with alligator clips on them and some resistors (or a pot) around the value. You can just copy the circuit as shown in Fig. 1-3 and you'll be in the ball park.

As mentioned earlier, you may wish to use your cordless iron in the car or when on vacation, which of course necessitates bringing along the charger stand, right? Wrong! Consider whipping up a charger cord that will allow you to charge your iron directly from your automobile, boat, plane, etc., 12 volt system. That tactic involves borrowing the cord from something that must go with you on every vacation (Unless you're going camping in the north woods)—your electric shaver. AC coiled cords for both the Schick Flexamatic and the Norelco Tripleheader shavers will mate with the Wahl 7500 iron with minimum modification. Shave about 1/32" off of the female cord receptacle end (completely around) with a sharp knife and then file it smooth to obtain a perfect fit in the iron. Don't get carried away—1/32 isn't very much. Yes, the cord still works in the shaver. Then wire the circuit shown in Fig. 1-4 into a standard female AC cord receptacle called a STA-TITE obtained from the local hardware emporium. Note the inclusion of the ever-present LED to show you when the iron is plugged into the shaver cord properly or when the shaver cord is plugged into the DC adapter cord properly. If the LED doesn't light, reverse either end of the shaver cord (no harm done because of that diode in the iron's case). The opposite end of the DC adapter cord is terminated in a standard cigar lighter plug. Note that this end is polarity conscious. The present standard seems to be negative ground (lighter plug shell), so wiring as shown will be correct. By the way, this DC cord is intended as a charger cord only. Attempting to solder a connection with it plugged into the vehicle's cigar lighter socket will only replace about 90 mA of the 5 or 6 amps needed to run the iron—and that won't buy much! The figures on this circuit are identical to the AC charger: 60 mA through the 150 Ohm, 1 watt resistor combination. These figures are based upon 13v DC input, which is an average lead-acid storage battery valve. You can purchase Wahl's plastic carrying tube for the iron, or you can

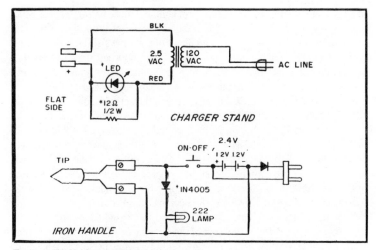

Fig. 1-3. Wahl Model 7500 iron and stand.

do just as well using a decent 1-⅜″ to 1½″ mailing or paper towel tube fitted with end caps. Make it long enough to house the DC charging adapter, too.

Wahl 7500 cordless draws 5 to 6 amps from the internal batteries once the element is hot. The initial current through a cold or cool tip is about 9 amps, tapering to the above after about 5 seconds. This suggests that the working time between charges is less when the iron is continually turned on and off. If you check the iron against the clock after a full charge, you should be able to get over 12-½ minutes of useful heat in a continuous ON condition. Presumably, off/on operation would yield something less than this. There are all kinds of ins and outs wiht something as simple as this seems to be, aren't there?

Before you relax for the evening, here's one additional easily accomplished modification. This time it is inside the iron itself. There's a very handy little light built into the business end of the Wahl 7500, with the nasty habit of burning out too fast. The type 222 bulb is rated at 2.25v and 250 mA, but the fully-charged nicads put out 2.4v (under load) and shorten the bulb life quite a bit. If you install a silicon diode (1N4005) in series with the bulb

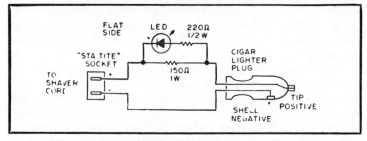

Fig. 1-4. DC charging adapter cable for Wahl Model 7500. 90 mA total: 30 mA through LED and 2200 ½s resistor, 60 mA through 1500 w resistor.

19

you'll get much better bulb life, plenty of light and more battery capacity for heat instead of wasted light (remember the energy crunch!). With this addition, the bulb has 1.5v across its filament and draws 200 mA due to the drop across the diode (junction potential). The important thing here is the well-documented fact that as you lower the voltage across a lamp from its design voltage, the lamp life increases dramatically. The reverse is also true. Now that you're convinced, just remove the screw holding the bar that makes pressure contact with the shell of the lamp socket and install the diode in its stead, wrapping the anode end around the screw and soldering the blanded cathode end of the lamp socket shell (use minimum heat and remove the lamp first). While you're inside the iron case, put a small dab of GC26-01 silicone lube on the switch contact. It's a good all-around keepercleaner. Note that these modifications directly apply to the model mentioned with non-rapid charge cells. However, they could be applied to other models and manufacturers and will hopefully act as food for thought for you.

Now you can relax for the rest of the evening or you could try out your newly modified cordless iron on some other interesting project or you could just admire the LED in the stand while the iron is charging.

The Solder Master

For contemporary assembly projects, use popular, screw-in element and tip soldering irons which, with tweezers and magnifying lenses, are tools of the microcomponent age. Two irons can be used at the same time—a 50 watt heavy duty for large joints and a tiny tiplet for compacted PG boards. Frustration is maintaining both irons at the proper temperature. The big element, left idle for any time, toasts its solder tinning and threatens self-destruction, requiring unplugging. The tiplet dissipates heat too fast for more than one connection and this involves impatient recovery time.

What is needed is a voltmeter for adjusting desired element temperature. The price of these commercial units is very high. The el cheapo answer is the commonly available light dimmer, rated at 600 watts and solid state at that. There is probably one in your junk box that will tame the heavy duty iron. But while lowering the voltage, why not also boost it briefly to help the tiplet along? A 24v, 2 amp filament transformer will solve this.

The major items can be squeezed into a 5″ × 3″ × 3″ aluminum utility box with two outlet sockets. An available miniature meter can be pressed into service, modified with bridge rectifier and series resistor to read zero to 1.0, for reference. This works great, but has some inadequacies such as how to control the irons separately. This can be taken care of with a few more parts.

The schematic of Fig. 1-5 gives the final story, including three small toggle switches. You could also splurge on a new meter of the same size, reading 0-150v AC. The transformer primary and secondary windings can be connected in series to make an auto-transformer, if the polarity connection with 110v AC across the primary, 130v AC is measured across the combined windings. The dimmer control adjusts line voltage to the primary so that regulated output to the sockets can be varied from 0 to 130 v AC with the front panel knob. S2 and S3, miniature single pole, triple throw (center position OFF) toggle switches are installed right below the meter in line with their respective sockets. These allow either socket to be switched to regulated or line voltage, or *off*, independently. S1 gilds the lily in switching the meter to read regulated supply, or doubles as a line voltage monitor. This miniature SPST switch is located immediately to the left of the meter. Due to the pulsating waveform output from the dimmer control, voltmeter readings are not really accurate rms, but close enough for reference adjustment.

Now keep the larger iron happily idling at 70v, while using the small one switched to "line." Or switch it off temporarily, while boosting voltage to the tiplet to solder multiple joints. Start up time is reduced by briefly applying full 130 volts, but boost voltage (over 110v) should be kept to the minimum to avoid premature element failure. And when the XYL sounds chow call, both irons can be left idling at approximately 60v, for fast reheat upon return. The neon pilot light, being connected across the regulated voltage, extinguishes at about this control setting. If you set the control where the pilot is flickering, this is just right for keeping the irons warm.

Finally, use a three wire power line input to provide ground to the metal case. This is not only a desirable safety-feature, but also provides shielding for noise spikes emitted from the dimmer control. Should you use grounded tip irons, you will also want to substitute 3-prong outlet sockets rather than the 2-prong types.

You may also find some other handy applications for this voltage controller around your workshop—but don't exceed the transformer wattage rating by plugging in the coffee pot. If all parts are bought new—and dimmer controls on sale for as low as $2.98—the total cost even in this age of inflation should not exceed $15.00.

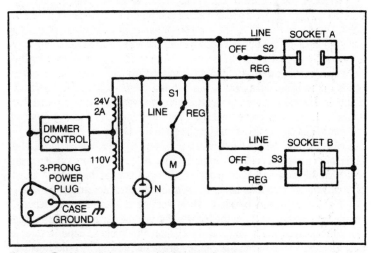

Fig. 1-5. Toggle switches are added to the dimmer.

One last warning. The push switch built into the dimmer controls serves to turn the regulated supply on and off. But, while plugged in, the unit will furnish voltage to the sockets when either S2 or S3 is switched to line. Inadvertently leaving the irons on is avoided by leaving S2 and S3 at center position OFF when not in use.

Shutting off the main bench safety switch, which controls all outlets, is double protection.

The Third Hand

When you're soldering small components, it can be a job to hold them, the soldering iron or gun and the solder. The easy way is with a vise made with two spring-type clothespins.

Cut one leg shorter than the other, and then use small brads to the side of a small wood block. Place the item within the jaws of the two clothespins while you solder.

How to Become a Troubleshooting Wizard

In times past just about every amateur acquired, if only out of sheer necessity, some degree of troubleshooting acumen. Of course, it might be argued that rigs of those bygone days were a lot less sophisticated than rigs of today. Is that single fact really good reason for the seeming decline in general amateur troubleshooting ability? It is contended here that anyone with a knowledge of theory deep enough to pass the General class license examination can make a reasonably good attempt at troubleshooting. Of course, years of experience are needed to make a commercial troubleshooter but we are only concerned with maintaining the usual hobbyist's station equipment and not how many high technology transmitters or receivers he can fix per day or week.

Crux of the problem. Lack of familiarity and zero technique are two reasons why, to many, troubleshooting is troublesome. Consider a case in point. One person bought a near-new condition Novice CW transmitter for a song($5) because the now-General former owner found out (after the new license arrived) that it wouldn't work on 20 meters. He even knew what the main symptom showed—no grid drive to the final on 20 meters. All other bands worked fine. This person was not an experienced troubleshooter at that time. He was merely a thoughtful teen-aged amateur. He reasoned that a loss of drive on any single band in a standard simple MOPA transmitter had to be either an open coil (on that band) or some defect in the band switch. In this case the wire on the band switch that went to the 20 meter tap on the grid tank coil was not soldered properly! This little story points out one main thing that will allow successful troubleshooting. Think out the problem ahead of time with schematic in hand, to ascertain all possible causes.

Test equipment needed. After working in a lot of service shops, people have grown weary of the fellows who use the excuse that they could "fix it themselves if only they had the equipment." While an experimenter's bench loaded with expensive laboratory grade test equipment is desirable, it is not strictly necessary for most troubleshooting. Sure if you were in the business of providing electronic service for a profit, it would be justified purchase because of the time savings on each job. The average amateur, though, only wants to get back on the air without having to wait weeks for a repair shop's turn-around time to elapse.

What constitutes "necessary" equipment? Consider, first, that the most valuable and versatile piece of test equipment is already in your possession—your head! The mind can only function, however, when given a data input concerning the problem to be solved. For this we need our learned observations about the performance of the rig vis-a-vis defect and some basic measurements, many of which can be taken with low cost or even simple home brew instruments.

Most basic in the troubleshooter's arsenal of instruments is some sort of multimeter. The traditional Volt-Ohm-Milliammeter (VOM) is often preferred over the Vacuum Tube Voltmeter (VTVM) in those cases where money or other considerations dictate that only one instrument be purchased. It is wise to purchase an instrument with a sensitivity of at least 20,000 ohms/volt (higher if possible). There are at least three reasons for this preference: portability, existence of a current range and insensitivity to rf fields. Many of the modern FET voltmeters fill the bill in the first two requirements (often better than the classic VOM) but most sadly fail in the third: They will read in error around rf fields such as exist in your transmitter.

The classic VOM will have AC and DC voltage scales, at least one current range (usually more) and at least one resistance range. Naturally, the more you can pay for a VOM the more features and ranges you can expect. You can extend the DC voltage scale of most instruments well into the kilovolt range by the use of an external high voltage probe of the type which has a built-in divider network. It is also possible to get at least a relative rf level indication through the use of a demodulator probe such as that shown in Fig. 1-6. Most manufacturers offering such probes claim operation well into VHF.

The VOM is one of those "every hobbyist should have…" instruments. Another in that category is the dip oscillator called *grid*, *base*, *gate* or

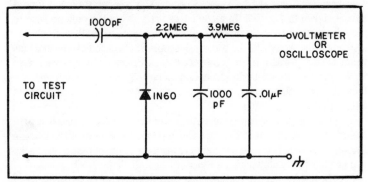

Fig. 1-6. Schematic of an rf demodulator probe for voltmeter or oscilloscope.

tunnel-dipper depending upon design. These instruments have an incredibly varied range of applications around an experimenter's workshop. They can be used to find the resonant points in tuned circuits, antennas and so forth. They can also be pressed into service as an absorption wavemeter, oscillating detector or impromptu signal generator. Best of all, they are simple to build and are low cost regardless of whether obtained in kit form or built from scratch.

In many troubleshooting procedures it might prove desirable to use a controlled, locally generated substitute signal. These are best supplied by a signal generator. Of course, a lab-grade instrument is the preferred type. It must be noted that many high grade signal generators are available on the used and surplus markets. While it might be nice to own a fine signal source, a more modest instrument will suffice for fixing the rig. It is necessary to keep in mind that troubleshooting and alignment are different procedures and that the signal generator requirements are vastly different. Commercial servicers have little trouble justifying a kilobuck signal generator for alignment purposes. They need the superior short term stability and a high quality attenuator; you don't. A *service grade signal generator* will be a very useful tool in troubleshooting. These are not up to the lab grade types in performance but they can be had for less than a month's rent.

A crude form of signal generator useful for troubleshooting is shown in Fig. 1-7. This is a square wave generator operating at about 1000 Hz. Such circuits are widely used in the service industry because they can help pin point a dead stage in moments. These instruments generate the fundamental square wave and a boat load of harmonics that can be used out to several MHz. In most amateur work this means that we can *quick check* the audio, i-f and rf (on 160 to maybe 40 meters).

Your crystal calibrator may also suffice as a troubleshooting signal generator. In Fig. 1-8 we have a circuit that can be home brewed from low cost TTL digital ICs. Although the possible choices of output frequency are limited in this case you can obtain almost any frequency by correct choice of oscillator crystal and division ratio. If zeroed against a frequecy standard such as WWV, this circuit can serve as an accurate method for frequency measurement. It can also be used as a regular signal generator or as a type of device shown in Fig. 1-7. In those latter applications it would be wise to provide an attenuator. If a smart troubleshooter were to be allowed only one instrument he would choose the oscilloscope. If triggered sweep and dual beam can be afforded, it would be very useful. Prices on 'scopes vary from $5/free to several kilobucks depending upon features, condition and vintage. Many oscilloscopes, some quite nice, remain UNSOLD for lack of interest at auctions!

Servicing typical ham transmitters requires several special instruments. Most of them are the sort which ought to be in every amateur's station. One item is the 50 ohm dummy load. At a cost of only a few bucks you can have a dummy antenna which will absorb all the power you have a right to be dishing out to a real antenna. Only the utterly irresponsible would attempt to service a transmitter connected to a *live* antenna!

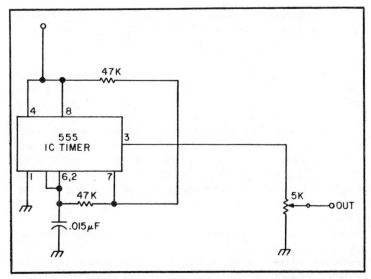

Fig. 1-7. 1000 Hertz square wave generator produces harmonics useful for signal tracing from audio to several MHz.

Rf wattmeters and relative field strength meters Fig. 1-9 are also amongst the highly desirable. Although the wattmeter instrument is sometimes preferred, make no mistake about it, the old fashioned swr bridge does have a lot of good mileage left.

No one seriously expects an amateur to be as well equipped with test instruments as a commercial or factory level shop. He should, however,

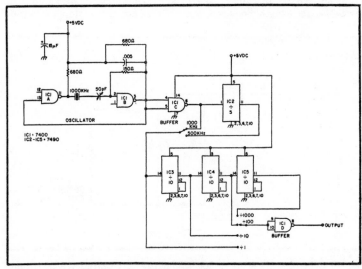

Fig. 1-8. Schematic of a TTL crystal calibrator.

attempt to obtain those considered as basic. For some of the others he might work through the local club to build a test gear collection of really nice pieces or arrange with certain locals to split the load by having each fellow buy certain instruments and then do a lot of lending.

Troubleshooting Procedure. Troubleshooting involves a logical, step-by-step procedure for determining a fault in a piece of equipment. Best results in a minimum of time are obtained by using a method formed from a logical analysis of the problem at hand.

One thing which makes troubleshooting easier is a service manual (or *at least* a schematic) on the equipment. It is the wise amateur who obtains (even at a ridiculous cost) the service manuals for all of his station equipment. In most instances new equipment comes with a manual; keep it. If you buy used or for any reason are without a manual, buy one and keep it on file. Do not expect your rig's manufacturer to be able to supply manuals indefinitely; get yours now.

One significant aid in troubleshooting is the block diagram. It is especially useful in rigs such as transceivers where a single stage might have different functions. An example of a block diagram, this one from the popular HW-101, is shown as Fig. 1-10. You can often cut the work of troubleshooting in half by doing a *desk check* of the service manual and block diagram before reaching for instruments to do the *bench check*. Note that many of the stages, shown here in dotted lines, are common to both transmit and receive functions. Although there are exceptions, you can usually overlook these stages when a defect affects only one one function or the other.

The element of logic in troubleshooting should be almost on a priori truth (despite the philosphers who say that such cannot exist!). After all, if the set is dead (no lamps or tube filaments lit) is it wise to worry about, say, the beat frequency oscillator? Fig. 1-11 shows what is affectionately called a *troubleshooting* tree. Although this is not offered as a universal approach you

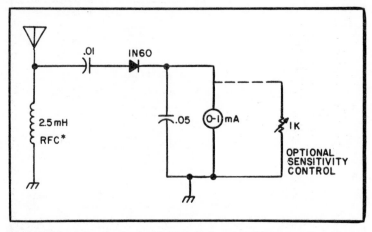

Fig. 1-9. Schematic of a simple rf field strength meter. (*Can be replaced by an **LC** tank circuit for greater sensitivity or use as an absorption meter.)

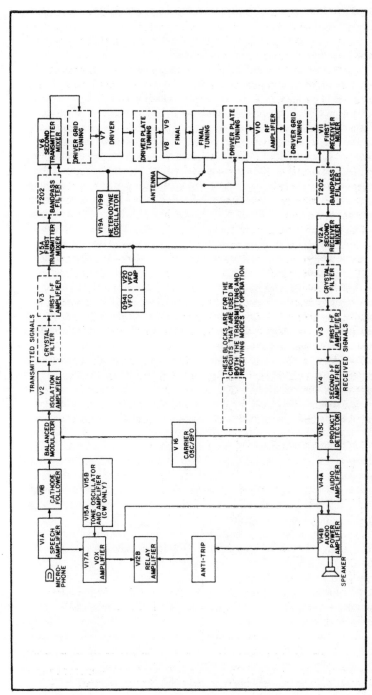

Fig. 1-10. Block diagram of a popular transceiver. (Courtesy of Heath)

27

can often at least get onto the right track by its use; it is intended only to be a *point of departure guide*.

Two terms pop up a dozen times in any servicing literature: *signal tracing* and *signal injection*. They are basically methods for locating a dead or otherwise defective stage. Before beginning one of these procedures, however, make certain preliminary observations. For example, can you hear noise in the speaker as you rotate the volume control? If so, the cause of the dead condition is probably *prior* to the volume control. What about dirty contact noise as you rotate the selectivity switch? This usually means that the fault is prior to that point. Band switch noise (again, as you change bands) could mean a fault in the rf or a dead local oscillator. What about the level of background hiss that sounds as if you were tuned to a dead band? High noise levels may tend to indicate troubles in the rf amplifier or antenna circuit. Here is one point where your calibrator is useful. Note how high the "S-meter" reads when the set is working normally. When a defect occurs, turn on the calibrator and see if it is lower. If not, suspect the antenna or feedline. These observations are of the sort that can save a lot of time. Most pros use them all day long even when unaware; it is the source of their supposed professional intuition.

Signal Tracing. The process of signal tracing involves injecting a known and controlled signal at the input to a chain of stages, then looking for it at the outputs of successive stages with a *demodulator equipment signal tracer* (high gain audio amplifier with a fancy name!) or *oscilloscope*. When you find the defective stage, you can then use the VOM or VTVM to locate the bad part. In some cases, you might want to inject an appropriate signal at the antenna terminals while in others it might be better to inject another sort of signal at the input of the i-f strip.

Signal Injection. Signal injection is a similar process. In this method you start at the output and inject an appropriate signal into the inputs of successive stages until you find one which no longer is capable of producing an output. In signal injection you begin at the output and work toward the input, while in signal tracing just the opposite situation occurs. In either event the procedure can usually be expected to ferret out the defective stage. Once found it is usually a simple matter to dig out the voltmeter and determine which component is at fault. Be sure not to overlook the obvious fault such as a bad tube or a loose connection.

When troubleshooting transmitters, you can use a demodulator probe to find relative rf levels on the 'scope or VOM. It is usually better, though, to read the DC voltage developed by the grid leak bias often used in transmitter stages. This voltage will be proportional to the rf level delivered by the preceding stage. This system works so well that it is the basis for the Motorola test set popular with FMers who run surplus commercial Motorola mobile equipment.

Alignment. The most exasperating mark of the neophyte troubleshooter is a wild *diddle stick* (alignment tool). It is a pretty good rule to assume that alignment points *never* shift enough to cause a gross fault in the receiver, especially a sudden fault. The only real exception is the local

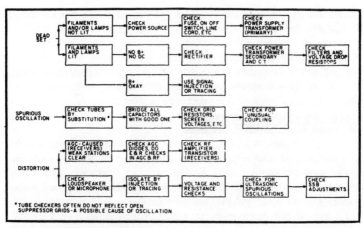

Fig. 1-11. Troubleshooting tree useful for finding a good jumping off point for troubleshooting specific defects.

oscillator. Even in that "sometimes" case there are other methods (i.e., measure the frequency) for determining its condition. One revealing way that shows whether or not the stage is actually oscillating is to measure either the grid bias or the voltage drop across the emitter resistor as the set is tuned through the range. In the case of crystal oscillators remove the rock. This will cause the bias to change suddenly for crystal pulling or smoothly in VFO type circuits. DO NOT EVER touch an internal alignment point without good reason, proper equipment and a knowledge of what it is' that you are doing. You can diddle stick a decent set to death that really only needed a small component. This is why the *screwdriver technician* is the bane of professional servicers everywhere.

Preventative Maintenance. There are a number of little tasks which will keep your rig in top shape for a lot of years. These generally fall under the heading of *preventative maintenance* (Fig. 1-12). One task would be to check all tubes on a periodic basis, say every six months or maybe once a year. Replace those which are defective or grossly weak. Some advocate doing a voltage check at critical points throughout the circuit especially in the power supply. It is recommended that you clean the inside of the rig, including all

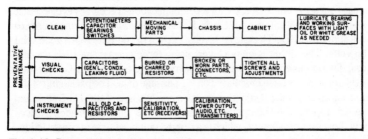

Fig. 1-12. Preventative maintenance tree.

ciruitry, with a paint brush and a mild solvent where needed. Use a switch/control, also called *tuner* cleaner spray on all band switches, function switches and controls. Clean capacitor and rotary switch bearings and relubricate using a white grease such as *Lubriplate*. *Peaking up* alignment is NOT part of this procedure unless you can perform factory-level service. Even in that case it should not be obligatory except every few years. Alignment is simply not all that common a problem.

A Look at Soviet Test Gear

One of the more enjoyable benefits of being an electronic engineer is the opportunity to meet engineers from other countries. While working on international programs, many hobbyists meet engineers and trade specialists from the Soviet Union. They are anxious to improve trade relations with our country to help with their balance of payment problem. So far, the bulk of trade has been in raw materials and commodities, very dull stuff to the electronics enthusiast. Many Americans do not know much about their equipment and are very curious about its design, the type of semiconductors used, how good it is and the prices. They have a state monopoly on all manufacturing industries, so all electronic equipment is sold through state trading organizations. Test equipment (VOMs, counters, scopes, etc.) is sold by Mashpriborintorg, electronic components by Eledtronorgteknica and communications equipment by Sudioimport. This approach results in a high degree of standardization throughout their country. It also results in fewer innovative designs by their enginners.

Through a few contacts, we received two pieces of test equipment for evaluation. Both of them had useful features not found in instruments currently on our market. The first was a small multimeter, the U-4323. It is designed for the hobbyist. It is similar in appearance and outward construction to many Japanese import instruments in its price class. Range selection is done by plugging the test prods into the appropriate tip jacks. Unpacking the instrument, we discovered a small package with spare diodes and a spare meter fuse. A comprehensive operating handbook—in English—complete with circuit diagrams and maintenance data was included.

A quick glance at the panel revealed that in addition to being a conventional VOM, it contained an audio oscillator and signal generator! The output of the audio oscillator is a 1000 Hz square wave. The rf output was fixed at 465 kHz, the standard i-f for their country. It was modulated by the 1000 Hz square wave. A quick check with a communications receiver revealed that the harmonic content was quite rich. It was usable as an alignment source to well above 30 MHz. The output level of each of the outputs was a good solid 0.5 volts. The circuit is quite clever, using three germanium PNP transistors. Two are used for the 1 kHz multivbrator, the third is an L-C controlled oscillator which is base modulated on the package data. The packaging of the transistors was quite distinctive, being a combination of the TO-5 style and the old "top hat." All components appeared to be very well made, particularly the meter movement, which is a rugged taut band type of construction. Diode protection is also provided.

The little handbook is quite comprehensive, going into the theory of operation for the oscillator circuit. The 40 microamp meter movement is described in such detail as to permit field repair and rewinding of the meter coil. A section is devoted to the variation of readings due to changes in ambient temperature, battery voltage, meter position and the frequency of AC voltages. All measurements are illustrated by simple one function drawings. It was evident that the manufacturer realizes his instruments will be used in a wide range of climates by people who may not be too well trained.

The U-4341 is a combination instrument that incorporates a transistor tester with a high quality conventional VOM. It is a quality instrument that would be classed as a bench test instrument adequate for commercial service work. It is housed in a black plastic case designed for tabletop use. Rated accuracy is 2.5 percent for DC and 4 percent for AC ranges. AC frequency response is from 45 Hz to 20 kHz. Both AC and DC currents are measurable. The instruction book is very comprehensive and, like the one for the U-4323, provides a lot of information for the user working in out-of-the-way places.

Another nice feature is the inclusion of special low range scales. The lowest AC current range is 300 microamps, AC is 60 microamps. The lowest DC voltage range is 300 millivolts full scale. The most interesting feature is the built-in transistor testing circuit. The meter will measure Iob, Ieb, and Ici (defined as initial collector current as measured in the common emitter configuration with zero emitter base voltage) with an accuracy of ±2.5 percent.

The quality of the instruments is quite good when the price class is considered. Both instruments are well sealed against dust and moisture. Battery replacement is accomplished by removing the back plates. All instruments have manufacturing seals placed by their quality control department to reveal any tampering that might void the warranty. A separate certificate of inspection giving the performance specifications and quality assurance sign-off is packed with each instrument.

Reading the manufacturer's literature provided insight into the type of equipment available to the Soviet hobbyist. There appears to be no production of strictly amateur radio equipment in the Soviet Union at this time. This means that the Soviet experiment has to build his own or modify military surplus equipment. This is made available to him through DOSAFF, an organization that has no American equivalent. It is a civilian auxiliary for supporting the armed forces, and would be roughly analagous to having MARS, civil defense, the National Rifle Association, sport parachuting and other paramilitary activities rolled into one super organization. Soviet defense policy places significant emphasis on having a trained cadre of civilian radio operators available to assist the military in the event of war, so they supply individuals and clubs with equipment much in the manner that our MARS program is supposed to operate.

There are a number of test instruments in their equipment catalogs that would be of interest to hobbyists in this country. Among these is a neat little

3 inch scope. It has a calibrated triggered sweep, 10 MHz bandwidth, built-in calibration and a number of other desirable features. Again, there is a design twist—this scope is designed for the hobbyist with limited facilities to maintain his test equipment. Although of recent design, it was designed around a series of high reliability vacuum tubes to minimize the number of components.

If and when these products appear on our market will depend on the political situation. One of the major stumbling blocks is the lack of a "most favored nation" treaty with the Soviet Union. Lack of such a treaty results in drastically higher import duties out of the market. There has been a great improvement in relations between our two countries in the past few years. Hopefully this will continue. With the ever-increasing prices of Japanese electronics, we may see a day when the USSR will be a major supplier of popular electronic equipment.

Chapter 2
Cheap-N-Easy
Handy-Dandy Testers

Just about everyone needs more test equipment, or easier ways to use the test equipment he already has. Just about everyone has an old radio sitting around and this can provide the needed test equipment. You merely need to make the proper modifications to the radio, then know how to apply the resultant piece of test gear. Here are some modifications which will allow you to add a mess of handy test equipment to your shop, and still leave your old transistor radio in working order. The finished device will be everything from a speaker subber to transistor checker. It will still provide your favorite news or music while just sitting on the bench.

The first consideration is what to do about packaging the radio. You may decide to leave the radio intact and add plugs and jacks of your choice in the radio cabinet. You could run leads out from the radio cabinet and put the plugs, jacks and switches on a mini-box. Or you could package the radio complete with modifications in the box of your choice. No matter what the package you choose, be sure to decide on plugs and jacks that you already have around so you don't make all your test leads obsolete. The diagrams that follow show the use of phone jacks, but you could just as well use banana jacks or whatever. Figure 2-1 shows a typical radio schematic. Some will differ, especially in part values, but the stages are essentially the same as the one shown. At least this diagram is helpful in locating where to put the modifications in your particular radio. You may add any or all the modifications to make a speaker subber, power supply, audio oscillator, signal injector, transistor checker, alignment generator or weather receiver.

Speaker Subber. This is the simplest modification. You may add a plug as shown in Fig. 2-2 or you may modify the earphone jack already on many radios. When a phone plug is plugged in, the speaker is automatically disconnected. A pair of alligator clips on the end of the leads will allow you to connect to almost any source.

Power Supply. Whether your radio is battery or AC power, it will give you an easy source of 9-12 volts DC depending on the radio. All this without tying up the main power supply in your shop! When you are using the other radio circuits to check out a radio or experimenting with a circuit on the bench, the radio power supply will run both the test equipment and the radio or circuit under test. Figure 2-3 shows some of the suggested modifications for both battery and AC power supplies. In Fig. 2-3C a transistor is used as a voltage source to provide variable voltage outputs from the radio supply.

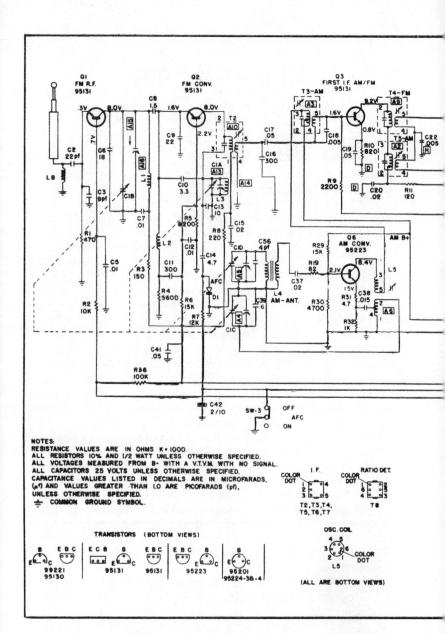

NOTES:
RESISTANCE VALUES ARE IN OHMS K = 1000.
ALL RESISTORS 10% AND 1/2 WATT UNLESS OTHERWISE SPECIFIED.
ALL VOLTAGES MEASURED FROM B- WITH A V.T.V.M. WITH NO SIGNAL.
ALL CAPACITORS 25 VOLTS UNLESS OTHERWISE SPECIFIED.
CAPACITANCE VALUES LISTED IN DECIMALS ARE IN MICROFARADS,
(μf) AND VALUES GREATER THAN 1.0 ARE PICOFARADS (pf),
UNLESS OTHERWISE SPECIFIED.
⏚ COMMON GROUND SYMBOL.

TRANSISTORS (BOTTOM VIEWS)

99221 95130 95131 95131 95223 95201 95224-38-4

(ALL ARE BOTTOM VIEWS)

You cannot ruin the pass transistor circuit. If you have a battery level indicator on your radio, it may be calibrated and used as a voltage indicator. By adding a resistor network and range switch in place of the potentiometer, you can have discrete voltage outputs such as 1.5, 5.9, etc.

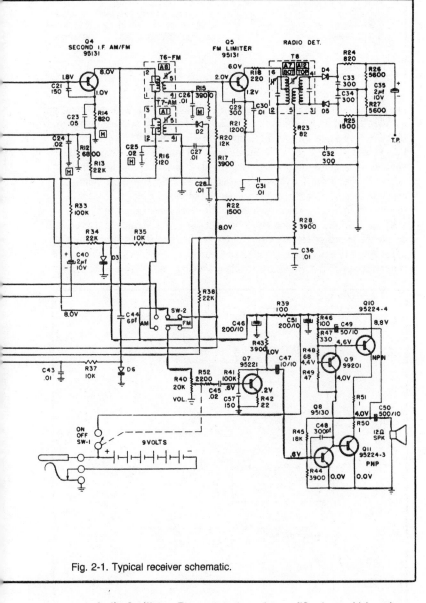

Fig. 2-1. Typical receiver schematic.

Audio Oscillator. Figure 2-4 gives the modifications which make an audio oscillator. The result can be used as a code practice oscillator, test oscillator, audible continuity checker or audio signal injector. The tone control provides some measure of control of the basic oscillator frequency.

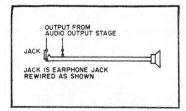

Fig. 2-2. Using the radio as a substitute speaker. You may wish to add a jack as shown while retaining the original earphone jack on the unit.

Signal Tracer. The radio will work as several different tracers. The most obvious use is shown in Fig. 2-5A, an *audio signaltracer*. Injecting a signal at the detector as shown in Fig. 2-5B yields an rf demodulator tracer. If you inject a signal into the i-f as shown in Fig. 2-5C, you have a 455 kHz tuned tracer (or 10.7 MHz if the radio has FM) for use with standard i-fs. If you allow disabling the oscillator as in Fig. 2-5D, you can inject another radio's oscillator signal to test the oscillator section of the other radio. Just inject the local oscillator from the radio being serviced, juggle the dial on both radios, and if you can pick up signals, the oscillator in the other radio is working. The same method could be used for FM systems. In that case, couple the oscillator signal to the FM converter stage through a gimmick of about seven twists of hook-up wire connected in the same place on the FM converter stage. You will have to add a switch to disable the oscillator as

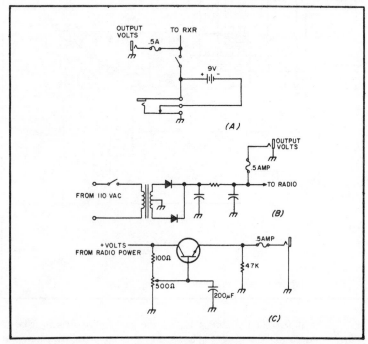

Fig. 2-3. Power supply circuits. (a) Simple battery supply. (b) Output is taken from an ac supply. (c) Continuously variable output power supply.

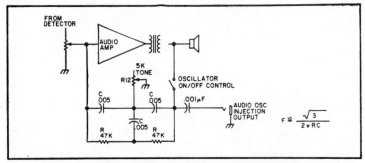

Fig. 2-4. Audio oscillator. May be used as an audio signal generator, code practice oscillator, etc. Values shown are for approximately 1 kHz; use formula for other frequencies.

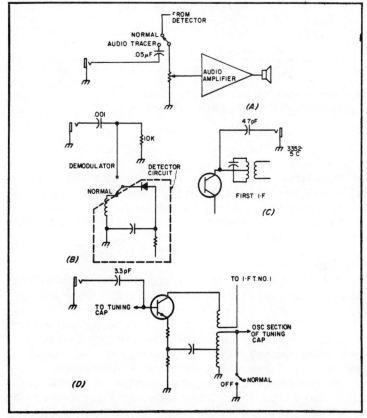

Fig. 2-5. Signal tracers. (a) Switch permits use as an audio signal tracer. (b) The input is applied to the detector for use as a demodulator. (c) I-f injection for 455 kHz. (d) Oscillator is disabled to allow injection of oscillator from another set to test its operation. With switch at normal, radio can be tuned to detect 1600kHz i-fs.

37

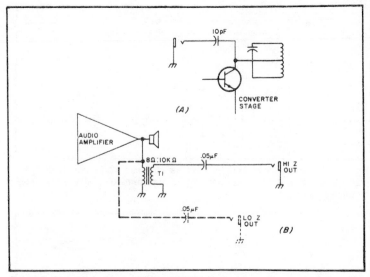

Fig. 2-6. Signal injectors. (A) The 455 kHz output from the converter can be injected in various i-f stages to check out their operation. (B) Addition of an impedance matching transformer for an audio signal injector with high output impedance. Dotted lines show alternate connection.

shown on the AM converter circuit. For the AM radio the oscillator injection point also serves as a point to inject rf from a system that uses 1600 kHz i-fs by letting the oscillator run and tuning the radio to 1600 kHz.

Signal Injector. The output of the converter taken from the point shown in Fig. 2-6A provides a 455 kHz signal for checking the i-f of a radio. An equivalent point on the FM converter provides 10.7 MHz signals.

Tune the radio to a station and the 455 kHz output will be the signal transmitted from that station. Figure 2-6B shows the connection for injecting audio signals; just tune in a station and use its audio to check out the audio circuits under test.

Alignment Generator. In Fig. 2-7 the output from the i-f is taken after it has gone through several stages of i-f amplification. Since this signal has the benefit of the several stages of selectivity, the i-f output signal can be used as a source of accurate 455 kHz (or 10.7 MHz) assuming the radio has been accurately aligned. Merely align the radio as one of the steps to making it into a piece of test equipment. Signals tuned in on the radio will be translated to their equivalent i-f frequency and can be used as alignment signals.

Transistor Checker. Figure 2-8 shows an outboard circuit added to the radio to allow in or out of circuit transistor checking. If the transistor under test will oscillate, it is generally good. This is a fairly accurate check in or out of circuit. Another simpler tester can be fabricated by merely replacing one of the transistors in the radio with a socket. When a transistor close to the type you are replacing it with is put in the socket, the radio should work. This test is only accurate for the NPN or PNP transistor type being replaced, and

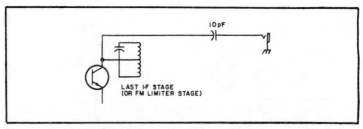

Fig. 2-7. This configuration will act as a fairly accurate i-f alignment generator. Since the output is from the last stage at the i-f, you have the advantage of the narrowing from all the tuned circuits at the intermediate frequency.

will only work with silicon or germanium types depending on what the radio had originally. This simple tester will not allow in-circuit checks at all, but the radio will operate normally as soon as the original transistor is placed in its socket.

Band Opening Monitor. Figure 2-9 shows the schematic of a citizens band converter that works into the 455 kHz i-f of an AM radio. You may shudder at the thought of listening to the citizens band, but it sure is a good band opening monitoring device. Just buy a crystal for channel 10 and you will rejoice when the DX comes rolling in. Crystals are available at your handy Radio Shack store for a few pennies. Just don't let them worry you with questions about what model transceiver the crystal is for . . . any one will do.

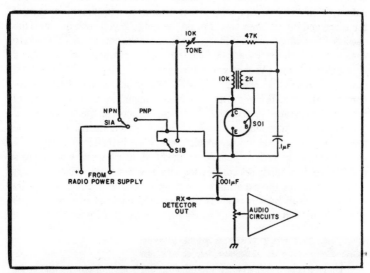

Fig. 2-8. Go-No Go transistor checker will work with silicon or germanium types with reasonable HFEs. Use a good unit for checkout; if circuit fails to oscillate, reverse either the primary or secondary connections. A good transistor will produce a tone in the speaker. You may wish to use clip leads in place of a socket.

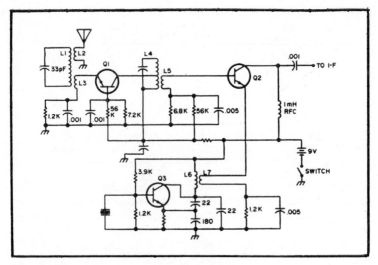

Fig. 2-9. CB converter for band opening monitor.

Commercial FM Receiver. If your radio has FM, it can be modified to make a receiver that will pick up commercial business-band FM signals such as police, fire, public service, weather, etc. Some find the idea of listening to these signals more appealing than music when the radio is resting on the bench. Begin the modification by changing the oscillator coil. Reduce the number of turns by ⅓, then bring it on frequency by adjusting the spacing of the turns and the trimmer capacitor on the radio. If you can't measure the oscillator frequency with a grid-dip meter or equivalent, use a TV tuned to channel 7. Watch the TV as you tune the oscillator; when you can see and hear a signal on the TV, the oscillator is set around 175 MHz. The radio will now tune around 185 MHz or 165 MHz, depending on the resonant frequency of the circuits in the rf amp and mixer stages. Reduce the number of turns in the rf stage and the mixer by the same factor as the oscillator coil. Tune them in by adjusting the spacing and by resetting the trimmers on the tuning capacitor. Final tuneup should come while listening to a signal.

Now that the radio is completely modified, it will serve you well as a radio as well as a handy test jig ideal for servicing transistor radios. The unit is also perfect for the experimenter who may need to use one or all of its functions in checking out breadboard circuits. Once you master the use of all its possibilities, it will prove as handy on your bench as your VOM.

World's Smallest Continuity Tester

Using a rubber-type two-pronged plug, the few components that make up this tester are mounted inside the plug (Fig. 2-10). The hole in the end of the plug through which the cord enters is used for the NE-2 neon bulb.

Two small holes opposite each other are made near the base of the plug for the probe wires to extend, and the two 100k, ½ watt resistors within the

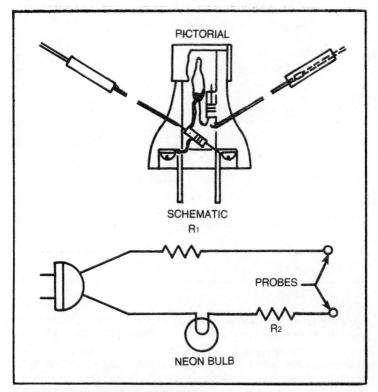

Fig. 2-10. Pictorial version and schematic.

plug cavity, being in series with the probe cords, prevent shock. A piece of
⅝" i.d. aluminum tubing ½" long, placed over the bulb end of the plug, with a
⅝" o.d. clear plastic lens pressed into the tube, protects the bulb and
enhances the appearance of the miniscule tester. Make the probe wires long
enough to suit your needs.

Photo Electric Bench Accessory

Combine the leftover power supply from an experiment that failed with
some 12 for a dollar CdS photocells purchased from S.D. Sales. Mix well
with a lull in regular hobbyist activities and the result is an interesting unit
with many uses.

The diagram in Fig. 2-11 shows the basic unit. The photocell is in series
with a pot. There is a voltage applied across this series combination to
ground. The op amp is used as nothing more than a high impedance driver for
the one mil meter used as an indicator of relative light flux impinging on the
cell. The word "relative" is important to note, as the meter is not calibrated
in any special units. Its reading is comparative only and its function is to tell
you that light has either increased or decreased at any specific moment. The

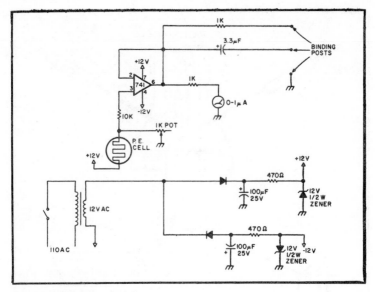

Fig. 2-11. All resistor ½W.

pot is used to control sensitivity. The higher the resistance, the greater the
sensitivity of the unit. The photocell is mounted on two back to back lids
from 35mm film containers of the plastic variety. One film can makes up the
body of the probe. This has a hole cut in the side to allow the cell leads to
exit. Exiting the leads from the side rather than through the bottom allows
the probe to be firmly positioned relative to a light source. A second
container has its bottom cut off and is used for a stray light shield around the
cell. These details are apparent in Fig. 2-12.

Notice that there are two outputs. One is DC coupled through an
isolating resistor and the other is AC coupled through a 3.3 uF capacitor.

With the values indicated, here is an idea of sensitivity for general use.
An LED energized from an audio oscillator and held next to the cell will give
about ½ volt of audio at the exciting frequency when the unit is at maximum
sensitivity. This makes a handy bench coupler into your counter. A 60-watt
bulb in a white glass shade will pin the meter from a distance of about 9 feet
as will an ordinary two cell flashlight.

If you play around with QRP rf levels and are addicted to using pilot
lamps as power indicators for tuneup, this unit will allow you to convert the
light into a meter reading that seems much more sensitive to slight changes
than the eye. When you are fighting for each milliwatt, this is very helpful.

The unit puts out a nice DC pulse with a flash of light hitting the cell.
Thus, the DC output can be used for triggering an SCR or used to bias the
base of a transistor used as a switch or some form of DC amplifier for control
purposes.

If you wish to raise the overall sensitivity of the unit, merely increase
the value of the pot to 5k or 10k. This will greatly raise the sensitivity but

Fig. 2-12. Completed project.

may create stray light problems. For general use, the indicated values work very well.

There is nothing magic about the voltages shown for the op amp—six volts or so would work as well.

Note that there is no need to use shielded cable for the cell leads.

As with most projects, just about the time you get the last screw in place, there is that little voice whispering in your ear, saying, "I wonder what would happen if . . . ?" Well, this project is no exception. Figure 2-13 shows what happens when you listen to little voices.

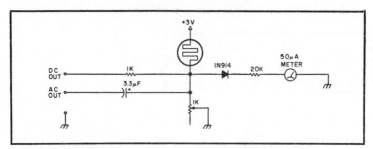

Fig. 2-13. Schematic.

The AC power supply has disappeared, replaced by two C cells in series. The op amp has vanished because a more sensitive (50 microamp) meter has been used. The diode in series with the meter is used to provide a hold-off threshold effect so a small steady meter reading is cancelled. Either unit does about the same job of providing your bench with a photocell dimension that will find many uses.

El Cheapo Signal Tracer

Every hobbyist needs a cheap signal tracer and audio amplifier at some time or another. If you try connecting a 117v AC to the audio output jack of a six meter receiver, you'll find you need one, too. Keep in mind that if you would try such a stunt, you'll have some painful memories.

One foolish experimenter tried this and when the smoke cleared, the two audio output transistors and their transformer were in such a mess that they were eligible for foreign aid (it's a Japanese receiver). Wanting to get back on the air fast, the hobbyist rummaged through the junk box until he found a cheap transistor radio. With this, a resistor, a capacitor and 10 minutes of work, he was back in business. Best of all, the transistor radio was still usable once the six meter receiver was fixed.

The modification to the pocket radio involves four steps.

- First, find the earphone jack and the earpiece and cord that plug into it. Inside, the jack will have three wires connected to it: a ground, a lead to the speaker and another one trailing off to the innards of the radio somewhere. This last wire actually goes to the secondary winding of the audio output transformer (Fig. 2-14).

    Leave the ground wire undisturbed. Unsolder the wire to the speaker and the one to the innards, both at the jack, and note which went where. Solder the ends of these two together and tape them. Now the radio is permanently connected to its built-in speaker.

- The second step involves finding the point where the diode detector connects to the volume control. This can be found by tracing back from the center pin of the volume control along the foil until you find the glass diode. Unsolder the end of this volume control, but leave the other end connected. Solder a piece of insulated hookup wire to the free end of the diode. The other end of this wire is soldered to the pin on the earphone jack that was formerly connected to the

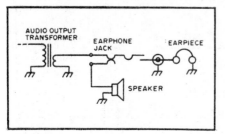

Fig. 2-14. The original circuit.

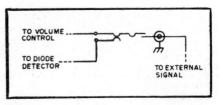

Fig. 2-15. The modified circuit.

speaker. Solder another piece of insulated wire to the point on the circuit board where the free end of the diode used to be. The other end of this wire is connected to the remaining pin on the earphone jack that used to be connected to the innards. Now, without a plug in the earphone jack, the pocket radio will play normally since the diode detector is connected to the volume control once again, although now through the contacts of the earphone jack (Fig. 2-15).

• For the third step, cut the earpiece off the end of its cord. Strip the ends of the wires, and with an ohmmeter or continuity checker, find out which of the wires goes to the inner pin of the jack and mark it. The other lead is the ground connection which can be connected to an alligator clip. Solder one lead of a 1 uF capacitor to the "hot"lead. This capacitor will keep stray DC voltages out of your pocket radio, thus preventing premature trauma. The free end of the capacitor is the probe tip, and is to be connected to the equipment under test, wherever you suspect audio should be. With the earphone plug inserted in the jack, and the probe connected to the circuit under test, you should now hear the desired signal, rather than Olivia Newton-John. The lead with the capacitor can be built into the plastic end of a discarded ballpoint pen, to make a neater probe tip. The voltage rating of this capacitor must be higher than any voltage you have in the equipment under test. For tube type receivers, 600 volts is usually adequate, while a 50 volt capacitor is adequate for transistor receivers and hi-fi gear.

If you're going to run the pocket radio from its own battery, this step may be omitted. If you would like to run the pocket radio from the voltage in the gear under test, this formula can be used to find the right value of dropping resistor:

$$\text{Resistance} = \frac{\text{(Available Voltage)} - \text{(Voltage Needed)}}{\text{Receiver Current}}$$

For example, if your pocket radio needs 9v to operate, and 12v is available, and the pocket radio draws an average of about .010a, then by plugging the numbers in:

$$\frac{12\text{-}9}{.01} = \frac{3}{.01} = 300\Omega \text{ resistance needed}$$

The wattage needed for the resistor can be figured by the formula I2R = P; that is, the current multiplied by itself, times the resistance, gives the

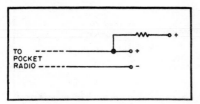

Fig. 2-16. Battery terminals with added dropping resistor.

TO
POCKET
RADIO

needed power rating in watts. In this example, it would be $(0.1) \times (.01) \times 300 = .03$ watts. A ¼ watt resistor would give a more than adequate safety margin. If you're planning to use a 150v supply to run the pocket radio, a 14.1 kΩ at 1.4 watts is the calculated value, and 15kΩ at 5 watts is adequate and a practical common value. This resistor is connected between the positive terminal of the battery holder and the supply voltage point, as shown in Fig. 2-16.

Once completed, the pocket radio can be used normally. By just plugging in the earphone plug/test probe, it becomes a signal tracer or audio amplifier. Total cost is less than 50 cents and you've still got your radio.

Test Instrument Saver

Delicate, costly test instruments may suffer severe damage when their conventional, too long leads snag on tools or on components spread out on the workbench and are inadvertently yanked off it to the floor.

A partial cure for this hazard is to replace the long leads with the much shorter coiled leads commonly used on cameras and flash units.

The coiled leads can be stretched out to where needed. When let go they retract out of harm's way. Should they lose their recoil power, rewinding the coils in the opposite direction helps restore their springiness.

Since each lead has two wires, their extra continuity gives them longer life.

The Violet and Other Houseplants Tester

The device we are about to describe might be classified as a "one evening kit" and does just as the title says—it tests violets. Not only does it test violets, but other miscellaneous *houseplants* as well. This gadget detects the presence (or absence) of moisture and informs people when it is time to water their plants.

The violet tester does not require exotic memory ICs, clock chips or printed circuit boards. One LED (garden variety), a single 330 ohm resistor and a 9 volt transistor radio battery are all the parts needed for the heart of the device. The testing probe, case, etc., may be constructed of miscellaneous junk that you have accumulated.

If you will notice Fig. 2-17 the 9 volt transistor radio battery is in series with the 330 ohm resistor and LED indicator. The 330 ohm resistor limits the current to the LED. Without the current limiting resistor, the LED would draw excess current and soon destroy itself. Also in series with this are test probes. These probes, when inserted in the soil, will provide a current flow or complete the series circuit if moisture is present. The amount of moisture and mineral content of the soil will determine the current available to light the LED. Therefore, when adequate moisture is present, the LED will come up to near full brillance. Varying lesser degrees of illumination will indicate less moisture.

At this point may we indicate that the violet tester is not a laboratory standard—it is merely a gadget in indicate relative moisture, and at that, serves its function well. What we are trying to say is—if some plant suddenly shrivels up, it is not the fault of the violet tester. On the other hand, if the vegetation involved suddenly tries to molest you some evening due to its fantastic state of health, it also is not the fault of the tester.

Anyway the device has proved to be much more sensitive than we had expected prior to construction. Common LEDs seem to be very responsive to minute current changes.

The moisture probe does not have to be a single chunk of copper tubing such as we have used. A pair of separate probes spaced 1" will work equally as well. The only requirement is that they be rigid enough to penetrate the soil.

Figure 2-18 illustrates construction of the violet tester. We used a small chunk of 1-7/8" plastic water pipe to enclose the main circuitry. The top and bottom plates are made from scrap ⅛" plastic. The top section (containing the LED) is cemented in place with epoxy. The bottom plate is held in place with 4-40 bolts. This allows the bottom to be removed for battery replacement. The 3/16′ copper tubing probe must be made secure in the bottom plastic plate as it will take quite a bit of stress when being inserted in the soil.

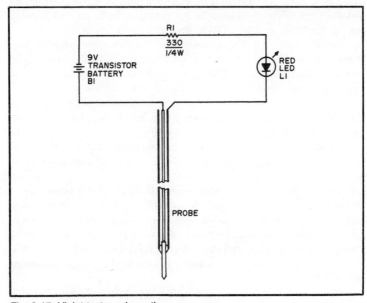

Fig. 2-17. Violet tester schematic.

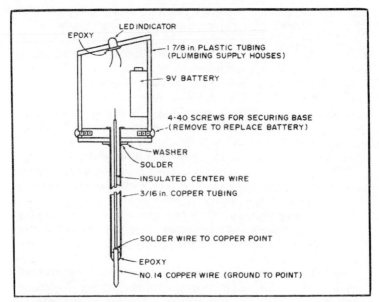

Fig. 2-18. Violet tester basic construction.

A hole is drilled at the center of the plastic bottom plate just a bit larger than the 3/16″ copper tube. A washer is then soldered about ½″ down on the tubing. After inserting the tubing through the hole, the washer will limit its travel. Split the tubing that protrudes through the hole with a hacksaw and fold down the halves flush with the bottom plate. Epoxy cement the *fold overs* to the plastic bottom. The washer acts as the stress point.

Grind a piece of #14 copper wire to a point and then solder a good length of insulated hookup wire to the non-pointed end. This assembly is then epoxied inside the 3/16″ tubing. A small piece of masking tape wrapped around the center probe will hold it in place while the epoxy dries.

The LED indicator is also held in place in the upper plate with epoxy glue. After the glue has set, the wiring is completed. Use a battery clip for easy replacement of the 9v battery when necessary.

Just about everything in this device is non-critical yet good results may be expected. The layout and case may be just about anything that is convenient.

Not only may the violet tester be used for plants, but it should serve well as a lawn, rose bush or garden tester. By attaching clip leads and a remote probe, it would tell you when your basement is flooding. It makes a good continuity tester also. Just think, if some clever individuals were to mass produce this gadget, it would probably sell for $5.99 at local discount stores. Therefore, before this happens, you can be the first on your block with a violet tester.

After completing this "one evening kit" you might really have fun with the XYL by telling her it is a tester for the presence of coffee in a cup. If the tester is inserted in a cup of coffee it will indicate with full brillance. With that, she will know for certain that you are completely "off your rocker" and really do need complete rest for six months in the Bahamas.

A $1 Strip Chart Recorder

In many instances a recording of a change against time can be of value in examining cause and effect. With recorders costing more than a 100 dollars, this advantage is not available to most amateurs.

This strip chart recorder, although of less than commercial stature, costs about a dollar and serves its purpose very well. It can be an interesting exercise in utilizing commonly possessed or available parts. The variations possible are myriad: direct clock motor drum drive, or through pulleys for mechanical reasons or to obtain the desired paper speed from an available motor. Similarly, the chart can be tensioned and paid off from its drum by simply attaching the loose end of the chart to a weight, or a more elaborate take-up may be devised by using another drum with an overriding rubber band slipclutch drive.

The choice in construction is facilitated by the wide range of drums available in practically every household—empty tin cans.

As an application example, a recorder is mounted above the main tuning dial of a receiver to record frequency deviations. The drum is an empty tomato can, 3¼" diameter, overhung from a 1 rph clock motor shaft, giving a chart speed of 10"/hr. The tuning dial motion is carried to the recorder pen by a hacksaw blade, which is ideally springy in the line of the pen, while being quite rigid laterally. The recording is made by constantly adjusting the receiver for zero beat, if the strength varies. If the signal strength is constant, a preferred method is to tune for a constant "S" meter reading on the side of the band pass skirt. The chart is attached to the drum with tape and prewrapped about six times. The loose end of the chart hangs over the back of the receiver and table, and is weighted to take up the slack as the chart is paid off the recorder drum

It is obvious that the receiver dial could be replaced by a variable resistor (pot) in a simple bridge circuit to record voltage or current. The XYL observes the bridge balance meter, preferably with a fixed magnifying glass and adjusts the dial on the resistor to hold the balance meter constant.

For convenience and a further exercise in applying junk box components, the bridge unbalance may be electrically sensed and used through an amplifier and small gearmotor to keep the bridge pot balanced and thus drive the recorder pen, as it is well known that many of the commercial recorders do.

The paper for the recorder may be cut from rolls of regular recording chart paper. If lines are not deemed necessary, plain paper or adding machine tape can be used. Pens can range from commercial recorder pens to fountain, ball or felt tip.

The little test instrument described here is something for the amateur who has nothing and something for the amateur who has everything. In the former case, it provides, very inexpensively, an instrument that can function as a continuity tester, transistor tester, diode tester, signal injection source, code practice oscillator, CW monitor, substitution microphone and substitution loudspeaker. In the latter case, it provides a very handy addition to a tool box, for quick continuity and relative resistance checks, without having to look at a meter.

The instrument is nothing more than an audio oscillator using a one transistor circuit. But the components are carefully chosen. A switching scheme is utilized so that a low current is passed through the circuit under test. The volume and pitch varies with the resistance placed across its test terminals, and maximum utilization is made of the circuit and its components for several modes of operation. Such basic testers, but without all the versatility of the one described, have been available commercially for years. They are popular with many service techniques, since one can visually concentrate on the circuit being tested or traced out without having to glance away to read a meter. This feature is particularly helpful when doing work on a detailed PC board, since one can lose one's place on the board in the time it takes to glance at a meter.

The circuit of the unit is shown in Fig. 2-19A. The oscillator circuit utilizes a transistor transformer, which has one or two center-tapped windings to form the equivalent of a transformer with three windings. One winding is used in the base circuit of the transistor, another as a feedback winding in the collector circuit and another as an output-coupling winding. Many of the usual miniature transistor transformers will work, aside from the TA-59 unit mentioned, such as the usual 10k ohm to 2k ohm CT or 1k ohm CT to 8 ohm units. One must be prepared to do a bit of experimenting to get the windings phased correctly and to get the output pitch desired. To achieve the latter with some transformers, it may be necessary to experiment with a small capacitor (.001 to .1 mF) across the base winding. The output "loudspeaker" should ideally be a unit such as a 600 ohm telephone receiver. But anything, from high impedance, miniature loudspeakers to cheap, dynamic-type microphones, can be used. Power is supplied by two 1½ volt batteries in series. No on/off switch is required, since no current can flow unless some resistance is placed across the test terminals.

The unit, as shown in Fig. 2-19A, can be used by itself, if desired. If the test leads are marked for polarity, one can test diodes and transistors and determine the direction of the junction involved. Resistance values, from a short to about 100k ohms, can be detected with the upper limit, depending on the specific oscillator components used. As the resistance value increases, the volume will decrease, but the pitch will tend to rise. This is a very handy feature, since, after a period of usage, one is not so aware of the volume changes as one is aware of associating higher pitch with higher

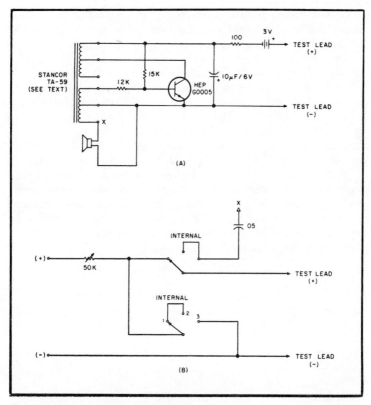

Fig. 2-19. Basic circuit of tester (a) and switching add-on for more versatility (b). See text for description of components not marked.

resistance. With usage, one can become familiar with the sound of at least the major steps in the output pitch, such as for resistance values of 1k and 50k.

By adding a few more components to the basic circuit, as shown in Fig. 2-19 B, more versatility can be gained from the unit. The addition of a series 50k potentiometer allows one to control the volume and also to limit the short circuit output current to less than 60 uA. The latter is useful as a safety feature, when testing some semiconductor devices, when one is unsure of the terminal markings. In the center position of the switch shown, the battery line is left floating and the positive test lead is connected to the speaker over a .05 mF capacitor. The speaker can then function as a replacement test speaker or as a dynamic microphone replacement. The reproduction quality is good enough to at least determine whether or not the speaker or microphone substituted for is basically defective. In the right-hand position of the switch, the battery circuit is completed to ground, and the internal speaker output remains connected to the positive test lead. In this mode, the circuit functions as an injection oscillator, the level of which can be controlled by the 50k potentiometer and monitored on the internal speaker. The output is quite harmonically rich, and it can be used to check amplifiers all the way from the audio range to the HF range.

One switch that could be used is a special miniature DPDT toggle switch with a center position. But, in the center position, instead of the usual "off" position, the poles still remain connected to opposite side terminals of the switch. The switch is available for $1 from Tri-Tek, 6522 North 43rd Ave., Glendale AZ 85301. The switch can, of course, be replaced by a regular 2P3T rotary switch, but then requires a larger enclosure. Using the miniature toggle and with the basic circuit wired on perfboard, the unit is assembled in a 3¼" × 2⅛" × 1⅝" Bakelite box, complete with batteries.

Probably some more uses can be found for the circuit, with a bit of imagination and a modified switching scheme. For instance, it would seem possible to rearrange things so that the circuit could also function as either a preamplifier or a low level audio amplifier complete with speaker. All in all, it is hard to find a more handy unit for general circuit or equipment checking, before one resorts to proper instruments for specific checks.

Chapter 3
Voltmeter and Ammeters

Build a Simplified, Sensitive Millivoltmeter

Here's a practical instrument suitable for home construction which is based upon an integrated circuit (NATIONAL type LM 4250 C). Switched ranges of 5 mv to 500v are provided (Fig. 3-1).

Construction. Layout has not proved to be critical in the prototype but input leads and associated switching have been kept compact; 2 percent tolerance resistors have been used throughout. Initial zero adjustment on the 5 mv range is set by selection of Rb in association with potentiometer R2 which is used for fine control. No zero drift has been observed on the prototype even on the most sensitive range.

An internal calibration check is provided on the 5v and 50v ranges by a pushbutton connection to the positive supply rail. This facility is used to adjust the meter deflection initially to 1.4v on the 5v range by means of pre-set resistor Rm. The value of the associated fixed resistor rf may also be varied if found necessary to suit the internal resistance of the meter used.

A miniature main switch serves to reverse the input polarity when required. The power supply comprises two mercury-button type cells (ER 675) contained within a small paxolin tube, spring loaded to maintain contact pressure. Other options are possible, of course. Consumption is minimal, being approximately 60 microamps for full-scale deflection. The unit is housed in an aluminum die-case box.

Input impedance is approximately the value of Rc for the range concerned, shunted by the input-lead capacitance (approximately 50k to 5 megohms, according to the range used). Overload protection is provided by two back to back signal diodes. Repeated application of 30v DC on the 5 mv range has had no detrimental effect on the prototype. External shunt resistors may be used to monitor DC currents down to quite low values. Figure 3-2 gives examples scaled for measuring 5 microamp and 0.5 microamp full scale deflection on the 5mv range.

Rf may be measured by the use of a suitable probe (Fig. 3-3). The approximate input impedance of this device is 6k at 2 MHz, falling to 1k at 25 MHz. The probe is usable to about 100 MHz.

Application. Since the instrument is sensitive and self-contained, it lends itself to a variety of tasks, some beyond the means of conventional multimeters. The very low power consumption means that battery life is very near shelf life.

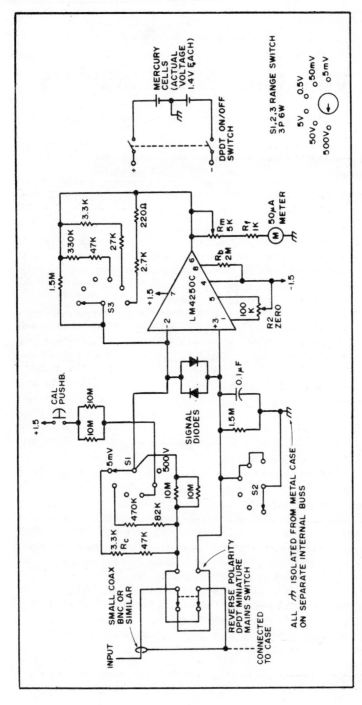

Fig. 3-1. Circuit diagram.

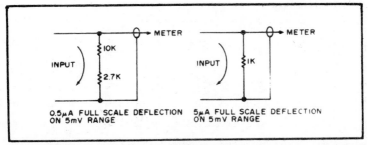

Fig. 3-2. External shunts.

The ability to read currents of less than a microamp enables diode and semiconductor device reverse and leakage currents to be read. With the millivoltmeter suitably insulated EHT tube currents can be measured (CRTs).

A Battery Voltage Monitor

A device introduced by Litronix, Inc., has wide application as a voltage monitor in all types of battery-operated equipment. The RCL-400 Battery Status Indicator is a current-controlled LED which has a voltage sensing integrated circuit incorporated into a small LED package.

The only additional circuit component necessary to build a voltage monitor is a suitable zener diode, or string of forward biased diodes, to bring the device into its normal operating range. The RCL-400 is designed to turn *on* at 3v and *off* at 2v; thus normal operation can be provided by selecting V2 = Vcc 3v (Fig. 3-4). When Vcc drops to Vz + 2v, the LED is switched off by the internal IC voltage sensing circuit to give a low voltage indication. Since the device has a relatively constant current demand in the *on* region (∼ 10 mA), the zener power rating need only be ¼w for most battery-powered equipment. One precaution is necessary: You must be sure that the voltage across the LED does not exceed 5v (its maximum rating).

For low voltage IC circuits using a nominal 4.5 v battery pack, the required value of Vz is only 1.5v. It is easy to obtain this value by simply substituting a pair of silicon diodes in series with the LED.

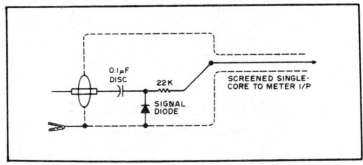

Fig. 3-3. Rf probe.

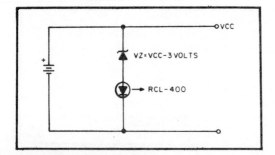

Fig. 3-4. Schematic.

The Easy Ammeter

Have you ever wanted to measure the current flow of a particular transistorized device that is sitting on your workbench?

Too many people sometimes bypass this important step due to the inconvenience of breaking the power lead and then having to resolder it again.

Here's a really easy way to measure current flow. Since most transistorized gear is powered from a battery pack of some sort, all you have to do is stick a piece of double-sided printed circuit board between any two batteries, as shown in Fig. 3-5.

Touch the two meter leads to each side of the PC board and there you have it—instant current reading without cutting and soldering any wires.

Adding An Ammeter to Your Car

In an effort to simplify the operation of automobiles, most car manufacturers have, for many years, incorporated an *alternator indicator light* on the instrument panel in lieu of the ammeter of days gone by. While this indicator light requires less attention while you are driving, it does not give you complete information on the state of your electrical system.

Perhaps your car should have an ammeter. Sometimes over a period of time, the output of an alternator steadily deteriorates and does not charge the battery properly. Eventually the battery can completely discharge and leave you stranded.

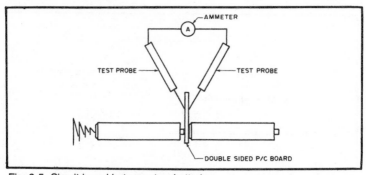

Fig. 3-5. Circuit board between two batteries.

You have to be very careful installing a new alternator because you could accidentally ruin a relatively new battery. For example, if during the entire time that the alternator was failing, the indicator light on the instrument panel would turn off during normal driving it would indicate that the alternator was "charging" and apparently, according to its indication, there would be no problem. It would be better to have some type indicating ammeter which would indicate the state of the system as to whether the battery was being charged or discharged. This section describes a method of installing an ammeter into your existing automobile system without any changes or modification to the existing system and at a very low cost. In fact, it is so easy and inexpensive to incorporate it into your system that you shouldn't be without one, especially if you operate mobile radio equipment which adds an additional drain to your car battery.

Advantages. This system has several advantages which make it attractive for an amateur to install in his car. These advantages are:

- All leads are at ground potential, so there are no problems with shorts in the system or the necessity of fusing the conductors.
- No heavy currents pass through the indicator, so small conductors may be used.
- Although the system makes use of shunt resistance to measure the current, a unique method is used so that no additional voltage drops are added into the present electrical system.
- It is very easy to add to your existing electrical system—no changes in existing wiring.
- The ammeter circuit can easily be adapted with a switch and resistor to read system voltage.
- It is inexpensive to install and depending on the type of meter movement you install, the cost can range anywhere from $1.50 to $7.00.

With these advantages, certainly many amateurs will want to install an ammeter to their car electrical system to monitor the battery charging rate.

Typical Electrical System. Referring to Fig. 3-6, a diagram for Ford Electrical System is presented. This system incorporates an electromechanical voltage regulator and is the type used in many cars (although other systems such as General Motors are similar). As you can see, the alternator consists of a stationary 30 stator winding in which the output goes to a set of diode rectifiers. These diodes are arranged in such a manner that the output at the battery terminal of the alternator is always positive. There is a rotating field which excites the stator winding during normal operation. As you can see, the alternator is always connected directly to the battery through a terminal block. It is at this point that the other electrical loads in the car are connected. The neutral from the stator winding is grounded through the field relay winding in the voltage regulator and its contacts are normally open. When the ignition switch is turned on, a certain amount of voltage from the battery is connected directly to the field through the charge indicator light on the instrument panel. After the engine is started, and the alternator is producing current, the stator current flowing through the field

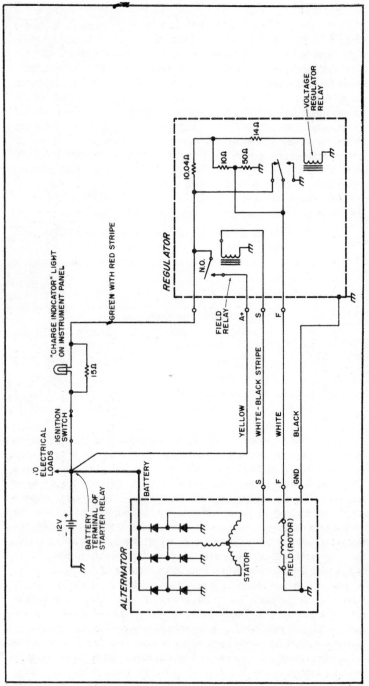

Fig. 3-6. Ford electrical system with electromechanical type regulator. (Other electrical systems are similar.)

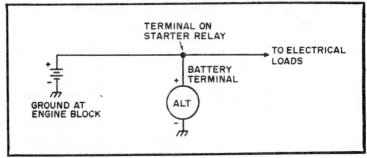

Fig. 3-7. Simplified diagram excluding the regulator.

relay winding pulls the contacts down which in turn bypasses the indicator light on the instrument panel. In other words, as long as the alternator is producing current of sufficient magnitude to hold the field relay contacts closed, the charge indicator light will be off, indicating that the alternator is producing current.

This system works fine as long as all elements in the system are operating under normal circumstances. However, as long as the alternator is capable of producing enough current to operate the field relay contacts, the charge indicator will be extinguished even though there may not be enough current produced to charge the battery. Under these circumstances, the charge indicator light does not give a true picture of what is happening in the system, and it also does not indicate the amount of discharge or charge condition of the battery.

Figure 3-7 is a simplified diagram of the battery-alternator connections without the regulator or field windings shown. As you can see, the alternator is continuously connected to the battery and to the other electrical loads usually at the terminal of the starter relay. During normal operation of a car, the alternator performs two functions. First, it produces enough current to operate the electrical loads in the car and second, it produces current to flow back into the battery to charge the battery during normal driving conditions. There may be a time when a faulty alternator does not produce enough current to do both, and it is under these circumstances that you may end up with a dead battery and no forewarning of that condition.

Also, from the diagram, it appears logical that the correct location for a charge indicator would be somewhere in the battery circuit, either between the positive terminal on the alternator and the positive terminal on the battery, or between the negative terminal on the battery and ground. This indicator should have a zero center scale so that it would indicate a discharge condition (battery drain) or indicate a charging condition from the alternator. The usual method of connection is to add some type of shunt into the system in order to sample the amount of current flowing in the conductors to provide an indication. However, if a shunt is added into the circuit (such as the ¼ ohm shunts which are used on most automotive ammeters you can purchase at a store) you incorporate an additional voltage drop into the system.

Ground Paths. Figure 3-8 indicates a simplified electrical diagram show-ing electrical loads connected to the system. This is a discharge condition in that the engine is stopped and there is no output from the alternator. Since the alternator remains connected to the system, it would appear that some drain would also result from the alternator. However, keep in mind that the alternator windings are connected to the positive terminal through diodes, and therefore, no reverse current can flow through the alternator. Under this condition, the alternator current $1_A = 0$. Applying Kirchhoff's Law which states the sum of the currents entering a junction is equal to the sum of the currents leaving the junction, we have the battery current 1_B entering the terminal and the load current 1_L leaving that junction. Since the alternator current is equal to zero, $1_B = 1_L$, simply stated, the battery is supplying all of the required power to operate the load.

Let's discuss for a minute the theoretical versus actual conditions. We normally think of the negative terminals of the battery, alternator and loads as all being grounded together so that all negative terminals are at the same potential. However, in most cases, we are depending on the frame of the car as a ground return path, and since steel is a mediocre conductor, there is always some inherent resistance incorporated into all automobile electrical systems. This resistance is represented as R1 between the battery negative terminal and the alternator negative terminal and R2 between the alternator and the electrical loads. Applying Ohm's Law, the voltage drop across R1 would be:

$$E = (I_B)(R1)$$

Granted, the resistance of R1 is small, in the neighborhood of 0.01 to 0.1 ohms, but the currents are very large—in the neighborhood of 10 to 30 amps

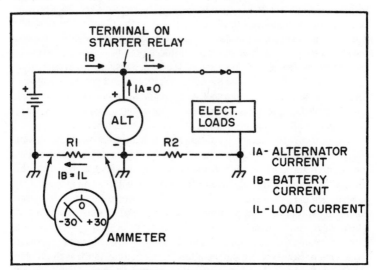

Fig. 3-8. Basic electrical system showing resistance in ground path circuit-discharge condition (engine stopped and electrical loads connected).

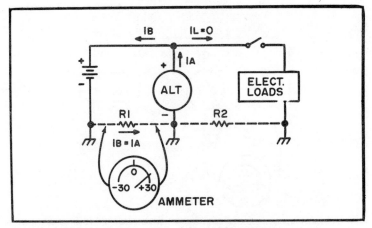

Fig. 3-9. Charging condition-engine running with electrical loads off.

(except during starts when it is much higher). Consequently, the voltage drop across R1 is a measurable quantity and can be used to indicate which way the current is flowing in the battery-alternator loop. This is shown on the meter in Fig. 3-8 as a discharge situation.

Figure 3-9 is a diagram of a basic charge condition in which the engine is running and no loads are connected. Again, applying Kirchhoff's Law with $1_L = 0$ we have: 1_B is equal to 1_A, and the voltage drop across R1 is $E = (1_B)(R1)$, but is of the opposite polarity of the discharge condition in Fig. 3-8. As the battery becomes charged, the alternator current charging the battery gradually tapers off and so will the voltage drop across R1. We get a direct indication of the condition of the battery and we know exactly when it has reached full charge condition.

In Fig. 3-10 there is a diagram showing normal operation of the vehicle in which the engine is running and certain electrical loads are connected. Under this situation, again applying Kirchhoff's Law, we have 1_B and 1_L leaving the terminal and the alternator current, 1_A, entering the terminal. Stated as an equation, $1_A = 1_B + 1_L$. This means that the alternator has to supply enough current to operate the electrical loads and charge the battery at the same time. Of course, as the battery becomes fully charged again, the battery current will gradually taper off, and the alternator will be operating mainly to supply current to the electrical loads connected to the system.

Ohm's Law Applied. It is apparent that there will be some voltage drop which can be measured across this ground path resistance, but just exactly how much will be measured and how can it be utilized in a charge indicator?

Let's take for example, that the ground path has a resistance of only 0.01 ohms and that a typical alternator is capable of producing at least 30 amps. The voltage drop across the ground path of resistance would then be:

$$E = IR = (30)(.01)$$

Therefore,

$$E = 0.3 \text{ volts}$$

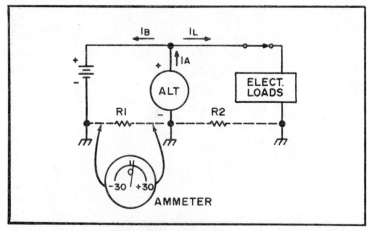

Fig. 3-10. Typical operating condition with battery charging current tapering off.

This voltage is easily measured and the instrument may be calibrated in terms of amps, that is, 0.1 volts would indicate a current of 10 amps.

If the normal charging rate is, for example, 10 amps, the voltage measured would be 0.1 volts. From this example, it is clearly evident that there is plenty of voltage available for a sensitive indicator and it can be used to advantage in a charge indicating instrument without adding additional shunt resistance into the electrical system. The next problem is to determine how and where to connect the meter in the circuit.

Basic Ammeter Circuit. Figure 3-11 shows the actual circuit used in a car with a 50 uA meter. This instrument, as with most instruments (even the very inexpensive tuning indicator instruments) can be converted to a zero center scale instrument so that the meter now has become a ±25 uA meter. A meter protector circuit was installed across the terminals to limit the current during starting, and a 2k resistor was added into the circuit to allow for calibration of the meter.

As mentioned, the basic movement that was used was a 0-50 uA meter manufactured by Midland, Model No. 23-206. However, any meter may be used, possibly up to a 0-1 mA meter, depending on your own particular

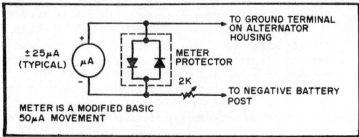

Fig. 3-11. Actual diagram for adding an ammeter to car electrical system—no changes are required to existing wiring.

situation. You may want to use a VOM to check your particular ground path circuit under normal conditions to find out how sensitive a meter movement will be required in your particular case. After this has been determined and you have a meter available, the next step is to carefully remove the plastic cover on the meter. Then, after removing the two retaining screws of the dial, it must be carefully slipped out from under the pointer. Then, all that is necessary is to erase the lettering not required and to add in new lettering according to your own particular circumstances.

The next step is to convert the instrument to a zero center scale type instrument. This is easily accomplished by moving the zero adjustment until the pointer is at mid-scale. It is possible to do this on almost all instruments, including the little horizontal movements that are available through surplus outlets for $1.50. The quality of the instrument is governed only by your own particular needs. The final step is to replace the dial and plastic cover and install the protective diodes across the meter terminals along with the multiplier resistor used for calibration.

Installation Details. The next step involves installation in the car and connecting the meter leads to the proper points in the electrical system. It is best to use solid conductor wire of approximately 18. gauge, and if you can find Teflon-coated conductors, so much the better. Since the leads will be near hot parts of the engine, Teflon-insulated conductors are more resistant to heat and grease in this type environment. The two leads from the instrument circuit should be routed through any available opening in the fire wall and around to the alternator and negative terminal of the battery. One conductor is connected to the ground terminal on the alternator housing. The other conductor is connected to the negative battery terminal right at the battery post. To check to see if you have the right polarity, leave the engine off and turn on some load such as headlights or press on the brake pedal. The instrument should swing from the zero center scale to the left indicating a discharge condition. If it swings the other way, the leads should be reversed. The next step is to get a rough calibration on the instrument. Turn on headlights or some other load with a known amp rating and roughly calibrate the instrument based on this amount of load. If you have another ammeter, you can use it to check and calibrate your new instrument.

It is imperative to connect the ammeter circuit into ground path R1 between the battery negative terminal and the alternator ground terminal. Otherwise, if by mistake it is connected between the battery negative terminal and ground near some of the electrical loads, it is easy to see from these basic diagrams that you would get roughly twice the indication on discharge which is the sum of the voltage drops across R1 and R2 in a discharge condition. In a charge condition, the indication will be the difference between the voltage drops across R1 and R2 since the voltage drops are in opposite directions. This obviously will give you false indications and will be useless as far as determining the state of charge of your battery.

Conclusion and Results. You are probably wondering at this point as to whether or not it would have been easier to buy one of the commercially available ammeters for 6 to 10 dollars and not have the problem of converting another meter for use in this project. It is true it may have been easier, but the meter movement in commercial automotive instruments is not known for its quality and there are usually no means of calibrating the commercial instrument. Also, most commercially available instruments make use of an additional meter shunt into an electrical system of approximately ¼ ohm. Admittedly, this is not very much resistance to add into a circuit, but a quarter ohm at 20 amps could introduce an additional voltage drop into the automobile electrical system.

Simple Remote Ammeter Using a Bus-Bar Shunt

This is really a simplified method for cutting ammeter shunts. There are two things that will not happen. It will not be necessary to do any deep math, and the meter will not get pinned during the cutting process.

The voltage drop along a piece of wire is proportional to its length and the current through it. If a piece of heavy wire is hooked up as shown in Fig. 3-12, all of the sport is taken out of making meter shunts.

A power supply, ammeter, resistor and the bus bar that is about to become a shunt are all connected in series. Long thin flexible leads from the 0-5 mA meter may be used. After all, they are going to carry no more than 5 mA. The negative lead is tied to the power supply and bus minus. The power is turned on and adjusted to give the required current. This is shown on the standard VOMQ. The plus lead of the 5 mA meter is touched to the bus *near* the minus lead. There should be a small deflection seen on the meter. The lead is then slid along the wire until full scale deflection is obtained. That point on the wire is marked.

If a single range is all that is required, it only remains to solder the leads in place. The resistance change at the point of soldering will be small in comparison to the resistance of the meter. Therefore, the soldering operation will not upset the calibration.

Leave a little more bus at each end than is needed. If the shunt is rather long, then slide insulation over it before the final soldering. The wire may be wound on a convenient coil form in order to make it somewhat more compact.

If a 0-20 amp range is needed and there aren't more than 2 amps available, or the shack VOM doesn't have a high amp range, all is not lost. Set things up as shown in Fig. 3-12 and adjust the power supply for, say, 2 amps. Then slide the plus lead for 20 divided by 2 or 1/10 of full scale. This

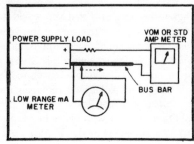

Fig. 3-12. Setup for shunting meter.

will generally prove to be adequate calibration if the standard is reasonably close. Later, when the high current supply is finished and a higher range standard is available, it will take almost no effort to touch up the calibration.

If the shunt turns out to be 10″ long for 500 mA, then halfway there, or 5 ″ up from the cold end will give a 1000 mA or 1 amp shunt.

A multi-range ammeter (Fig. 3-13) was constructed with this method in less time than it used to take to make a single shunt. Again, since the meter and its leads are carrying only 50 uA or 5 mA (whatever the basic movement is), then a cheap multi-position switch and light hook-up wire will do the job. Number 22 wire may be used for shunts up to about 1a.

This should get some of those low range mA meters out of the junk box and into circulation. And isn't that like getting a new meter for an old one? Although a 5 mA meter was used in this example, other meters may be used.

Find That Meter Resistance with This Simple Bridge

There comes a time in every hobbyist's life when he must seek that unknown meter resistance. Here's a simple solution to that age-old problem. The schematic is shown in Fig. 3-14. It's equivalent circuit is shown in Figs. 3-15A and 3-15B.

In Fig. 3-15B, R2 is equal to R2, and R$_{BP}$ is the equivalent parallel resistance of branch 1 and branch 2. Neglecting R$_{BP}$, the current through R2 would be 1.5 (E)/1500 (R) = .001 A or 1 mA, the full-scale reading of most meters. Thus, when we reinsert R$_{BP}$, we know that the current is less than 1 mA. This keeps the current through each branch (Fig. 3-14) less than 1 mA, protecting both meters.

In Figs. 3-14 and 3-15A, when the resistance of branch 1 is equal to the resistance of branch 2, the currents through both are equal. Thus, you know that when the reading on the meter under test and the current reading on your meter are equal, the resistances of the two branches are equal.

The resistance of branch 1 is equal to the resistance of M1 (which *must* be known) plus R1, a potentiometer with a calibrated dial. If we select R$_{M1}$, then, when R1 is equal to R$_{M-test}$, the resistances of the branches are equal. If the resistances of each branch are equal, the currents through them are equal.

To find the meter resistance, one must plug in the meter under test and rotate R1 until the currents through both meters are equal. Then we know that R1 = R$_{M-test}$ and its resistance can be read directly off the calibrated dial.

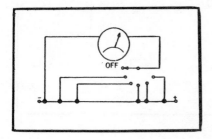

Fig. 3-13. Multi-range ammeter.

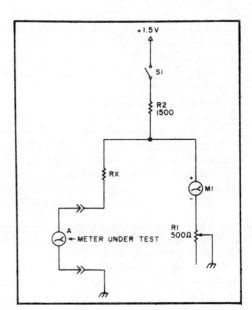

Fig. 3-14. Schematic.

The smaller the value of potentiometer R1, the more accurate is the measurement of R$_{M\text{-test}}$. This is because the dial can be calibrated in smaller units.

As an option, a more accurate circuit is shown in Fig. 3-16. A rotary switch can select different values of resistance to be added to R1. Thus, R1 can be made as small as you wish. R$_{M\text{-test}}$ is now equal to R1 plus the switched-in resistance.

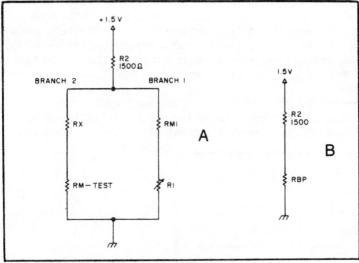

Fig. 3-15. Equivalent circuit.

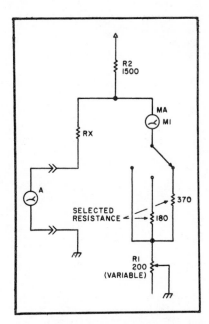

Fig. 3-16. More accurate circuit.

Let's say you wanted to measure a meter's resistance using only R1 (Fig. 3-14). If your dial was calibrated with 100 notches, the result would be 5 ohms per notch. If we use the circuit in Fig. 3-16, the potentiometer is only 200 ohms, leaving 2 ohms per notch on the same calibrated dial. Thus, we see how there is more accuracy in a circuit such as the one shown in Fig. 3-16.

Choose a meter with a low resistance. Also, if you prefer, you can use an ohmmeter to read the resistance of R1, thus saving yourself the trouble of finding a calibrated dial.

As you can see, the circuit is a flexible one and can be customized by the builder. All that is needed is a pen, paper and E = 1R.

Chapter 4
Capacitance Meters

A time-honored and very practical supplier of components for the latest experimenter's project continues to be the junk box. A well-stocked junk box not only reduces the cash outlay of a project, but also serves as a source of comfort and inspiration to the happy owner.

A major factor limiting the utility of these readily available goodies is the difficulty of identifying them properly. The number of books describing transistor checkers, IC probes and programs for identifying and checking ICs points not only to the popularity of the junk box source of supply, but also to the difficulty mentioned above.

Capacitors are a part of this problem. It seems that the original equipment manufacturers, the source of many of these components, delight in concealing the true value with an esoteric part number.

Cheer up! All is not lost. This is an unbelievably uncomplicated and cheap device that will go far toward blowing the cover of all those mysterious micas, discs and ceramics smirking at you from your hoard.

The Circuit. This little device is ridiculously simple and inexpensive. It consists of an oscillator and a rectifier, with a meter to indicate the value of the rectified current. Let's refer to Fig. 4-1 and be a bit more specific. U1A and U1B are two NAND gates of a CMOS quad two-input NAND gate. You can get one for $.29 if you don't have one. These two gates are cascaded, biased in the linear mode by the resistor network and caused to oscillate by capacitive feedback around the two gates. The output is a square wave which is buffered by the third gate U1C to insure an output of constant amplitude. We apply the output of the buffer to the diode rectifer through our unknown capacitor. If the unknown capacitor is small (has high reactance) in comparison with the frequency of the oscillator, it will pass a small pulse on each cycle which will be rectified by the diodes. The meter will read the sum, or integral, of these pulses. The larger the value of the capacitor, the larger the pulses, the greater the sum and the higher the meter reading. Simple, isn't it?

This is a basic counter circuit. With a given value of capacitor, if the oscillator frequency is increased, there will be more pulses per second, thus a higher integrated meter reading. This forms the basis for a very simple and useful counter which is linear over quite a few octaves. We are simply turning the circuit around to measure capacity instead of frequency.

Circuit Details. The circuit was set up to use a 1 mA meter movement. The values shown in Table 4-1 list the oscillator frequency, capacitor and resistor values for a 1 mA meter movement. There are five ranges arranged as decades. The lowest range is 0-100 pF, the next 0-1000 pF, etc. Each range is linear, so it is quite possible to read a 5 pF capacitor on the lowest range. The highest range then reads full scale on a 1 uF capacitor.

It would be inadvisable to use a meter with less sensitivity than 1 mA if the 0-100 pF range is desired. Note that with the 1 mA meter, the oscillator is running at a frequency slightly greater that 1 MHz. This is approaching the top frequency for this chip in this circuit. Conversely, if a sensitive meter such as 50 uA is available for this use, it should be desensitized to some extent if the 1 uF range is desired. With a 1 mA meter, the oscillator frequency is only 109 Hz on the 1 uF range. The 50 uA meter would require this frequency to be reduced to 5 or 6 Hz for this range, which would require awkward capacitor values in the oscillator and an excessively large smoothing capacitor to provide a steady reading on the meter. Approximate oscillator frequencies for meter movements from 50 uA to 1 mA may be interpolated from Table 4-1.

Construction. Table 4-2 lists the necessary parts. The instrument was housed in a discarded multimeter cabinet. One of the many miniboxes available would be entirely satisfactory. The circuit was constructed on a piece of perforated board which was mounted on the meter studs. Trimpots were used to set the battery check and individual capacitance ranges. The 5 pF capacitor for the 0-100 pF range was soldered directly across the IC

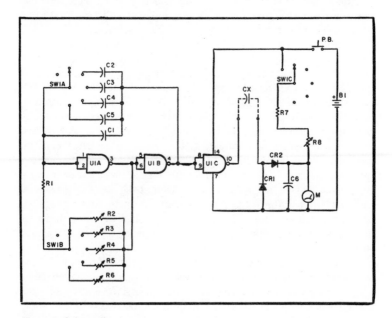

Fig. 4-1. Schematic.

Table 4-1. Values for a 1 mA Meter Movement.

Range	Total R	C	Frequency
0-100 pF	15k	5 pF	1100 kHz
0-1000 pF	31k	100 pF	112 kHz
.01 uF	36k	1500 pF	11.2 kHz
.1 uF	45k	.012 uF	1.170 kHz
1 uF	45k	.1 uF	109 Hz

socket and is in the circuit at all times. A five-position two-pole rotary switch is required. We added a battery check position on the switch, which required a third pole and sixth position on the switch. A push-button for the battery check function would do as well. The two leads to the binding posts of the unknown capacitor should be routed in the clear and *not* twisted together for neatness; their mutual capacity will show on the low range!

Calibration and Operation. When the unit is completed, it should work just fine, but it probably will not. First check to make sure the gates are oscillating. Since they are biased in the linear mode, all gates and outputs should be at about half the battery voltage. If they are at zero of full voltage, they cannot oscillate. If they are at about half-voltage, they just about have to oscillate if the circuit is wired correctly. When oscillation has been proven, the diode circuit will rectify and indicate on the meter, unless a diode is reversed or there are other wiring errors. This is not a cranky circuit.

After you find the wiring error and get the thing working, the rest is a snap. With a fresh battery, set the battery check trimpot for a full-scale reading on the check position of the switch. In spite of the junk box complex, most of us have *some* capacitors that are marked. Calibration was begun on the most sensitive range. Select three or four 100 pF capacitors and set the trimpot for a full-scale reading. Micas are usually pretty close and very little discrepancy will be observed.

Also, check some smaller values, such as 50 pF, 10 pF and 5 pF to make sure that the scale is linear and all is well. Using the 100 pF capacitor, decade

Table 4-2. Parts List.

B1	9 volt transistor battery
C1	5 pF mica
C2	100 pF mica
C3	1500 pF mica or mylar
C4	.012 uF paper or mylar
C5	.1 uF paper or mylar
C6	.1 uF
IC1	CD 4011 quad NAND gate
CR1	1N34 or equiv.
CR2	1N34 or equiv.
M	1 mA meter
R1	12k ¼ or ½ W.
R2	50k trimpot
R3	50k trimpot
R4	50k trimpot
R5	50k trimpot
R6	50k trimpot
R7	5k ½ W.
R8	10k trimpot

up to the next range and set the trimpot for 10% full scale. Now try a 1000 pF. It should read pretty close to full scale.

In short, if you have a selection of known capacitors, you will be able to set each range without difficulty. Note that once you have proven the scale is linear, you can use intermediate values to calibrate. A .5 uF capacitor can be used to set the 1 uF range, for example. It is best to try three or four different capacitors of different types for each range. You will find that consistency can be achieved and the calibration is sufficiently accurate for all practical purposes. If you are a nitpicker and have a friend with a good bridge, you can achieve good accuracy.

Operation is very simple. Place the capacitor across the binding posts and depress the button. Read the capacity on the meter. If the meter tries to go beyond full scale, go to a higher range. If the meter reads less than 10% full scale, turn to a lower range. That's all there is to it! The circuit is self-limiting; the meter may bang over full scale if the capacitor is too large for the range selected, but the current supplied will *not* burn out the meter. No current is drawn from the battery unless you depress the push-button, so the battery cannot be run down because you forgot to turn the thing off. Current drain while checking is less than 5mA.

Build This Digital Capacity Meter

One of the common problems any constructor faces is determining the exact capacity of a capacitor from his junk box or for use in af filters, etc. Small (and cheap) RLC bridges are not accurate enough and if, for example, you try to measure 100 mixed values of capacitors from a surplus pack, you may well run out of patience. Furthermore, in the age of integrated circuits and frequency counters, everybody is getting lazy. Well, here is an instrument which eliminates all these troubles, presents the value of capacity instantly in digital form, is highly accurate and is fairly simple to construct.

First, some information about the C-meter:

- It measures capacitors from 1 pF to 1 uF in two ranges—9999 pF and 999.9 nF.
- Display is four digits in these ranges, with leading zero suppression and overflow indicator.
- Accuracy is better than ± 1% of full range ± 1 digit for higher values of capacitance in both ranges; for lower values of capacitance it is still very good (i.e., it is possible to determine if the measured capacity is 1 or 2 pF). This accuracy is of course in relation to the standard used for calibration of the C-meter and applies only to capacitors with a good Q.
- No warm-up period is required. It measures immediately after switching on, with full accuracy.
- Operation is extremely simple, with only two controls: zero adjustment and range switch.
- With the exception of a power supply, the whole unit fits on two small printed circuit boards.

- The price of the complete unit should be around $50, depending on how familiar you are with cheap sources of TTL integrated circuits. Maybe most of them are in your junk box.

The basic principle is very simple and well known; the only difference is *how* it has been used. There is a similar unit, but it measures only higher values of capacitance and uses different devices.

Figure 4-2 shows the block diagram of the C-meter. The heart of the unit is the monostable multivibrator (MMV), which produces gating pulses. The length of these pulses is directly proportional to the value of capacitor C_x. Consider the practical limits of your MMV integrated circuit, the crystal oscillator and the frequency counter. The MMV integrated circuit used in this case is number SN74121. Using only C_x and no C_F, with the highest permissible value of the timing resistor, will result in a pulse length of about 25 usec. The crystal oscillator frequency must then be 40 MHz in order to obtain "1000" on the display. But, if due to the long leads to the terminals for C_x, hum, etc., the display is not stable, especially for values of C_x under 100 pF, start with a value for C_F of 1000 pF. This should cure the problem. With no C_x on the terminals there will be "1000" on the display. Instead of resetting to "0000," reset the counter to "9000;" then with no C_x it will count to "10000," but the first digit will not be displayed and the resultant display is "0000." Similarly, for higher range, the counter can be reset to "9990." Reset circuits are then a bit more complicated and the overflow indicator must be a two stage counter, but this is not a serious complication.

Capacitance ratio between range 1 and 2 is 1:100. To obtain a correct reading on range 2, the frequency of 40 MHz (for range 1) must be divided by 100, resulting in a frequency of 400 kHz.

The last part of the C-meter is the timing circuit. It generates trigger pulses for the MMV, strobe pulses for latches and reset pulses for counters. The last ones are distributed with the help of a few TTL gates.

Flicking of the least significant digit is not suppressed for two reasons: to simplify construction and to enable recognition of differences between, say, 17.0 and 17.5 pF. In the latter case, the display would change from 17 to

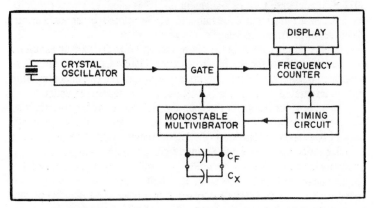

Fig. 4-2. Block diagram.

18. Using cheap Minitrons for display determines the frequency of the timing circuit; it must be low enough to read both values comfortably, e.g., 17 and 18. After a few tests leave it at about 2 Hz. The timing oscillator is then running at 20 Hz and is divided by 10 with the SN7490. Pulse "4" is used for triggering the MMV, "6" is used as strobe pulse for latches and "8" is used for reset pulses. This system works very well and leaves about 100 ms between trigger and strobe. With 1 uF and 40k ohm, the SN74121 generates about 25 ms pulses, so that it is well within the above 100 ms. Any other combination of the outputs from the SN7490 is, of course, possible (Fig. 4-3).

The instrument can also be divided into two parts:

- Crystal oscillator, MMV, timing circuits and gate.
- Frequency counter and display with overflow indicator.

This arrangement reduces the size of the printed circuit boards which are mounted in parallel and interconnected by means of a few wires. Both parts can be tested separately before final assembly of the whole C-meter.

Frequency Counter and Display with Overflow Indicator. This unit is mounted on a single-side printed circuit board 110 by 80 mm. Using a single-sided printed circuit board for such a complex circuit results in a few jumpers being necessary. However, the use of double-sided printed circuit board would not be fully justified—it would be much more difficult and expensive.

This unit is basically a four-stage frequency counter with overflow indicator, which reacts only to every second pulse from the last decimal counter. Decade counters used are SN7490, with the exception of the first one, which must handle frequencies around 40 MHz. An SN74196 is used here instead, as it can handle frequencies over 50 MHz. Latches SN7475 and BCD-to-seven-segment decoders SN7447A are used to drive 7-segment incandescent displays (Minitrons) of a cheap foreign make, type 3015-F. If any other type of 7-segment display is used, you must check to see if the pin arrangement is the same. If not, the printed circuit board must, of course, be modified accordingly. The overflow indicator uses an SN7473, one surplus plastic switching transistor and an LED (TIL209 or similar).

Use IC sockets for the Minitrons only (MOLEX type); all the other integrated circuits are soldered directly into the printed circuit board. With surplus integrated circuits this is a bit risky. The decision is yours.

There are four resets for the counters, as well as one for the overflow indicator and strobe for the latches. Two wires are for switching the decimal point of the second Minitron for the higher range.

Oscillator, Gate, MMV and Timing Circuits. One of the most successful of harmonic oscillators is shown in Fig. 4-4. It is reasonably stable and the output voltage is high enough to drive a buffer stage, which drives high speed gate IC14A (SN74H00). Crystal X is a harmonic type (3rd or 5th harmonic) and *any* frequency between 35 and 45 MHz is suitable. Coil L1 has

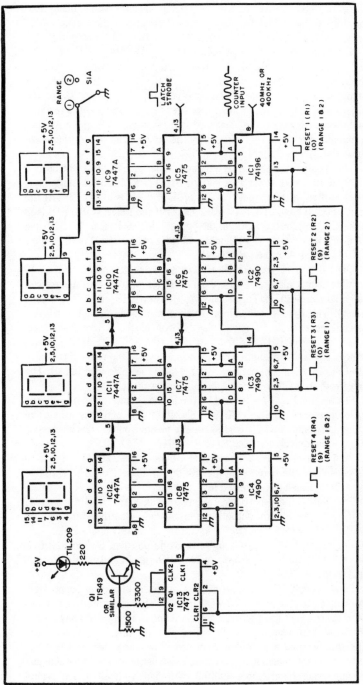

Fig. 4-3. Note: Cut off pin 4 of ICs 2 and 3 (SN7490).

10 turns of 28 SWG enabeled wire on a 5 mm diameter with a tuning slug. Capacitor Co tunes with L1 to the harmonic frequency of the crystal X. The coupling coil is 2 turns of insulated wire (24 SWG) over the "cold" end of L1.

The buffer stage Q3 must safely drive the buffer gate IC14A. The easiest way to check if IC14A is driven enough is to measure the voltage with an rf diode voltmeter at pin 11 of IC14. If the rf voltage is low or unstable, change the value of the base resistor of transistor Q3 or the transistor itself. Unstable or low rf voltages (e.g., under 2v) will result in unstable or no display of the measured value of capacity. So get it to work one hundred per cent. This is probably the only difficult part of the whole instrument.

The rf output from IC14A is fed into another gate IC (14B) and divided by 100 (IC15 and IC16). Range switch S1B blocks IC16 for range 1 via IC20E, so that at the output (pin 11) of IC16 there is logical "1." At the output of IC14C (pin 3) is a 40 MHz signal. If the range switch S1B is in position 2, gate IC14B is blocked, the output (pin 8) is logical "1," IC15 and IC16 divide the 40 MHz by 100 and the 400 kHz output (pin 11 of IC16) goes via IC14C to pin 4 of the gate IC14D.

One half of IC17 (dual Schmitt trigger) works as an oscillator; a 47 uF 10 V solid tantalum capacitor and a 470 ohm resistor make up the timing circuit for approximately 20 Hz. This frequency is fed into IC18 (an SN7490 decade counter). The B, C and D outputs of IC18 are used to produce trigger, strobe and reset pulses. Trigger pulse "4" from pin 8 of IC18 is reshaped in the second half of IC17 and fed into IC21 (the SN74121 MMV), which produces a pulse with length directly proportional to the value of capacitor Cx (+CF). This pulse (at pin 6 of IC21) is used to enable gate IC14D. The output from IC14D (pin 6) goes to the counter input.

Strobe impulse "6" is formed by the B and C outputs from IC18 (the decade counter) in the gate IC19B (SN7400). The output is a negative impulse and, as the SN7475 latches require positive pulses, it is inverted in one of the hex inverters (IC20B) of SN7404.

Reset 1 and 4 pulses are common for both ranges. The D output from IC18 goes via IC19A to Reset 1 (negative pulse). Reset 2 must be a positive pulse—thus, Reset 1 is inverted in IC20A.

For range 1, the Reset 3 pulse must be positive. With S1B in position 1, gate IC19C is open but the output pulse is negative —it is therefore inverted in IC20C. The Reset 2 output is logical "0", as IC19D is blocked.

For range 2, the Reset 2 pulse must be positive. Switch S1B is in position 2 and gate IC19D is open, while IC19C is blocked and the Reset 3 output is logical "O."

Power Supply. The last part of the C-meter is the power supply. It is very simple, thanks to IC22 (MC7805CP). This integrated circuit has built-in overcurrent and thermal protection and is a very good buy for the money (Fig. 4-5). The output voltage should be very close to +5v (according to the specifications ± 0.2v, but there are a few worse than ± 0.1v). It is rated for 750 mA minimum (short circuit current limit) and this value just suits our requirement (about 700 mA with display "8888'"). The typical

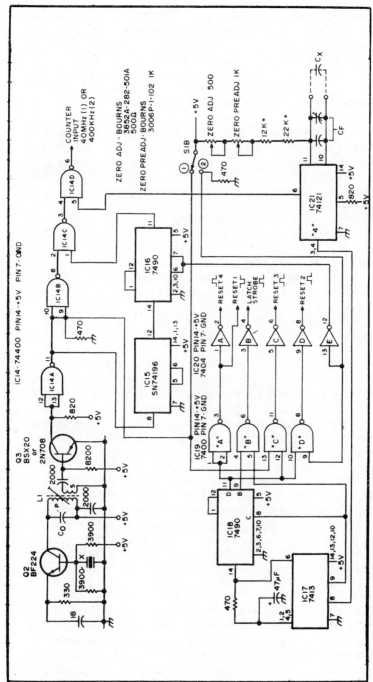

Fig. 4-4. Schematic

77

input-output voltage differential of the MC7805CP is 2v, but to be on the safe side use 8v AC. With a large smoothing capacitor the input voltage is over 9v DC. Do not try to go under this value (at nominal mains voltage); otherwise, when the mains voltage drops, you are in trouble.

A small capacitor across the output improves transient response. The connection between the 5000 uF and the input of IC22 should be as short as possible. Use heavy wires—they eliminate voltage drops and improve stability.

To suppress spikes from TTL integrated circuits, it is essential to connect a few capacitors of 0.1 uF and 50 or 100 uF tantalum at intervals on both printed circuit boards. The 0.1 uF capacitors are Siemens, type B32540-A3104-J metalized polyester. Their advantage is very low inductance, but even the ceramic disc capacitors will do the job. Be careful about the polarity of tantalum electrolytics.

By the way—the IC22 regulator needs a heat sink to dissipate about 3w. A small piece (100 x 100 mm) of 2 mm (approximately 14 gauge) aluminum is more than adequate; if a metal box is used, just mount the regulator on the rear side of the box.

Construction. Mechanical construction is probably the weakest point of most of the equipment. Not everybody has a well-equipped mechanical workshop.

But C-meters can be built in your home. One can be built into a homemade cabinet. The basic dimensions and the general idea are shown in Fig. 4-6.

U-shaped parts *a* and *d* can be bent in a big vise. The other parts are very simple. Aluminum (2mm, 14 gauge) can be used for all parts of the cabinet. Start with parts *a*, *b*, *c* and *d*, and then measure the inside dimensions and cut front and back panels *e* and *f* accordingly.

The front and back panels are held 94 mm (inside dimension) apart by means of four 4 BA screws 102 mm long. This sub-assembly is then inserted between the two U-shaped parts *a* and *d*. Parts *b* and *c* prevent the front and back panels from moving and hold parts *a* and *d* together.

After drilling all necessary holes into the front and back panels, clean the front panel and mark all holes with a pencil on a piece of thick white paper which does not change its dimensions with moisture. A white plastic sheet would be even more suitable. Cut off all the marked holes with a sharp knife. Use transfer letters (Letraset or similar) for descriptions of the functions on the panel. Then cut a 1.5 mm thick piece of clear Perspex (Plexiglas) and drill all holes with the exception of the ones for display LEDs and overrange LED. Cut a piece of red Perspex (3 mm thick) which will fit exactly into the hole for the LED display in the front metal panel *e*.

Perspex is supplied with paper sheets glued on both sides. Do not remove these; mark all the holes and outside lines with a pencil and then make cuts with a sharp edge along the outside lines, about 0.2 mm deep. Use the sharp edge of a small chisel (along a steel ruler) with success. Do only one cut at a time. Break both ends of the cut slightly with your fingers. Lay the Perspex over the edge of the bench and break carefully. Try this

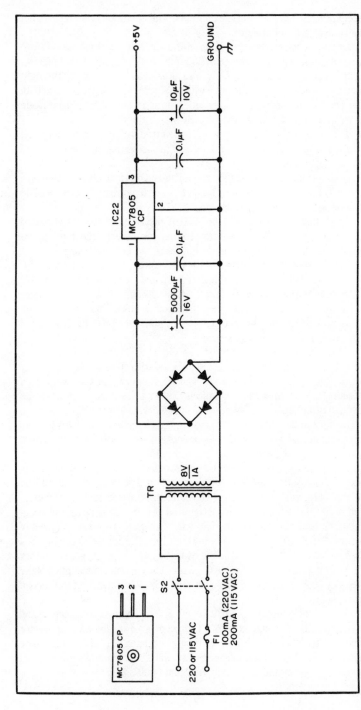

Fig. 4-5. Note: MC7805CP mounted on a heat sink.

79

procedure a few times on Perspex off-cuts. The operation is not difficult but does require a bit of practice.

Drilling of holes in Perspex is also a little difficult. The best method is to use a drill press with a maximum of 500 rpm. A bit of practice on a piece of off-cut is not a waste of time. Only after drilling all holes, cutting the right dimensions and smoothing the edge with sandpaper should you remove both paper covers and wash the Perspex with soap and water. A front panel made this way looks professional and is really worth the effort.

The printed circuit board with LEDs is fitted parallel to the front panel by means of four screws with countersink heads so that they are covered with the paper and not visible.

Both terminals of Cx must be made from a good insulator or fitted on a piece of Teflon (PTFE). Perspex tends to absorb moisture from air—a bad insulator will cause erratic readings.

The transformer, IC22 and the 5000 uF electrolytic capacitor are fitted onto the rear panel. Use a small transformer (or increase the depth of the box), so that it does not interfere with the printed circuit boards.

For the second construction of the instrument, you could use a ready-to-use plastic cabinet with a metal front panel. This is the lazy man's construction and it is, of course, less mechanically difficult. Due to the bigger size of the box, it does not need the cooling holes that proved to be necessary in the previous construction. The construction of the front panel is the same as in the previous case.

Fiberglass printed circuit boards should be well washed with soap and water after etching, rinsed, dried and polished with steel wool. Then spray both boards with soldering varnish and let them dry in the oven. This protection is very important. Otherwise, after a few years you might be surprised to find that your C-meter does not work. Only after this procedure can you start drilling the holes into the printed circuit boards. Holes 0.8 mm in diameter are adequate for all components and wires, except the five thick interconnecting wires for the ground and +5v.

Soldering should be done with a small soldering iron—preferably temperature controlled. Start with the jumpers; a few long ones should be insulated wire, while the others can be bare wire 0.6 to 0.8 mm in diameter. Then solder the components and integrated circuits. Be sure that all integrated circuits are positioned the right way. Do not forget to cut off pin 4 of ICs 2 and 3 (SN7490), before soldering them onto the printed circuit board.

You can test each printed circuit board separately. Use your laboratory 5v power supply or the one for the C-meter. The current consumption of the frequency counter part with display is approximately 500mA; of the second part, around 200 mA.

To test the counter sub-unit, Resets 2, 3 and 4 must be grounded and Reset 1 must be connected to +5v. Then connect a square wave generator to the counter input. A circuit of the timing oscillator IC17 (SN7413) can be used (use pin 6 as output). The display will start counting at a random number and will continue to over 9999 and 0000 again.

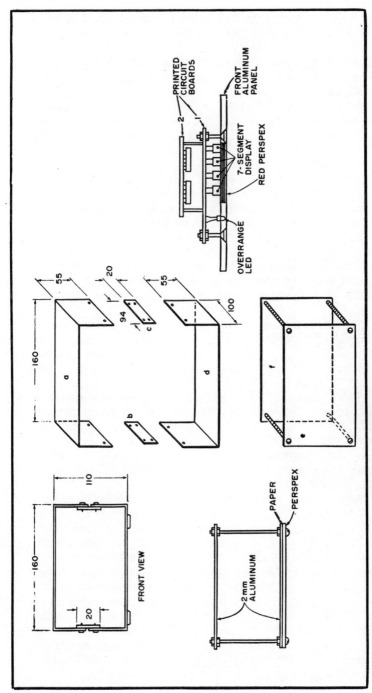

Fig. 4-6. A C-meter can be built into a homemade cabinet.

The second part of the C-meter can be checked with a frequency counter (output from IC14D pin 6). You must, of course, simulate the function of the range switch S1B. Output frequencies (at pin 3 of IC14C) must be exactly at the ratio of 1:100. The latch and reset pulses can be checked with an oscilloscope or, preferably, with a logic probe.

Alignment. If both parts of the instrument work satisfactorily, solder the two printed circuit boards together. Five thick wires should be soldered onto the counter board. To make this operation easier, use differently colored insulation on the interconnecting wires. The same applies to the range switch wiring.

Check the output voltage of the power supply. If possible, load it with a 6.8 ohm resistor to check if the output voltage is stable. This checking is important, as any voltage over 5.25v can damage some integrated circuits. Then check the current consumption of the whole unit. It should be under 0.75 A. If everything is alright, you can start with alignment.

Set "Zero Adj" (ZA) and "Zero Preadj" (ZP) to the center positions. Solder a few (two or three) capacitors, which are stable with temperature, across the Cx terminals (inside the cabinet). Used plastic molded silver mica and polystyrene. In parallel, solder a small mica trimming capacitor (20 to 50 pF). The total value of all of them should be about 980 pF. They form capacitor Cf.

The difficult problem here is to get hold of a very accurate capacitor whose value is between 5000 and 10,000 pF. The future accuracy of the instrument depends upon the accuracy of this capacitor. So have a good look around for it. Note that either input terminal is grounded. Of course, the one marked "+" is not very sensitive to hum, etc.

With no capacitor Cx you should be able to adjust the instrument to show zero on the display, with the values of resistors 12k and 22k next to ZA and ZP. But if you used a crystal of a frequency lower than 40 MHz, these resistors will be slightly higher in value (say 33 and 10k) and vice versa. By the way—they must again (like ZA, ZP and Cf) have a very low temperature coefficient. It is best to start with ordinary resistors and replace them, after finding the right values, with high stability ones.

Anyway, adjust zero with ZP and then connect the precise capacitor Cx. If it reads low, adjust capacitor Cf to the lower value. Remove Cx and adjust zero with ZP (changing 12k and 22k if necessary). Try it again with Cx and adjust Cf. Remove Cx and adjust the zero with ZP. (Basically, when the display shows a low value, you must lower the value of Cf, and vice versa.) Continue in this way until the display shows zero without Cx and the right value with Cx. During this procedure ZA is adjusted to the middle position. And that's it. Now you can take, say, 20 mica or polystyrene capacitors 200 to 500 pF, measure each one of them, mark the value and add them one after another in parallel on the terminals of Cx. The difference in total readings should at any point be not more than ±2 digits. If more, then the insulation between terminals 10 and 11 of IC21 (MMV) or between terminals Cx is bad and must be improved.

This procedure of adding measured capacitors in parallel, demonstrated the excellent linearity of the SN74121 MMV. It is really amazing how these small plastic 14- and 16-legged spiders perform.

The higher range 2 does not need any alignment, as it is automatically alright when the lower range 1 is adjusted.

Conclusion. Inside the cabinet, the integrated circuits, display, IC22, etc., dissipate over 5w. Certain crystals tend to be unstable at the high temperature which can develop when the instrument is switched on for a long period of time. In this case, drill a lot of small holes in the top and bottom of the cabinet in order to allow air to circulate along the printed circuit boards.

Take advantage of this instrument—it measures with practically full accuracy within a second of being switched on. Just adjust the zero by means of ZA (with range switch in position "pF," even when measuring in the nF range) and connect the capacitor across the terminals Cx. If you measure a lot of capacitors, check the zero from time to time, as it drifts slightly with temperature.

The overrange flickers when you exceed the range. This is much better than a steady signal.

If you are measuring solid tantalum capacitors, connect them with the correct polarity at terminals Cx. The polarizing voltage across these terminals is about 3v DC. This low voltage limits accurate measurement to only low voltage electrolytic capacitors.

If the reading is not stable and changes by more than ± 1 digit, then there is something wrong with the capacitor and you would do best to throw it away.

How Many pF Is That Capacitor, Really?

Tucked away in our junk box someplace, most of us have jars, boxes, paper bags or whatever of surplus capacitors—capacitors that are perfectly usable but not particularly useful because we can't decipher the color code or numbering system that designates their value. Many of us have also had occasions when we wished to twist up a "gimmick," or check the value of an unmarked air-variable or determine the exact value of a ceramic disc before we install it in a particularly critical circuit. The more sophisticated tinkers may recall a time when they wanted to measure the stray capacitance between the leads in a wire bundle or between the foils on a printed circuit board. If any of these situations sound familiar, read on, because here is a project to help you out of all those dilemmas. When completed, it will provide you with the capability to meesure any value of capacitance between 0 pF and 1 uF by merely pushing a button and reading a meter. Not only that, the cost is reasonable, construction is quick and easy and the parts are available from your nearest Radio Shack or mail-order parts supplier if they aren't already in your junk box.

The circuit is an adaptation of an old principle which, simply stated, says that the AC current flow through a capacitor is dependent upon the voltage applied, the frequency of the waveforms and the value of the capacitance. In this case, use of a square wave allows the capacitor to charge to its full potential and permits measurement of current flow as a linear function of capacitance.

In operation, depressing the read button applies the regulated 15-volt output of the LM78L15 to the LM566 phase locked loop voltage controlled oscillator (VCO). Regulating the Vcc input to the VCO insures that the calibration of the capacitance meter will be stable as long as the battery voltage remains above approximately 16 volts. The frequency at which the VCO oscillates is determined by selection of the proper RC combination (R1/C1, R2/C2, etc.) with the rotary switch. The square wave output from pin 3 is fed directly to the unknown capacitor whose value determines the current flow and subsequent meter reading. Rectification and filtering are provided by diodes D1 and D2 and the 30 uF electrolytic capacitor. The switch selected RC combinations provide six linear scales: 0-10 pF, 10-100 pF, 100-1000 pF, 1000 pF-.01 uF, .01-.1 uF, and 1.1 uF. Accuracy, dependent primarily upon the tolerance of the meter and the capacitors used for calibration is approximately ±5 per cent. If a high quality meter is used and 1 per cent tolerance capacitors are available for calibration, an accuracy ±2 per cent—or better is possible.

For the most part, component selection is noncritical. The values of frequency-determining capacitors C1 through C6 may deviate from those specified by as much as 50 per cent, but should be as stable as you can find to insure that the accuracy of the instrument is maintained under changing environmental conditions. The prototype used an aluminum electrolytic for C6 and ceramic discs for the remainder. After a night in the car at 15°F, the .1-1 uF scale read 10 per cent low and the other scales were in error to some lesser degree until the instrument warmed up to room temperature. Diodes D1 and D2 can be either silicon or germanium, and the value of the 30 uF electrolytic can vary from 22 to 50 uF. The higher value will provide better filtering action at the low operating frequency of the .1-1 uF range, but slows the meter response when the circuit is activated. For R1 through R6, the fancier multiturn trimpots are nice and ease calibration, but the cheaper single-turn variety are more than adequate and are what the circuit board was designed for. If you're going to splurge anywhere, capacitors C1 through C6 and the meter movement are where quality counts most in this project.

Construction is accomplished by referring to Figs. 4-7 through 4-10 and Table 4-3. Begin by drilling and lettering the front panel. For professional-looking results, the panel can be roughed with a stainless steel pad, or, if you prefer, sprayed with the background color of your choice. Lettering is most easily accomplished using "Dri-Transfer" letters available from Radio Shack, stationery stores and other sources. For durability, the lettering should be sprayed with a clear protective coating such as Krylon. After lettering, install the meter, binding posts and push-button switch, and then put the case aside for the moment.

Next install the remaining components, including the rotary switch, on the circuit board. R1 through R6 and C1 through C6 mount on the foil side,

Table 4-3. Parts List.

Bakelite box—6-¼″ × 3-¾″ × 2″
Rotary switch—2-pole, 6-position
Momentary switch—SPST, NO
*Precision panel meter—100 microamps
Knob
Binding posts—red & black
LM566—8-pin DIP VCQ IC
LM78L15—15 V regulator, TO-92
8-pin IC socket
Small signal doides (2)
Printed circuit trimposts, 10k (6)

100 pF		
.001 uF	C1	Mylar, polystyrene, or
.01 uF	through	silver mica preferred; however,
.1 uF	C6	good quality ceramic disc will
1 uF		suffice
10 uF		

1 uF ceramic disc/electrolytic
.001 uF ceramic disc capacitor
30 uF, 16 V electrolytic capacitor (20 uF-50 uF suitable)
1.5 ¼ W resistor
10 k ¼ W resistors (2)
Circuit board (available from author for $5.00)
9 V batteries (2)
9 V battery clips (2)
9 V battery holders (2)/double-sided tape/Super Stuff/etc.
*Note: A 50 or 150 microamp meter can be substituted if desired.

the remaining components on the reverse. The circuit board is supported by the switches and the leads from the binding posts. Solder a short length of solid conductor (a clipped-off resistor lead will do fine) directly to the tip of the screw end of each five-way binding post. Fit the board into position over the binding posts and push-button switch terminals and install the nut on the shaft of the rotary switch. Solder the leads from the binding posts and lugs from the push-button switch to the board. Connect the meter leads to the points indicated and connect and install the batteries using double-sided tape or "Super Stuff" to hold them in position next to the meter.

After admiring your handiwork, you're ready for calibration. Use the most accurate capacitors available and begin with the x1 scale (0-10 pF). Connect the calibration capacitor to the teminals and adjust R1 for a meter reading that matches its value. When you have completed this procedure for the other five scales, install the completed instrument in the case and grab a handful of those previously unusable capacitors out of your junk box.

When measuring values of less than 10 pF, stray capacitance can be a problem, so connect the capacitor leads directly to the terminals if possible. If not, plug in a set of extension leads with banana plugs on one end and alligator clips on the other. Position them as they will be when you are making the measurement and check the stray capacitance. Subtract this reading from the indicated value of the unknown capacitor to obtain the

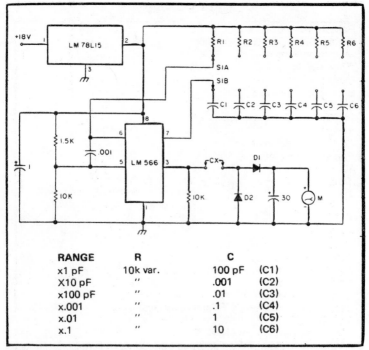

RANGE	R	C	
x1 pF	10k var.	100 pF	(C1)
X10 pF	"	.001	(C2)
x100 pF	"	.01	(C3)
x.001	"	.1	(C4)
x.01	"	1	(C5)
x.1	"	10	(C6)

Fig. 4-7. Note: All capacitance in microfarads unless otherwise annotated.

actual value. If extension leads of more than 2″ are required, the same procedure should also be used on the x10 (10-100 pF) scale.

A Digital Capacity Meter Simple Construction Project

There are many means of taking the measure of a capacitor. For the range of capacitors that the instrument to be described can handle (approximately one microfarad to 99,900 microfarads), the best method to refer to is time versus voltage. Mother Nature and science reached a detente relating the charge or discharge to reach some specific limit. The charge limit reaches 63 per cent of the applied voltage in RC seconds, where R is in megohms and C is in microfarads. Thus, if you were to apply exactly 10 volts to a capacitor which you made sure was totally discharged, through a resistor of 1 megohm, and you monitored the voltage rise across the capacitor with a voltmeter that did not load down the circuit, you could use a stopwatch to time the number of seconds it took for the voltage to hit 63 per cent of 10 volts, or 6.3 volts. It is easy to see that a 1-microfarad capacitor would time out at 1 second and that a 100-microfarad condenser would time out at 100 seconds.

You could work the same general method using the discharge curve, but your point of measurement would be when the voltage had fallen 37 per cent from a fully charged capacitor. Herein lies the rub, for it is much harder

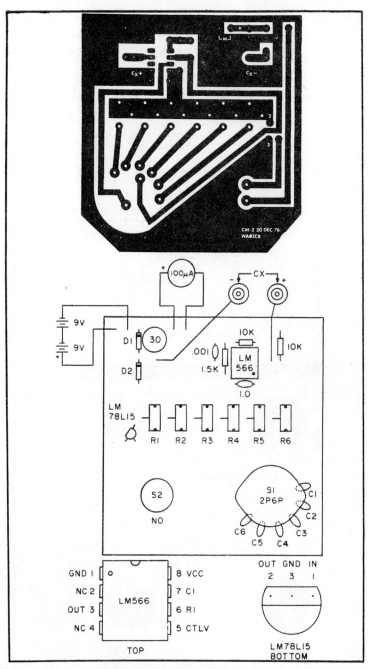

Fig. 4-8. Component placement. R1 through R6 and C1 through C6 mount on foil side of board, all other components on reverse.

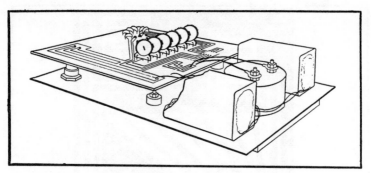

Fig. 4-9. Panel.

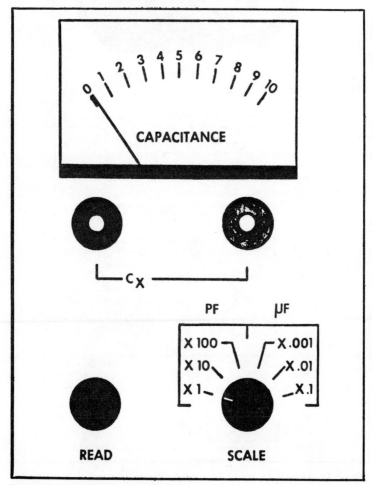

Fig. 4-10. Suggested front panel layout.

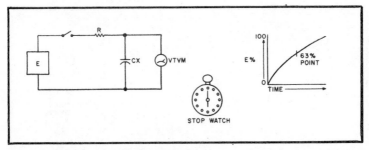

Fig. 4-11. Basic concept of capacitor measurement using time versus charging voltage rise.

to tell when a capacitor is fully charged than fully discharged. It takes about six RC time constants for a capacitor to charge to about 99 per cent of final full charge, so you would have time to read the paper if the capacitor under discussion was a 92,000-microfarad unit from your favorite computer.

Figures 4-11 and 4-12 diagrams the basics required to translate the time-versus-voltage method of capacitor measurement into terms of electronic hardware.

Figure 4-13 is the schematic diagram of the actual unit, which works as follows:

Note the "Function" switch S-1-1a, 1-b, 1-c (three pole, double-throw). In the OFF position, this switch performs the following three tasks:

- It provides a short across the capacitor connected to the test leads so that you start out with a fully discharged capacitor
- It blocks the flow of 60 Hz timing pulses to the counting system, which consists of a squaring circuit followed by a divide-by-six counter which produces 10-Hertz pulses.
- It makes the reset terminals of the three 7490 decade counters HIGH, which is the condition required to make the three-digit display show all zeros prior to making a count.

In the Test position of this switch, the following conditions prevail:
- The short is removed from the capacitor under test and it is connected to the measuring circuitry which starts out basically with the range switch S-2 (one-pole, four-positions) and the 741 op amp used as a comparator.
- It connects the 60 Hz timing waveform to the sine wave squaring circuit which uses two sections of the 7408 and gate package.
- It puts a ground on the 7490 reset line so that the counters will now be enabled to count.

Switch S-2 is the range switch, giving scaling factors of 1, 10, 100 and 1,000. The zener-regulated nine volts positive is applied to the capacitor under test through one of these range resistors. Notice that the inverting input of the op amp is connected to a positive voltage through a voltage divider. Under conditions where the positive voltage to this input is greater

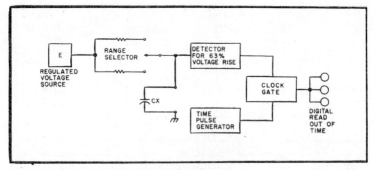

Fig. 4-12. Basic translation into electronic hardware.

in magnitude than the positive voltage applied to the noninverting input, the output of the op amp (pin 6) is highly negative. When the charging voltage of the capacitor under test reaches and just slightly exceeds the reference voltage on the inverting input, then pin 6 (output) goes highly positive.

In this fashion, by changing output polarity, the comparator gives a fixed point in time when the charging voltage just exceeds the reference voltage applied to the inverting input of the op amp.

Now all you have to do is provide a means of automatically starting a "clock" coincident with the start of the charging cycle and use the flip-flop of the comparator to stop the clock. Then the capacity of the unit under test is merely the multiplier of the range switch times the number of full and fractional seconds shown on the three-digit readout.

Tackling the bits and pieces of how this all happens, consider the output swing of pin 6 of the comparator. It goes from approximately minus 8 to plus 8 volts in the course of normal operation. The function of Q1 (2N3904) is to convert this voltage swing to standard voltage levels acceptable to the TTL logic blocks used in the unit. In addition to this interfacing function, the output from the collector is inverted in polarity, which you will see is needed to fit the rest of the circuit functions.

Since the timing chain starts in the power supplies, just a word should be said about that here. The 5-volt supply is run of the mill with a 1-amp capacity regulated by an LM309K. You can lash up any kind of a plus and minus 9-volt supply you care to for the op amp, but it must be zener regulated, at the least. From the transformer for this split supply, you need to provide a source of 60 Hz voltage from a voltage divider (1.3 to 1.5 volts AC), the only proviso being that the leg of the divider to ground should be about 1500 ohms or less. This is so the squaring circuit which is next in line sees a reasonably low impedance.

The low-voltage AC goes to the two sections of the 7408 through a diode. The output amplitude is a rather decent 60 Hz square wave, about four volts in amplitude. This square wave is applied to the divide-by-six section of a 7492, which results in a 10 Hz output.

The digital readout section consists of three 7490 decade dividers, each of which is connected to a 7447 seven-segment decoder driver. If you skip

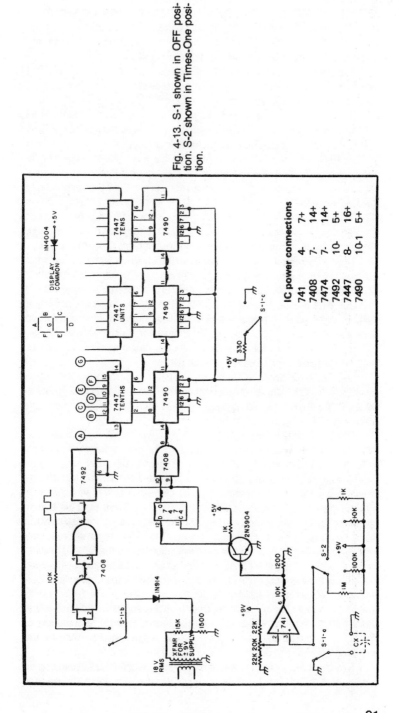

Fig. 4-13. S-1 shown in OFF position. S-2 shown in Times-One position.

91

.over some of the intermediate control circuitry and merely connect the 10 Hz to this three-digit divider-display chain, the display would consist of the right-hand digit showing tenths of seconds, the middle digit showing unit seconds and the left-hand digit showing tens of seconds.

You now have the comparator, the range switching and the 2N3904 interface. All you have to do is use it to stop the clock when the charge on the capacitor under test reaches the comparator reference voltage.

Interpose some logic circuitry between the 10 Hz pulses from the 7492 and the three-digit counter display as follows. A third section of the 7408 AND gate package is used as a gate. One side of this gate is fed by the 7492 with its 10 Hz output. The second output to this gate is a series chain of the output of the 2N3904 feeding into a 7474 D-type flip-flop used as a synchronizer. The output, A, of this flip-flop then goes to the 7408 section being used as a gate.

The 7474 makes sure that you always get a full final pulse or count through the gate no matter when the comparator triggers at the end of the measuring cycle. Note that its clock input is fed from the 10 Hz source.

Now let's take a quick trip through the whole shebang to see what happens when you measure a capacitor.

Suppose you have an electrolytic which is marked 6 uF and you want to check it. With the function switch in the OFF position, you connect the test leads to the capacitor, being careful to observe polarity. The test voltage does not exceed 5 volts. In passing, if the capacitor has not been in service, take the time to form it for a few short minutes at near its rated voltage, or your reading can be way out of the ball park.

Set the range switch to Times One, as this will basically measure up to 10 uF, which will be shown by a readout of 10.0 (this range will really go up to 99.9 uF, but it will take 99.9 seconds to do so, which makes another range more logical for a larger capacitor).

When you throw the function switch to Test, things begin to happen. The short is removed from the capacitor and it begins to charge. The 60 Hz is squared up and divided by six, feeding the gate and the synchronizer. The 7490 counter chain has had its reset bus grounded so that it can count any pulses coming its way. The output of the op amp is now highly negative, making the collector of the 2N3904 high (positive 5 volts). This high is applied to the D-input of the 7474, and the first low-to-high transistion of the 10 Hz signal applied to its clock input causes its Q-output to go high (and stay high). This TTL high is passed to the clock pulse gate, causing 10 Hz pulses to be passed to the 7490 divider-readout chain, and the readout begins.

When the charging cycle finishes, tripping the comparator output positive, this makes the collector of the 2N3904 go negative. On the very next low-to-high transition of the 10 Hz clock which is applied to 7474, the Q of this flip-flop goes low, which shuts off the time pulses to the 7490 count-display chain. Now you merely multiply the range switch setting by the indicated time on the display and you have your capacitor measured.

Now for a few notes on calibration: The range multipliers are in decade ranges of 1, 10, 100 and 1,000. As noted on the schematic, the resistor values for these ranges are one megohm, 0.1 meg 10k, and 1k. Five per cent resistors will do a decent job, but 1 per cent resistors are preferred for aesthetic reasons if not for practical ones. Electrolytics are generally anything but what is marked on the pretty package, generally erring heavily on the high side of what you think you bought. Oil-filled or large paper cups are generally truer to the mark, but, like all generalities, this can lead you astray. To start, let's assume that you have one favorite 5 uF oil-filled capacitor that you know is on the money (you checked it in the well-equipped CB shack of the guy next door). Connect it to the machine (ignoring polarity, as it is not electrolytic) and start a testing cycle naturally using the Times-One range (one meg). If you have started out with the variable element of the voltage divider feeding the inverting input of the op amp set to the middle of its range, you will probably be close to the mark. Adjust this variable trimmer resistor on subsequent timing cycles until the display agrees with your known value of capacitor. Be sure that you have the function switch in the OFF position at least 30 seconds between successive measurements on the same cap to guarantee that it is once again discharged. If you do not, then your readings will vary. This is the entire calibration effort, for, if your range resistors are on the money, then the other ranges should be in good shape.

A good quality multiturn trimmer resistor is a must for the calibrating pot. Anything else will lead to frustration. There is nothing magic about the plus and minus 9 volts. My particular zeners came out at 8.8 volts, which worked fine.

The choice of readouts is optional. I used RCA DR2000 incandescent units, as Herback and Rademan was selling out some readout kits from RCA at an unreal $2.00 per digit, including PC board and decoder-driver IC. As shown in the diagram, the segments are fed through a diode to lower the 5 volts to about 4.2 under load. This way, the life is extended. LED readouts are perfectly acceptable according to personal taste; it just means that you have to add the usual current-limiting resistors to the circuitry.

An open capacitor will show no count; a shorted capacitor will make the display count forever. This is handy, as it gives you a clock that counts up to 99.9 seconds for timing any event in that range. It could even have a darkroom spin-off for photo buffs.

Adrift Over Your Cs?

Once upon a time capacitors were nicely marked 0.1 uF or even 0.05 MFD. But now it is necessary to deduce their value from a rainbow of bands or spots, not to mention such numberings as 4k7, 5n, 5pj. Is that last colored band a figure in the value or a tolerance? Not to mention the useful collection of mica and other capacitors with obliterated markings and those with only a maker's number. Having some quite large tubular capacitors marked 5kp, in the expectation that they were 5,000 pF, or 0.005 uF, only to find they were in fact 50 pF, is quite frustrating. Here is a simple means of checking the values of doubtful capacitors.

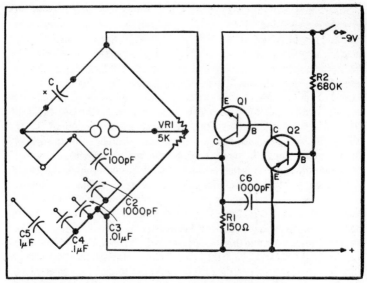

Fig. 4-14. This is the whole simple circuit, which isn't much.

The circuit of the result is shown in Fig. 4-14. Despite its simplicity it works nicely for values from 25 pF to 10 uF. Q1 and Q2 are combined to form an audio oscillator, output from Q1 being applied across the bridge.

The bridge has five capacitors for five ranges. Cx is the unknown capacitor. VR1 is a linear pot (it must be linear) for balancing, and high resistance phones indicate the null, so that the capacitor value can be read from the setting of VR1.

Oscillator. Assembly details will probably not matter much, but the audio oscillator can be wired on a tag strip. Both the transistors are audio or small output types, such as are available in great number and it is difficult to find two transistors which will not work here. Q1 must, however, be NPN, while Q2 is a PNP type.

A quite high tone is best. If necessary juggle with C6 or R2 or both to obtain this. The supply need not be 9v. The phones put across R1 will show how this works and two flying leads are soldered on to go to the bridge part of the unit.

Bridge. The five-way switch selects any capacitor C1 to C5. Without using the extreme settings of VR1, where accuracy falls, the center setting of VR1 is obtained when Cx is the same as C1, C2 or so on. As an example, if C3, 0.01 uF is in circuit, the middle setting of VR1 balances the bridge at 0.01 uF for Cx. From here, the swing of VR1 goes from 1/10 to 10 times, so that this range is 0.001 uF to 0.1 uF.

In the same way, C2 (1000 pF) gives a range of 100-10,000 pF, while C4 gives 0.01-1 uF and C5 gives 0.1-10 uF. C1 would by the same token provide 10-1000 pF, but the null or balancing point for VR1, easily audible with larger values, grows a little difficult to hear at the extreme low capacitance end of this scale.

It will be seen that the overlap is such that the same total coverage would be achieved with only C1, C3 and C5. These three ranges would be 10-1000 pF, 0.001-0.1 uF and 0.1-10 uF. However, the extra capacitors C2 and C4 are well worthwhile to fill in for easy checking.

C1 and C2 should be silver mica 1 per cent items. For C3, a 2 per cent or 5 per cent item will probably have to be adopted. C4 may be difficult to get with better accuracy than 10 per cent, unless costly, while C5 is actually two 0.5 uF tubulars in parallel. This means that calibration can be accurate for the lower values, but is less so for high values. In any case much accuracy is often not necessary with large values, except possibly in audio filters and some other applications.

These items are grouped mainly around the switch and the layout shown (Fig. 4-14) allows convenient wiring with the capacitors supported by a near tag of the audio oscillator section.

Case. A plastic box about 6" x 4" x 2" with insulated panel is most suitable. It carries an outlet for the phones, on-off switch, two terminals for Cx and a bracket to clamp the battery in position.

The scale for VR1 is marked 100 pF to 10k pF for C2 range and 0.01 uF to 1 uF for C4 range. The switch is marked pF and uF for direct reading of these ranges, and for uF x 10, uF ÷ 10 and pF ÷ 10 for the other ranges.

Calibration is carried out on the C2 range, with several known capacitors, preferably 1 per cent, such as 100 pF, 250 pF, 500 pF and a few others. Values can be obtained by paralleling some, such as 250 and 100 for 350 pF, while switching to C1 range allows the 100 pF, 250 pF and other low values to be used again for what will be 1000 pF, 2500 pF and similar 10x values on the C2 range.

No calibration is made for the large capacitor ranges, as these will be progressively 10x the existing ranges (assuming C3, C4 and C5 are accurate enough).

A clip or two on short leads from the terminals will be handy for some capacitors. Simply rotate VR1 for the null and read the value on the scale. High impedance phones, in the 2k to 4k range, will be best. A dip but no real null shows leakage and is to be expected with electrolytic capacitors.

The Capacitor Comparator

If your junk box is tucked way back in the corner in a spot where you throw all those unmarked, unidentified capacitors that you just KNOW will come in handy some day, then this project is for you. Round ones, square ones, flat ones, fat ones, piston and bypass ones and those a guy could have a great time screw-drivering if he just knew where to use them are probably all in your box.

This simple circuit is an easy one-evening project that when completed will provide an audio tone comparison of a built-in reference capacitor to an unknown capacitor connected to the test clips. Bearing in mind that the

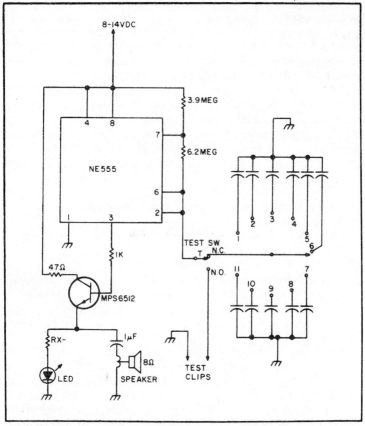

Fig. 4-15. LED indicator and RX-see LED specs. Capacitor bank: 1 = .7 pF; 2 = 3 pF; 3 = 5 pF; 4 = 10 pF; 5 = 25 pF; 6 = 50 pF; 7 = 100 pF; 8 = 330 pF; 9 = 470 pF; 10 = 680 pF; 11 = 820 pF. Test switch-SPDT push-button.

larger the capacitor the lower the tone, it is a simple matter to establish the value of unmarked units. The circuit (Fig. 4-15) will identify caps between .5 pF and .001 uF by providing tones between 8 kHz and 100 Hz. The heart of the tester is a 555 timing IC and may be operated from any DC voltage source between 8 and 14 volts.

An LED indicator is provided for testing values larger than .001 uF which do not produce a tone but merely turn the LED off and on. A .1 uF unit will trigger the indicator at approximately 5 Hz.

Piston, compression and rotary trimmers may be identified by first making a comparison fully closed and then fully opened. Small gimmick caps made from twisted leads are also easily sized.

The LED is, of course, optional, as well as the number of reference capacitors. Any NPN switching or audio transistor may be used in place of the MPS6512. If the LED is not needed, the transistor may be eliminated and the speaker wi ⸻ connected directly to pin 3 of the IC. Why no ⸻ sortment at 100 caps for $1.98?

Computerized Capa⸻

Caught up in the ⸻ the design of some higher-order active filters? If you have ever spent more than a few hours programming an SR-52 to solve cubic and quartic equations, you have realized that some heavier-duty number-crunching was called for if you really wanted to obtain those steep skirts. If you head to your local friendly computer store, you will return several hours later with several heavy boxes and one extremely light wallet.

The kits will go together surprisingly well and produced only a couple of minor bugs.[1]

You may now also need several close-tolerance capacitors. But if you do not have any kind of a bridge for measurement and if the state of your wallet precludes the purchase of such an instrument (or even the cost of several 1 per cent capacitors for a one-off filter), you can make use of items you find in your junk box to build something that would do the job with a little assistance from the computer.

Overall Design Considerations. The design considerations are simple:

- It has to be extremely cheap to build.
- It has to be extremely easy to build.
- It has to use a minimum of both hardware and software.
- It has to be reasonably accurate.

The resulting circuit uses only three readily-available components, a small machine language routine (which can be called from BASIC) and the accuracy is limited only by the clock speed of the computer system being used (Table 4-4). Figure 4-16 offers a thorough description of the method with a flowchart. This enables the system to be tailored to any computer.

Circuit Design. The circuit consists of a 555 timer chip wired in a standard monostable configuration (Fig. 4-17).

Circuit Operation. The monostable is triggered under control of an output port bit. Triggering the 555 requires a trigger signal going from a high (+5) level to a low (0) level for a brief period and then returning the trigger signal to a high level (Fig. 4-18). The monostable now switches its output to a high level and this level is sampled at an input port until the monostable times out. While the monostable is timing out, a count is made. This count is

Capacitance in pF	Accuracy
10	40%
100	4%
1000	0.4%
10000	0.04%
100000	0.004%

Table 4-4. Assuming count loop for one count is approximately 20 microseconds.

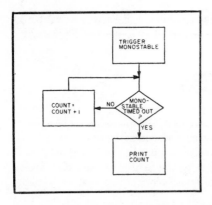

Fig. 4-16. Flow chart.

software scaled to equal the value of capacitance that determined the length of the monostable (Table 4-5).

Software. The whole program can be written in machine language using very little memory, or the count routine can be inserted in memory and called from BASIC via a USR or CALL statement. The example shown in Table 4-6 uses a simple BASIC routine which can be expanded to the desired esoteric level. The count from the monostable is averaged over ten triggerings and the resultant count is then multiplied by a calibration factor to give the value of capacitance used. This count may then be output in whatever form you may choose.

Calibration. Insert a known value of capacitance in the circuit and run the calibration program (Table 4-7). Enter the result from this program into

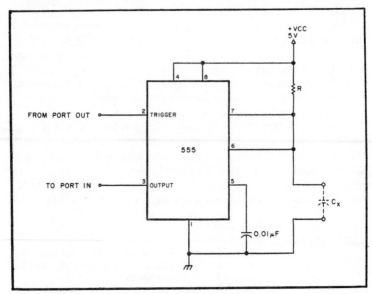

Fig. 4-17. Standard monostable configuration.

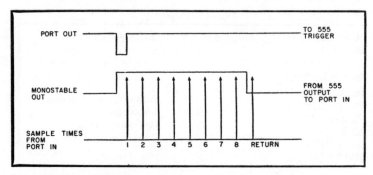

Fig. 4-18. Trigger Signals.

statement 80 of the main program (Table 4-6). Stray capacitance in the test leads can be calculated by running the program without a capacitor in the test leads and changing the BASIC program to compensate. (Don't forget that you are averaging 10 triggerings!)

The representative listing (Table 4-5) is from a hobbyist's read-boarded setup attached to a parallel port, using the high-order data in/out bits. For breadboard purposes, a line was run from the bit 7 data out to the trigger input and fed the monostable output to the bit 7 data in. The other lines were left floating, which accounts for the bit structures you see addressed to the port in the machine language listing.

Using a 4.7 megohm resistor, the breadboard setup measures from less than 100 picofarads to better than 0.1 microfarads. If you need to cover a larger range, you need only change the value of the resistor and recalibrate. Out-of-range detection can, of course, be accomplished by inserting a software routine to check for a carry during the count and return you a message to change ranges when this occurs. Again, be cautioned, any

Table 4-5. Program A.

Top	3E FF	MVI A,0FFH	Make sure trigger
	D3 xx	OUT PORT	is not active.
	21 00 00	LXI H,00	Clear counter.
	3E 00	MVI A,00	Send bit to
	D3 xx	OUT PORT	trigger monostable.
	3E FF	MVI A,0FFH	Reset the
	D3 xx	OUT PORT	trigger.
LOOP	DB xx	IN PORT	Watch the monostable.
	FE FF	CPI 0FFH	Is it active?
	C2 xx xx	JNZ OUT	No — go end routine.
	23	INX H	Yes — increment counter.
	C3 xx xx	JMP LOOP	Loop some more
OUT	7C	MOV A,H	routine to pass
	45	MOV B, L	value in H, L back
	CD 9C 0C	CALL 0C9C	to BASIC (MITS).
	C9	RET	Return to BASIC.

Table 4-6. Program B.

```
10 PRINT"CAPACITANCE METER"      Initialize count.
20 A = 0                         Set up for 10 triggerings.
30 FOR I = 1 TO 10               Link to user routine here.
40 X = USR(0)                    X = returned count value.
50 A = A+X
60 NEXT I                        Add this count to total.
70 B = A/10                      Loop again.
80 B = B*C                       Calculate average.
90 PRINT B                       C is calibration factor.
100 END
```

increase in the size of the count loop will affect the accuracy of measurement.

Timing considerations for the 555 in a monostable configuration are given by $T = 1.1\,RC$. Because of this relationship, this circuit can also be used to measure resistance, though it then needs to be calibrated with a fixed value of capacitance.

Table 4-7. Program C.

```
10 X = USR(0)                    Link to user routine
20 INPUT"ENTER KNOWN CAPACITANCE IN pF" ;Z
30 PRINT"CALIBRATION FACTOR C =";Z/X
40 END
```

Reincarnating Old Test Equipment

(A 1942 capacitance meter is born again in this project.) Purchased in 1942, the model BN cost about $12.00. Eighteen years later, the 12A7 failed and was replaced. In 1972, a resistor failed, and then, in May 1977, the new 12A7 failed. Obviously such an unreliable piece of gear was begging for replacement—until the price of an equivalent unit revealed the true extent of inflation. The only item needed to bring the unit back to life was a bridge balance amplifier and an indicator of balance. Some people have never been thrilled with the magic eye indicator, as it has to be shaded from ambient light to really see when balance is achieved. New housing was provided by a cabinet from an old 5" Sony TV. It was mated to a panel made out of a formica cutoff from a woodworking project. Figure 4-19 shows the original schematic.

Electrically (Fig. 4-20), all parts associated with the old bridge amplifier and magic eye indicator were discarded, keeping the basic bridge components intact. The voltage divider, consisting of a 10-megohm resistor and a one-megohm resistor (across the output of the bridge), serves two purposes. First, it keeps the impedance high at this point and allows the bridge to function. Secondly, it controls the amplitude of the input to the op amp used as a bridge amplifier.

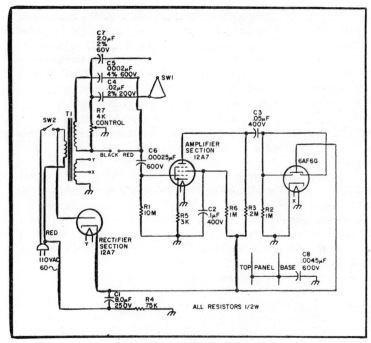

Fig. 4-19. Circuit diagram for capacitor bridge model BN.

The output of the bridge when unbalanced can run as high as 30 volts, and at balance, depending on the range in use, can be as low as a few tenths of a volt. The divider reduces this variation of amplitudes by a factor of 11 to avoid real problems with the op amp, which is set up for a gain of 10.

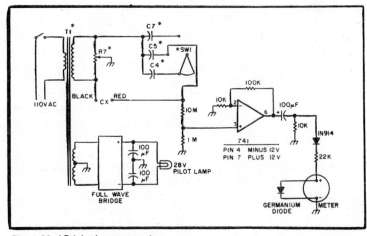

Fig. 4-20. *Original components.

The balance indicator is a meter salvaged from an old tape machine and serves quite nicely in this function with the advantage over the old magic eye in that it does not need to be shielded from ambient light. Here is a simple way to set up the proper series resistor to scale the meter. Unbalance the bridge and insert just enough series resistance so that the meter pins full scale. Then shunt the meter with the germanium signal diode, as shown in Fig. 4-20, and the meter indication should drop about one or two scale divisions. This simple method obviates any fuss and feathers log amp problems to handle the signal variations that the op amp handles. It allows for a very nice null reading to show balance, while saving the meter from overload.

The original transformer has two windings—one driving the bridge and the other a center-tapped winding which formerly lit the filaments of the tubes. This latter winding is used to feed a full-wave bridge, producing the DC voltages needed for the 741 op amp. Putting a 28-volt pilot lamp across the output of the supply serves the second purpose of acting as a bleeder.

Be sure to ground the dial plate of the bridge to the ground terminal of the supply. If you use a metal cabinet, keep the bridge circuitry isolated from it, as it may upset bridge calibration on the low capacity scale.

Chapter 5
Resistance and Impedance Measurers

Build A Bridge to Measure Transformer Impedances

This project is for those who are scared silly at anything that even remotely resembles an algebraic formula. If you have several transformers around your workshop that are unmarked, you'll enjoy this project. Three pieces of equipment are needed, two of which you probably have. The first is a signal source at about a 1,000 cycles (you can use 60 cycles in a pinch), since most audio transformers are measured at this frequency. The source should have an output of about 5 to 10 volts, although it can be less if you use a sensitive oscilloscope for the null detecter, which is the next piece of gear that you need. You can use a VTVM, a scope, a simple old 6E5 magic eye null detector, or (if you have enough signal) an AC micro-ammeter (not recommended). The third and last thing that you will need is an instrument improperly called a Wheatstone Bridge.

Please don't rush for your pen to defend old Sir Charles Wheatstone. We're not trying to deprive him of his place in history, but the fact is that he just didn't invent the darn thing. True, he publicized it, but the device was invented by an English scientist named Samuel H. Christie in 1833. At any rate, this simple piece of gear is the solution to not only this problem, but to so many others on the workbench. Considering its simplicity (Fig. 5-1) it is perhaps the most valuable instrument that you will own.

Now for the construction data, which will be meager because no one will build a carbon copy anyway. First, don't use a standard volume control for your calibrated pot. They are just too small to permit any degree of accuracy. The potentiometer is a 5½" diameter, wire wound, 10k ohm log taper job. You don't have to have one that big, but try to get a pot over 2" in diameter.

Wire with heavy leads, the heavier the better (#14 wire) and make sure that you use well-insulated binding posts. Also, be sure that your signal input connector is well insulated from the chassis. Plug-in resistor jacks are made using banana plugs on pieces of Lucite. The only other components that you will need are some precision resistors and capacitors, preferably 1 per cent (Table 5-1). The accuracy of the instrument depends on two things: the precision with which you calibrate your dial and the accuracy of the standard resistors and capacitors that you use.

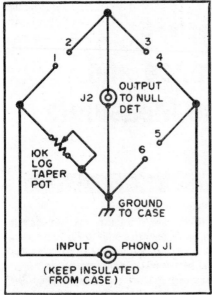

Fig. 5-1. Schematic.

You can make the dial 6″ in diameter using a large skirt knob and a flange from a daylight load spool from 16mm movie film. File a flat on the shaft of the pot so that the dial will never slip and it will retain its calibration if it should ever have to be removed. After the instrument is assembled sit down with a handful of precision resistors and a VOM to calibrate. Hook the ohmmeter across terminals #1 and #6 and mark the dial carefully with a pencil, checking every reading against a known 1 per cent precision resistor wherever possible. You can also apply white press-on numbers and dots and spray them with clear lacquer to protect them. The calibration procedure is not difficult, just tedious. Calibrate the major 1k ohm points first then the 100 or 200 points in between. The more careful you are, the more accurate your bridge will be.

Use standard RCA phone jacks for input and output. Using them whenever possible on projects is a good idea because of the wide variety of shielded cables available that match them.

Just one more thing on construction before we get to the good part. Space the banana plug posts a standard ¾″ apart. That way you can use standard dual banana jack instrument plugs. It doesn't affect anything; it's just a matter of looks.

It's difficult to believe that such a simple device can really work so well. Don't be disappointed if at first you are testing an audio transformer

1	1 Ohm res.	1	.0001 uF cap.
1	10 Ohm res.	1	.001 uF cap.
2	100 Ohm res.	1	.01 uF cap.
2	1k Ohm res.	1	1 uF cap.
2	10k Ohm res.		

Table 5-1. List of Required Standard Resistors and Capacitors, 1 per cent.

with a known impedance of 10k ohms and you don't get the correct reading with your new tool. The other winding also should be loaded to its proper impedance in order to get an accurate reading on the winding under test. Hook up a four ohm resistor to the output winding and it will measure 10k ohms on the primary. Since four ohms is a pretty low resistance, what do you think would happen to the impedance of the other winding if you were to short the four ohm winding? Try it. You'll now measure approximately 1,800 ohms across the primary. Could it be that shorting the low impedance winding gives you a reading of about one-fifth of the true impedance? Subsequent tests with other transformers show that, although this is not 100 per cent accurate, as a general rule of thumb you might use this as a ball park figure to start.

Start by measuring the DC resistance of all windings with an ohmmeter (or with the bridge using the described procedure for DC resistance measurements) and marking them. If you are sure that it is an output transformer, hook a 4 or 5 ohm resistor across the lowest resistance output winding. Connect the high resistance winding across the #5 and #6 terminals of the bridge, a 1 kHz signal to the input and a suitable detector to the output. Measure the resistance and hook a carbon resistor of this value across the winding you have just measured. Reverse the transformer and measure the previously shunted low impedance winding. If it was an output transformer and you had used the 4-5 ohm shunt, you will probably find that you are within 5 to 7 per cent of the true value of the transformer. Close enough for all practical purposes.

If you are not sure of the type of transformer you are working with, then proceed as follows. Short the leads of the lowest resistance winding together and connect the highest resistance winding across the bridge. Measure and multiply the reading you get by six and connect a resistance of that you just measured. Now reverse the transformer, measure the previously shorted winding, connect a resistor of that value across it and repeat the procedure on the high impedance winding. It only takes about two or three measurements, going back and forth from one winding to another, to achieve a remarkable degree of accuracy. Practice on a few transformers of known value and if you find your readings coming out wrong, look for shorted clip leads or open connections. With a little practice and care you can solve your unknown transformer problems.

Connect unknown resistor between 5 & 6		
Range (Ohms)	Posts 1 & 2 Std. Res.	Posts 3 & 4 Std. Res.
.1—1	10,000	1
0.1—10	10,000	10
1—100	10,000	100
10—1000	10,000	1000
100—10,000	10,000	10,000
1000—1,000,000	1000	10,000
10,000—1 Meg	100	10,000

Table 5-2. Resistance Measurements.

Connect unknown capacitor between 3 & 4		
Range	Std. Res. Posts 1 & 2	Std. Cap. Posts 5 & 6
1—100 pF	10k	.0001 uF
10—1000 pF	10k	.001 uF
100 pF—.01 uF	10k	.01 uF
.001—.1 uF	1k	.01 uF
.01—1 uF	10k	1 uF
.1—10 uF	1k	1 uF
1—100 uF	100 Ohms	1 uF

Table 5-3. Capacitance Measurements.

It will interest you to know that you can also measure carbon and wire wound resistors to a very high degree of accuracy using DC current. First connect a source of DC, a battery or other supply of from 1½ to 4½ volts to the input jack with a push-button in series and a 0-50 or 0-100 DC center scale micro-ammeter to the output jack. Hook the standards and the unknown up in accordance with the resistance chart of Table 5-2. Using the push-button for momentary contact, start rotating the dial until you start to get a null, after which you can hold the button down to finalize the null.

Using an audio oscillator you can also measure capacitance (Table 5-3) and inductance (Table 5-4). Just remember to keep all test leads as short as possible, make tight connections. Also, be careful of shorts and open connections.

A Simple RC Substitution Box Using a Matrix

Those experimenters who are constantly engaged in circuit experimentation or development work have probably purchased or built elaborate R and C substitution boxes. Such substitution boxes greatly simplify the problem of finding just the right component value to use to optimize circuit performance for any desired condition (maximum gain, bandwidth, stability, etc.). For those who just occasionally build some accessory circuits, substitution boxes would also be very useful when a given circuit doesn't work exactly as it should or when one wants to determine if an available component will do in place of a specified value. Unfortunately, cheap RC substitution boxes don't buy you too much. The increments in which they cover RC

Connect unknown inductor between 1 & 2		
Range	Posts 3 & 4 Std. Res.	Posts 5 & 6 Std. Cap.
1—100 uH	1	.01 uF
10—1000 uH	10	.01 uF
100 uH—10 mH	100	.01 uF
1—100 mH	1k	.01 uF
10—1,000 mH	10k	.01 uF
0.1—10 H	1k	1 uF
1w8100 H	10k	1 uF

Table 5-4. Inductance Measurements.

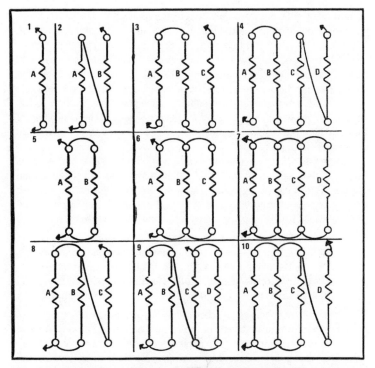

Fig. 5-2. Interconnection diagrams for four resistors in substitution matrix.

values are usually too large and calibration can be poor. One can substitute potentiometers for fixed resistors in a circuit, but their calibration is poor and one has to constantly check the value of a setting with an ohmmeter. The result of all this is that most occasional circuit builders end up soldering components of a fixed value into a circuit and by trial and error arrive at a successful result or a hopeless jungle of tack-soldered components.

This simple matrix can solve those problems for the occasional circuit builder at a very low cost and still achieve accurate results.

Most RC substitution boxes, whether they be for resistance or capacitance substitution, make use of switches to select components of various values. However, another approach would be to take a fixed number of components (R or C) of different values and hook them up in as many ways as possible (series, parallel, series/parallel, etc.) to obtain a succession of different overall values. However, how many fixed values should be chosen to keep the circuit complexity within bounds? What should their individual values be both to obtain a good overall substitution range and have a smooth progression of resultant substitution values?

We'll let a computer solve some of the latter questions. It's not surprising that a computer comes up with a "binary" type answer. We'll use resistors to demonstrate the idea, but it is equally applicable to capacitors. Take four components (resistors) of value 10, 20, 40 and 80 ohms and let's

107

see how many different ways they can be interconnected. The interconnection possibilities are illustrated in Fig. 5-2, starting with the simple use of one of the resistors alone and then various series and series/parallel connections. There is nothing new about these circuits, but what is interesting is what happens if one tries all the circuit possibilities and then arranges the resultant values that can be achieved in order. The result, as shown in Table 5-5, is a wide, smooth range or resultant values going from 5 ohms to 150 ohms. Four properly chosen resistors have resulted in the equivalent of almost 40 different resistor values progressing in very even steps over a 30 to 1 range!

Table 5-5 is, in fact, a universal value table. It shows resistance values of 5 to 150 ohms being achieved by four individual resistors of 10, 20, 40 and 80 ohms. But, it can be scaled up or down to achieve different ranges with other, similarly ordered values of individual resistors. For instance, the following would be the range of values achievable with the resistors shown:

- 0.5 to 15 ohms—use 1, 2, 4 and 8 ohm resistors
- 50 to 1500 ohms—use 100, 200, 400 and 800 ohm resistors
- 0.5k to 15k ohms—use 1k, 2k, 4k and 8k ohm resistors
- 15k to 150k ohm—use 10k, 20k, 40k and 80k ohm resistors
- 150k to 1.5 M ohm—use 100k, 200k, 400k and 800k ohm resistors

The resultant values can be checked by anyone using the usual series/parallel resistor formulas for any of the circuits shown in Fig. 5-2. A more practical question might be how to implement the idea in a simple, reliable form. One could, of course, devise a special switch to achieve all the hookups shown in Fig. 5-2. Some manufacturers capitalized on this idea. However, a simple home brew way to arrange the four resistors in any of the circuits shown in Fig. 5-2 is to use a simple plug and jack switch matrix as shown in Fig. 5-3.

Four columns of six plugs each are used with the individual resistors inserted in the positions shown. The lower "output" terminal is fixed at the bottom of the "A" column. The upper "output" terminal is connected via a flexible wire so it can reach any of the plugs on the upper row of the field. By moving the latter jack wire, plus the use of a few jumper wires between columns, any of the circuits of Fig. 5-2 can be wired. With a little practice, one quickly develops a system to go from one circuit configuration to the next.

The matrix of Fig. 5-3 can be built on any insulating material (perforated board stock or plexiglass), using inexpensive noninsulated plugs and jacks. One could permanently solder in the resistors, but if they are made "plug-in", the same matrix can be used for different resistance ranges by plugging in different sets of four resistors. The matrix board is a neat and orderly way of interconnecting the four resistors. But if one had to do it, they could also be interconnected using simple jumper leads by keeping Fig. 5-2 and Table 5-5 handy for reference.

Standard value resistors do not come in all the desired values (4k and 8k, for instance). Choose the closest available values to those shown (3.9

and 8.2k, for instance), but measure them once with a good ohmmeter. Usually resistance values, unlike capacitor values, are well within the manufacturer's stated tolerance. Standard, inexpensive 10 per cent tolerance resistors should work fine for most circuit work. But, again, check them first before relying upon them. The tolerances can either build up or cancel out in the various series/parallel circuits. The power rating of the resistors used depends upon the circuit applications intended. For most low level transistor circuits, 1 or 2 watt resistors will certainly suffice. If one obtains the four resistors for the 0.5 to 15 ohm range, they should be of the 10 watt size since they will most likely be used for power supply or audio power amplifier circuit applications.

What about using the matrix for capacitor substitution? Of course, series and parallel resistor circuits have their equivalent in parallel and series

Ohms	Circuit	A	B	C	D
5	7	10	20	40	80
5.6	6	10	20	40	
6.2	6	10	20	80	
6.8	5	10	20		
7.5	6	10	40	80	
8.0	5	10	40		
9.0	5	10	80		
10	1	10			
11	6	20	40	80	
13	5	20	40		
16	5	20	80		
20	1	20			
22	9	10	80	20	40
24	9	10	40	20	80
26	8	20	80	10	
27	5	40	80		
28	8	10	40	20	
29	8	10	80	20	
30	2	10	20		
33	9	10	20	40	80
36	8	40	80	10	
40	1	40			
45	10	10	20	80	40
46	8	40	80	20	
48	8	10	20	40	
50	2	10	40		
56	8	20	80	40	
60	2	20	40		
70	3	10	20	40	
80	1	80			
90	2	10	80		
91	8	20	40	80	
100	2	20	80		
110	3	10	20	80	
120	2	40	80		
130	3	10	40	80	
140	3	20	40	80	
150	4	10	20	40	80

Table 5-5. Circuit Numbers Refer to Those Shown in Fig. 5-2.

Ohms	Circuit	A	B	C	D
5	7	10	20	40	80
5.6	6	10	20	40	
6.2	6	10	20	80	
6.8	5	10	20		
7.5	6	10	40	80	
8.0	5	10	40		
9.0	5	10	80		
10	1	10			
11	6	20	40	80	
13	5	20	40		
16	5	20	80		
20	1	20			
22	9	10	80	20	40
24	9	10	40	20	80
26	8	20	80	10	
27	5	40	80		
28	8	10	40	20	
29	8	10	80	20	
30	2	10	20		
33	9	10	20	40	80
36	8	40	80	10	
40	1	40			
45	10	10	20	80	40
46	8	40	80	20	
48	8	10	20	40	
50	2	10	40		
56	8	20	80	40	
60	2	20	40		
70	3	10	20	40	
80	1	80			
90	2	10	80		
91	8	20	40	80	
100	2	20	80		
110	3	10	20	80	
120	2	40	80		
130	3	10	40	80	
140	3	20	40	80	
150	4	10	20	40	80

Fig. 5-3. Matrix panel suitable for either resistor or capacitor usage.

capacitor circuits. The circuits of Fig. 5-2 and Table 5-5 can be used if one remembers the equivalent circuits. For instance, for 150 ohms Table 5-5 calls for the use of circuit 4 in Fig. 5-2, which is all resistors in series. If one wanted 150pF (using individual capacitors of 10, 20, 40 and 80 pF), the equivalent capacitor circuit is used. Connect all capacitors in parallel. For 5 ohms, Table 5-5 uses circuit 7 and connects all resistors in parallel. For 5 pF, connect all capacitors in series. For 30 ohms, Table 5-5 uses circuit 9 or AB parallel in series with CD parallel. For 30 pF, use AB in series and then parallel with CD in series. Table 5-5 can be scaled up or down for capacitors the same as it can be for resistors. The total range for each set of four capacitors will remain a 30 to 1 range: 5 to 150 pF for fixed capacitors of 10, 20, 40 and 80 pF, 0.1 to 3 uF for fixed capacitors of 0.2, 0.4, 0.8 and 1.6 uF, etc.

Unfortunately, capacitors, and especially electrolytic types, vary widely from their marked value. Tolerances of up to −20 and +40 per cent for electrolytics are common. One can build a substitution matrix using disc ceramic or mylar capacitor and rely upon the capacitor values to be close enough to their nominal values for general substitution work. In practice, this means a substitution matrix going up to about 1 uF maximum. Beyond this value, when one has to use electrolytic capacitors to build the substitution matrix, it is imperative that the actual capacitor values be measured if the substitution matrix is to have any reasonable accuracy.

This describes a simple individual R or C substitution matrix. If the resistors or capacitor sets are made for plug-in use, the matrix of Fig. 5-3 can be used for either type of substitution as required by an individual circuit. With a few matrix boards and several sets of resistor and capacitor plug-in sets, a handful of components can be used to replace several hundred for substitution, testing and experimental purposes.

Finally, how about RC combinations such as time constant circuits, low pass and high pass filters? These can also be built for various values and frequencies by combining R and C elements on one or more matrix boards. The variety of values here is so broad that it is impossible to define any table for them. However, by using the known R and C values for the individual RC substitution boards and standard formulas, one can easily develop a duplicatable set of response configurations usable for a wide variety of circuit experimentation.

Megger for Peanuts

During WW II days, a certain landlady named Gladys had goldfish in a big aquarium, had a husband Al, a two-year-old Carol and a cat. She had noticed an attrition in the goldfish to the feline. She was, however, realist enough to suspect her baby daughter too. As the resident electronics expert, could I invent something that would determine guilt or repel—without harm—the culprit?

The basic scheme was to connect the light line between a ground screen (you could get galvanized iron screen in those days) and the water itself. The fish and the pipes would probably ionize the water enough.

Obvious enough, but there remained safety considerations. The kid was a card, and definitely not expendable. Limiting resistances, sure, but how big? I had heard 15 mA put forth as the least amount of current, applied directly across the heart, that would be fatal. Since this stated a unique value, not a range, I had little faith in it. Human beings vary at least three to one in susceptibility to inimical influences, whether electric, ballistic, chemical or a knock in the head. Taking 15 mA as a point to depart a long way from, a megohm in one side of the line seemed like a conservative place to start. But the direct other side—how could I make sure the other side stayed grounded? Simple: duck the problem, put a meg in each side.

I tested the system while Gladys and the goldfish watched curiously.

I didn't expect to feel much, considering the *decoupling* resistors. But I did expect to feel something, if only a tiny tingle. But still thinking of the baby, I decided to leave the setup alone for a few days before dropping resistance values.

Gladys counted her herd every day and gave me the unchanging total each day at the supper table. One evening we discussed the goldfish and decided the cat was making fools of us. Al caught the cat and gave her to me. Kitty didn't know what was coming, but she suspected she wasn't going to like it.

Taking her firmly by the neck, I pushed her head so that her nose would barely touch the surface of the water. At the exact instant of contact, she jerked. I tried it several times. It seemed to me that here was biological-electrical rather than nerve response.

Gladys said, "Aw, the poor kitty—," took her from me and shoved her whole face under the water. The cat exploded out of her arms and escaped. Puzzled, I dipped my finger in again, contacting the ground screen with my other hand. If I got hit, I'd at least fall clear. Nothing. I was conscious of my audience—Al, Gladys and the fish. I stuck my own nose into the water. Nothing. I stuck my tongue in the water and the fish backed away. Nothing. Not even the coppery taste you get from a dry cell.

We concluded that the setup was working, though it was hard to see how. Obviously, the cat had a horror of that tank. Cats smell very well, though their noses are isotropic. They often miss a bit of food within reach, while a dog scans, goes into a hunt and locks on in practically one motion. But maybe we had an entirely new biological phenomenon here. Was a cat's nose a supersensitive indicator of an electrical charge? It certainly seemed so.

The limiting effect of a resistance is instantaneous. The system was a big voltage divider and the better the contact, the less voltage drop.

I forgot all this for 20 years. I wound up on a microwave test bench and found out a lot of other things. You've heard of the Miller Effect? Ever see any evidence of it? I did. The 70 meg i-fs would occasionally exhibit it and until I caught wise, drove me up the wall. The cause was defective bypass capacitors which could fail in a variety of ways. They could go open or shortened, or just lose their capatance. These were button-mica feed-through types. We were furnished an excellent GR electronic megger. It would measure megohms up in the hundreds or thousands. When a bypass had as little as one meg resistance to ground, it was time to get rid of it. The 250 volts and the heat would break it down in a couple of weeks and you'd get the amplifier back from the field. This was not good. The best solution was to replace every cap after the amplifier was in service for a couple of years, but there simply wasn't enough stock to permit this.

Good as it was, the GR megger was too much instrument for the job. What I needed was something simple, something cheap. It should be small—the GR took up too much bench and was heavy and clumsy.

The first design worked very well in practice. The basic standard was my wristwatch, which gave me better than one part in 720 accuracy. Lighting juice went into the little plastic box to a DPDT switch. Thrown one way, it merely lighted a red pilot light (neon). The other way, it connected it through a diode to a filter capacitor for a primitive power supply. A pot regulated the voltage. The slider went through a one meg limiting resistor to a binding post. The ground side of the pot went through a timing capacitor and another one meg resistor to the other binding post. Across the capacitor, an NE-2 bulb.

After the first few days, I forgot all about calibration, except for the most basic adjustment. What I needed was a go-no-go instrument and I really had one in this gadget. All I really wanted to know was, would the cap work for a few months, or should I replace it? Twenty megs or 200 would be fine. Ten wouldn't.

Of course, theory-wise, you are way ahead of me. It is a time-constant deal; when the voltage rises to 63 per cent of the input voltage, you have one time constant, whether it is one Farad and one ohm, or a megohm and a microfarad, right? There is something about that 63 per cent. Maybe here it suddenly becomes more horizontal than vertical—the charging curve. Anyway, an oscillator will come out to pretty close to the calculated time-constant value. And the NE-2 is both a relaxation oscillator and a timing indicator, right?

That's what I thought. There is a time constant that uses 63 per cent right enough, but the time constant this gadget uses is very far from the classic one. Instead, it is time taken for the NE-2 to fall from the firing voltage of maybe 70 volts to the extinguishing voltage of about 69. You get approximately a one volt spread between these values. Of course, the timing capacitor size, the limiting resistor sizes and the charging voltage are all factors. If you are looking for resistance, then you juggle voltage and timing capacitor size until you come up with the proper value. Of course, it is most convenient to vary the voltage with a pot.

It surprised me that the ionize-non-ionize range was so small. I expected the voltage to drop to 15 or so. The CRO sweep tubes. 884/885/6Q5 will, but remember, they all have hot cathodes and this makes all the difference.

In actual use, I discovered that the safety switch was as useless as the well-known complementary accessories on the bull. Knowing that I couldn't possibly be hit, I was unable to remember the switch and found my fingers all over the external wiring. It was time for a new model. Perhaps a true go-no-go model without any adjustments? No, this was too radical. A pot and a knob and a hole to look through—that was the idea! And while I was about it, why not get rid of those silly binding posts? The other change was to use standard commercial 8-lug strips rather than individual miniature standoffs (Fig. 5-4). Still another was to use shallow grooves in the plastic box ends for the AC input cord and for the output clip-cords. I once knew a man who claimed that if you wanted to blow fuses, the best gadget for this was the standard alligator clip. He is right, beyond question, but in this case, with the load dead, it doesn't matter if the clip slips. And it saves a lot of bother with binding posts, which will eventually have clips leads on them anyway.

One thing I learned from commercial equipment: a ¼ watt, 100 ohm resistor makes an excellent fuse and protective device. Sometimes they shatter explosively, leaving only the leads. Sometimes they crack invisibly and you find blown ones by twisting them with flats. Or, they may cook and spray the whole inside with bakelite varnish. No question what happened, the stink tells you. You can remove varnish and stink with mineral spirits (keresene). They serve another purpose—they limit the peak charging current through the diode.

The model illustrated in Fig. 5-5 is the NE-2 (actually Japanese). It was originally taped to the paper clip, to a ground lug, which cannot be used for active connections. There are two of them. The panel has a ½″ hole for viewing the neon. The two straps short out padding resistors in the voltage-adjustment circuit which were not needed.

Calibration is a cinch. Short the clips—with your fingers if you like—and adjust the pot for the firing rate you want. Say, 10 flashes in 20 seconds. Put you wristwatch next to the gadget and count flashes seen in your peripheral vision. This is a lot easier than it sounds, because every flash is equally bright, no matter how long you have to wait for it.

I have waited for minutes at a time to measure really high values, but for my purposes, this is time wasted. Who cares about hundreds of megohms? Now, with 10 flashes in 20 seconds you are timing out at one second per megohm, and actually accounting for the safety, or limiting resistors in the box. Naturally, you start your count with a flash, and then count 10 more, excluding the start, or zero flash. Now if you have one, stick in a ten meg resistor. What would your timing be now? Twelve seconds between

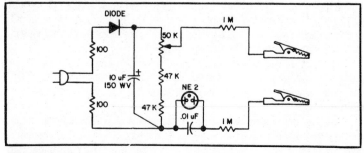

Fig. 5-4. Schematic.

Fig. 5-5. Completed project.

flashes? That's 10 on the outside and two inside. How about one flash after 19 seconds?

Maybe you don't want to wait 19 seconds. Then calibrate ½ second per megohm. Or any other value you want. The instrument is surprisingly linear over a reasonable range, mostly because of its limited voltage excursions as it oscillates. In theory, when the external resistance being measured exceeds the leakage resistance of the calibrating capacitor, and it's got some, the said capacitor never will charge up. It will just sit there, discharging as fast as it charges. Don't worry about it—you are not likely to expereince this effect.

All right, you built yourself a megger. What good is it? The first thing to do is pick a holiday—your birthday, 4th of July, whatever. On this day you measure the insulation of your refrigerator, your airconditioner, furnace motors and don't forget your electric drill—they are killers because you are wrapped up around them in use. Don't forget the wife's kitchen either—stove, mixmaster and iron. Write down all the values you can read. Next year, do the same thing or sooner if you like. Does any of them show a marked drop in the winding-to-ground value? Dump it. Get a new one. It could save a life. The box probably will save you money, maybe 50 times its cost. If it does, you'll be a believer from then on out.

A VOM or a bridge is a multi-purpose instrument. I dislike other examples; however, too often an additional function just complicates an instrument and makes it expensive. For instance, you could set up the megger to measure small capacitors. Please don't. Leave the "high current" high voltage in that box where it belongs.

But I did have a very odd use for the megger. I was investigating hum sources in a Kellogg 40.1 vacuum tube. I tried various connections and discovered that the insulation in these was very good, whether the tube was lighted or cold. The bright ends of the heater wires looked interesting, so I measured and got a nice reading—50 megs—in the conducting direction. Assuming 60 volts, this gave me 1.4 microamperes between heater end and the plate support rods. This is a significant source of hum.

As far as I know, there are only two of these gadgets in the world. If they became popular, there might be two or three in a large city. Why wouldn't these little boxes be a perfect present for the man who has everything else?

If he fixes TVs as a hobby and likes to measure leaks in coupling capacitors—yes. He'd appreciate one—any real technican would. But never think you could persuade a non-technical type to measure the electrical equipment around his house. To him, it would be pointless. To gift such a man with one of these is to provide a source of embarrassment to both of you.

Chapter 6
Testers for Solid-State Components

Ultra Simple Diode Checker

This simple diode checker is an up-to-date version of an idea that has been around for a number of years. It can be built in one or two evenings from the parts in most experimenter's junk boxes. The parts required are one resistor, two LEDs, and any 117v AC transformer that will provide from 3 to 25v AC. Discarded audio interstage transformers from old tube-type radios and TVs can be used. If all new parts are bought, the cost will be about $5.00, including the small aluminum box. The small cost can be recovered many times over by buying unmarked, untested, manufacturer's closeouts, diodes by the pound, etc., available from most discount mailorder houses (like Poly Paks). Bad diodes can cause disastrous results in some circuits and can be difficult to detect and locate in other circuits. It is a wise precaution to check them all before installation. This simple diode checker was conceived and built for just such purposes.

The Circuit. The simple schematic is shown in Fig. 6-1. The transformer provides a low AC voltage, through the current limiting resistor, to two LEDs connected back-to-back. The diode to be tested is connected in series with this combination and the return side of the transformer. The LEDs will respond to the four possible conditions of the diode under test. If the diode is open, no current flows and neither LED will light. If the diode is shorted, one half cycle of the AC voltage will light LED1 and the other half cycle will light LED2. Since each LED is lit 60 times per second, a shorted diode will cause both LEDs to appear lit continuously. If the diode is good, LED1 will light when the diode's anode is toward the return side of the transformer and LED2 will light when the diode's cathode is toward the transformer return side. By proper physical arrangement of the LEDs and diode, the LED near the diode's cathode will always light.

The resistor should be sized to limit the current through the LEDs to about 10 mA. Most LEDs will have a voltage drop of about 1.5 volts across them and most signal-type diodes will have from 0.1 (germanium) to 0.5 (silicon) volts drop across them. The resistor value can then be found by subtracting these two voltages (say 1.5 and 0.5) from the transformer voltage and dividing by 10 mA:

$$R = \frac{VXFMR - 1.5 - 0.5}{.010}$$

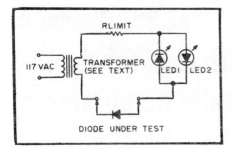

Fig. 6-1. Schematic.

For a 6.3v AC transformer, the resistor value is 430 ohms. A 330 or 470 ohm resistor will do. Its value is not critical.

Construction. The simple circuit lends itself well to point-to-point wiring, which is probably the quickest way·to build the checker. If the builder prefers a neater appearance, the printed circuit board layout shown in Fig. 6-2 can be used. It is easy to duplicate with an etch-resist pen, or, if a professional look is desired, by photographic means. Figure 6-3 shows the parts placement for the circuit board and Fig. 6-4 shows the matching hole locations for mounting it in a 4 × 2-1/8 × 1-5/8 box. A Radio Shack 6.3 volt transformer, stock number 273-1384, was used for the circuit board layout and hole patterns. No on/off switch is used. The unit is simply plugged in for use. A TV cheater cord plug and socket are used so that the diode checker is easy to store without dangling AC cords everywhere.

Pin jacks, banana jacks or five-way binding posts can be used for connecting to the diode to be tested. The binding posts allow for a variety of connections to diodes that cannot be brought directly to the checker. Adapters with a V-notch are used so that loose diodes can easily be dropped into place for testing. Figure 6-5 shows two easy methods of making such adapters.

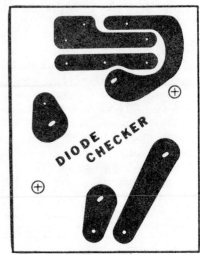

Fig. 6-2. PC board layout.

The holes should be cut in the box, and it should be painted and labeled to suit the builder's taste. The current limiting resistor and transformer should be mounted to the circuit board next. The AC plug and pin or banana jacks should then be mounted in the box, with short lengths of wire soldered to them. Next, insert the LEDs and bend their leads so that they will not fall out. Do not solder them yet. The LED leads must be long enough for the LEDs to protrude through the box, so the builder may have to add short lengths of wire. Now feed the four wires from the AC plug and the jacks through the proper holes in the circuit board, check for proper fit and clearance and solder the four wires. Position the LEDs and solder them to the circuit board. Put insulation on the inside of the box bottom to prevent any possibility of shorts to the circuit board. Don't forget that it has 117v AC on it! Make adapters to fit your jacks and you are ready to check out the unit.

Test the checker by applying power to the 117v AC plug. Neither LED should light. Now put a short circuit across the jack terminals and both LEDs should light. A diode that is known to be good should now be connected across the jack terminals. The LED closest to the diode's cathode should light. Try it both ways to make sure the LEDs are oriented properly.

After building and using this simple diode checker, the owner will find a desire to also know if the diode under test is silicon or germanium. Since a germanium diode will develop about 0.1 volts across itself and a silicon diode about 0.5 volts, it seems that some very simple circuit might be devised that would light an LED if the diode under test were silicon.

The Amazing Zener Sweeper Tests Zener Diodes

Have you ever acquired a packet of unmarked, untested zener diodes from one of the surplus dealers? Perhaps you have discovered that sorting and testing 100 such zener diodes is a rather formidable task.

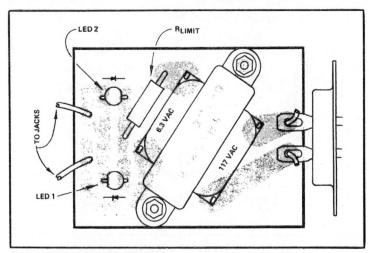

Fig. 6-3. Parts placement drawing.

The common method of testing a zener diode is by connecting the zener in series with a limiting resistor across a variable voltage source and metering the zener voltage (Fig. 6-6A). As the supply voltage is increased, the voltage across the zener diode will be equal to the supply voltage until the zener point is reached. At that point, the voltage across the zener will cease to rise with further increase in the supply voltage, and the meter will indicate the zener voltage. This test, however, does not make it readily apparent whether or not the zener under test is leaky. In Fig. 6-6B, this zener diode leakage is represented by RZ in parallel with the diode. The applied voltage will divide across RL and RZ until the zener point is reached, at which time it will stabilize. In this case, the voltage across the zener diode increased with increasing supply voltage, as it should, but it was less than the supply voltage.

The shortcomings of this method can be overcome by using a sawtooth test voltage and monitoring the zener voltage with an oscilloscope whose horizontal sweep is driven by that sawtooth. This is possible with an oscilloscope that provides front-panel access to its internally-generated sweep voltage, as shown in Fig. 6-7. The displayed ramp will rise to the zener point, beyond which it will be a horizontal line. If the zener passes current before the zener point is reached (is leaky), the ramp will be curved. If the zener does not go into complete conduction at its zener point (is resistive), the horizontal trace will continue to rise. This would seem the ideal method, but it does have several disadvantages:

- Not all oscilloscopes provide front-panel access to the sweep voltage.

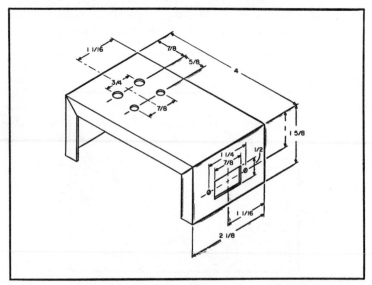

Fig. 6-4. Hole pattern for a 4 × 2-1/8 × 1-5/8 inch box. Cut holes to fit parts on hand. Holes are centered on centerline of box.

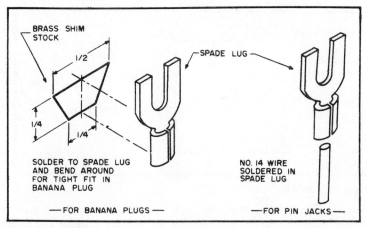

Fig. 6-5. Adapters.

- The trace on the screen often begins after the voltage on the sweep output terminals has risen to some value, making it impossible to test zeners of voltages lower than that value.
- It is somewhat time-consuming for use in testing any great number of devices. (Remember that packet of 100 unmarked, untested zeners?)

Perhaps you have anticipated the next step. A sawtooth generator whose output has a zero baseline and sufficient amplitude to exceed the voltage of the zeners to be tested will do nicely. Described hereafter is a tester that provides these features. The circuits, far from original, have been borrowed from various sources and modified as necessary to utilize the contents of a junk box.

Figure 6-8 is a block diagram of the tester. A sawtooth, controlled by a multivibrator, is amplified and applied to the test circuit. Jacks are provided for connections to the oscilloscope inputs and for leads to connect to the zener under test.

The heart of the tester is an integrator with a clamping transistor to reset the timing capacitor at the end of its timing cycle, which produces a

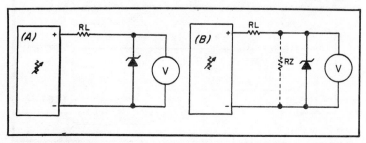

Fig. 6-6. Schematic.

121

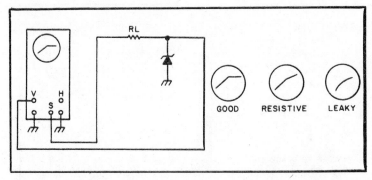

Fig. 6-7. Schematic.

sawtooth (Fig. 6-9). The required control voltage, EC, is obtained from an astable multivibrator. When EC is high, D9 is reverse biased and the positive voltage, applied through R8, holds Q3 in saturation. R12 is thus in parallel with C7, the timing capacitor. EO is then essentially zero. When EC goes low, D9 conducts and D10 and D11 are reverse biased. Q3 turns off and the circuit begins to integrate at the rate of $I_{R10} \times C7$ volts per second. This rise continues until the control voltage again goes high, driving Q3 to saturation and discharging C7. With the components shown, the rise is approximately 560 volts per second. The amplitude of the ramp produced is dependent on the timing rate, limited by the voltage applied to the integrated circuit.

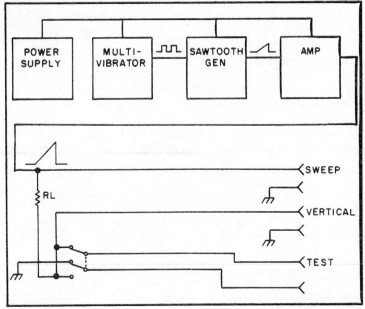

Fig. 6-8. Schematic.

122

Figure 6-10 shows the astable multivibrator that produces the control voltage. The timing components are C5, R5 and C6, R6. The time interval of conduction of the transistors is given by: $T = 0.692 \times C \times R$.

With the components shown, pulses of approximately 7 milliseconds duration are obtained with an amplitude of 18 volts. This is quite adequate to control the sawtooth generator.

The sawtooth from IC1, approximately 4 volts in amplitude and of 7 milliseconds duration, is applied to a two-stage direct-coupled amplifier (Fig. 6-11). Output from the amplifier section is a 40-volt sawtooth whose duration is somewhat less than 7 milliseconds, having lost a bit due to the bias on transistors Q4 and Q5. Q5, a GE152, is the only critical component in the tester. It must have good linearity, low leakage and an adequate voltage rating.

The power supply (Fig. 6-12) is designed to provide both positive and negative 6 volts from a single low-voltage secondary. A bridge rectifier is connected with equal loads on its positive and negative outputs. Equal current flows in both loads; hence, equal voltages of opposite polarity are developed. These power the LM741. The high-voltage secondary provides +40 volts for the output amplifier, which ensures that the test sawtooth will be sufficient amplitude for the range of zener diodes normally encountered in the shop where this tester is in use. R3 is used to drop the +40 volts to +20 volts to power the multivibrator.

The zener diode tester is quite simple to use. Connections are made from the tester to the vertical and horizontal inputs of an oscilloscope whose vertical amplifier is set for DC and whose horizontal amplifier is set for *external*. The zener diode is connected to the test terminals and a graphic representation of the zener characteristics is displayed on the oscilloscope screen (Fig. 6-13). The zener voltage may be read on the screen if the

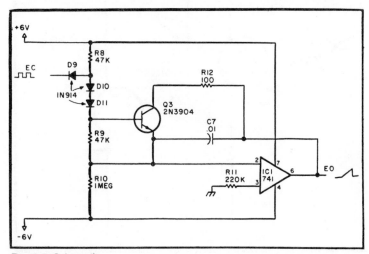

Fig. 6-9. Schematic.

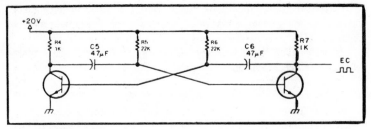

Fig. 6-10. Schematic.

oscilloscope is calibrated. A polarity-reversing switch is provided for convenience in switching the polarities of the test terminals. A normal zener will produce a distinctive trace, while a defective device will be readily appparent (Figs. 6-14 and 6-15), thus facilitating rapid testing. Figures 6-16 and 6-17 show the layout used in the protoype of this tester. Nothing in the layout is critical, which means a great deal of variation is possible, so you can make use of any available parts.

Though constructed only recently, this tester already has proven to be of tremendous value. Several previously inexplicable power supply prob-

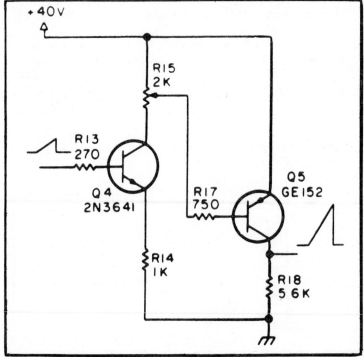

Fig. 6-11. Schematic.

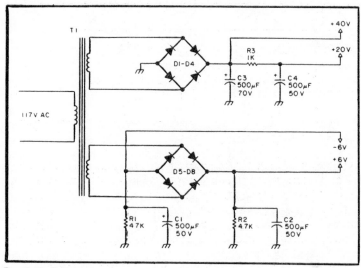

Fig. 6-12. Schematic.

lems have been resolved by demonstrating that a zener knee was not sharp. Additionally, more than 100 unmarked, untested zener diodes, purchased from a surplus dealer has been tested and sorted.

Presto Transistor Checker from VOM

Vacuum tubes and transistors have to be tested in special testers, usually in the shop or lab, to determine their condition or defect. This still holds true for vacuum tubes but, because the transistor has advanced so rapidly and become so popular in the electronic field, it has been difficult to test.

Many good transistor testers have been designed. But, they still require that they be used in the shop or lab. They may even be tested on an oscilloscope for performance to determine their operative ability or leakage condition.

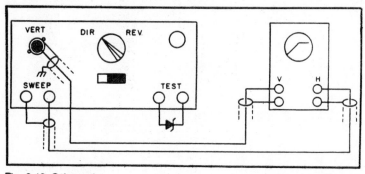

Fig. 6-13. Schematic.

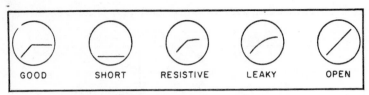

Fig. 6-14. Reactions of tester.

The Multimeter Test Meter, Simpson or Triplet type, has become the technician's and serviceman's most handy and important tool for servicing on the spot today. This Simpson test meter made possible the fast checking of tube circuits and other electronic parts on the spot on complex systems and important electrical and electronic devices. Today this same meter is being used on the new state-of-the-art printed circuit cards, solid state circuitry, transistors, diodes, etc.

This *adapter/conversion board* makes possible the simple change over of a common test instrument (Simpson or Triplet) into a transistor tester for on the spot testing of normal defects of transistors, diodes or other solid state devices. For a complete parts list, see Table 6-1.

Using the standard Simpson 260 meter, a printed circuit card adapter, with a circuit for testing general type NPN and PNP transistors and solid state components for defects, is "piggy-back, plugged-in" over the existing controls on the meter. The settings are then set to 50μA and DC on the meter controls. This results in now having a full transistor tester with new controls on the printed circuit adapter (Fig. 6-18).

Connect the PC board to the meter by way of the mounting banana male studs. This makes the necessary connection to the Simpson 50 microammeter direct. Add the clip-on battery, to complete the conversion of the

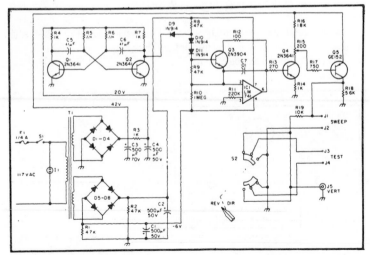

Fig. 6-15. Zener diode tester.

126

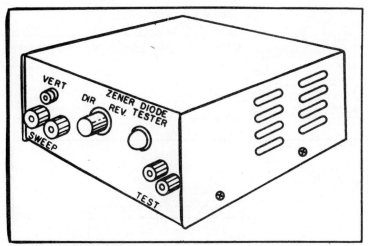

Fig. 6-16. Panel layout.

multimeter to a transistor tester. The controls on the adapter enable the operator to adjust the 50 microammeter for full scale. Or, scale 1.

This full scale of 1 can be used for the gain measurement of DC or alpha, which is less than unity. By using the selector switch on the adapter board, you can check the DC gain—Iceo, -Icbo, Iebo—the last three being diode action, reverse current and action test for front-back leakage. This takes

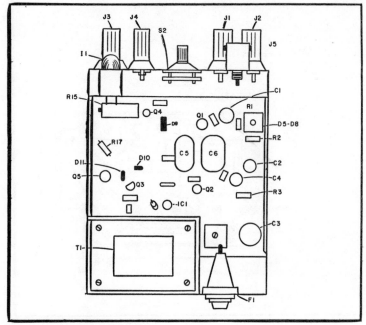

Fig. 6-17. Component location.

Table 6-1. Parts List.

No.	Name	Description
1	Switch—2 pole-5 pos. wafer type ¾" diameter non-shorting (calectro EE-163) or equal.	
1	Adj. pot—¼" mount—½" diameter case 50k (IRC).	
3	P.B.SW.—Mini-type, N.O. contacts-¼" mount (ALCO).	
2	DPDT SW.—¼" mount (ALCO-MST 215) or equal.	
2	Sockets—transistor chassis mount type—any type.	
2	Resistor—150 Ohm & 120k ¼ Watt.	
3	Phono jacks—chassis mount—¼" type (external).	
1	Bat. clip—(Keystone Enterprises) single #139.	
6	Banana plugs—male test type plug—⅛" or ¼" mount. (Note: May take apart a test lead-male plug and use tip, or banana).	
1	Battery—(Eveready type #523) or equal. Misc. decals, 4-40 & 6-32 nuts and bolts, etc.	

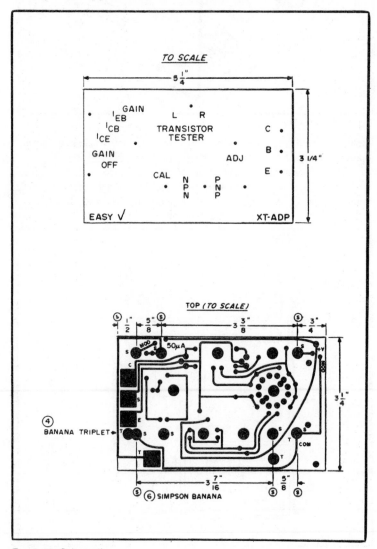

Fig. 6-18. Schematic.

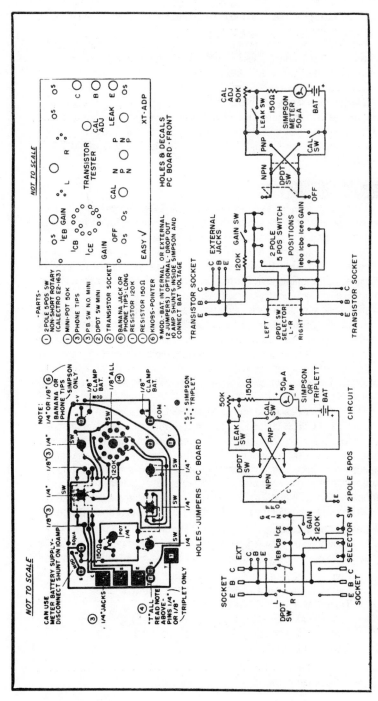

Fig. 6-19. Transistor adapter circuit for Simpson 260.

care of the leakage currents in the base, emitter and collector, and makes a quick gain check on the tested transistor.

For diodes, external pin connections "C and E" and left position on the selector, make this check possible. Transistor comparison testing is made possible with the "L and R" selector switch. Leakage switch is used to making the 50 microammeter more sensitive for lower leakage readings.

This tester adapter can be hand wired on a small board and adapted to other meters. This PC board has been designed for either the Simpson 260 or 270 and the Triplet 630 series. Mounting and drilling holes for the two types are marked as "S" and "T" Fig. 6-19. The board is called EASY XT-ADP.

Test Those ICs

There is a definite need for an integrated circuit tester with much greater flexibility and capability than those commonly found in current literature or available commercially within reasonable cost.

Many IC testers which have been described recently are designed to use either a dedicated IC socket for each IC tested, or some form of hardware programming device to interconnect Vcc, GND, logic in, logic out and other necessary control signals. Other types of testers commercially available employ multiple-pole slide or rotary switches for programming. Although both approaches are satisfactory for testing a limited number of elementary gates, counters and flip-flops, the cost to dedicate the very large number of IC sockets or other hardware programming devices necessary to test a great variety of chips generally has been considered too prohibitive; as a result, the testers previously described have a limited number of ICs they can test.

The IC tester described here is capable of testing a very wide range of DTL, TTL and CMOS digital logic functions, including single and multiple-input gates, inverters and buffers, flip-flops, counters, shift registers, latches, one-shots, pulse synchronizers, multiplexers/encoders, demultiplexers/decoders, arithmetic logic functions, switch debouncers, priority encoders, true/complement elements, parity generators and many others. Specific TTL and CMOS ICs which are testable are listed in Table 6-2. Others may be added to the list, but have not been investigated.

Design Considerations. The primary objective of this project was to design a quality tester which maximizes the number and types of digital ICs testable and to do so with minimum cost. Other design goals achieved were:

- Unit can act as an in-circuit, logic-state monitor
- Tester tests over 450 different TTL digital ICs
- Tests CMOS digital ICs (directly compatible)
- Tests 7400, 74H00, 74LOO, 74LSOO and 74SOO series
- Tests open-collector and totem-pole output TTL ICs
- Full CMOS input gate protection
- Internal three speed clock
- Vcc applied before any logic signals applied

Table 6-2. ICs Known to be Testable.

TTL (Includes N, H, L, LS & S series of 5400 and 7400 family ICs)
00,01,02,03,04,05,06,07,08,09,10,11,12,13,14,15,16,17,
20,21,22,25,26,27,28,30,32,33,37,38,40,46,47,51,54,55,
64,65,71,72,73,74,75,76,78,80,86,87,90,92,93,102,103,
106,107,108,109,110,111,112,113,114,120,121,122,123,128,
132,133,135,136,140,153,155,156,157,158,164,165,174,175,
176,177,183,190,191,192,193,196,197,260,266,278,279,280,
290,293,298,386,390,393,490.

CMOS (4000 family)
00,01,02,03,04,05,06,07,08,09,10,11,12,13,14,15,16,17,
.18,19,20,21,22,23,24,25,26,27,28,30,31,33,37,40,41,42,
49,50,66.

- Icc monitor provided
- Rapid, easy programming
- Tests both 14- and 16-pin dual in-line packages
- Six separate logic input and six separate logic output lines
- Rapidly tests many ICs of same type by using IC clip
- Each logic output line monitored independently
- Built-in 110v AC power supply
- +9v or +12v DC input capability for portability
- Convenient size, attractive case

The above features, and the logical design of the IC tester circuit, were arrived at by seriously examining the required parameters to be tested on the various IC chips. Two basic categories of parameters are normally specified for digital ICs—*dynamic* and *static*. Table 6-3 summarizes the basic differences between static and dynamic testing.

Dynamic parameters of digital logic chips include gate propagation delay, maximum clock toggling frequency, output rise and fall times and other properties designed into the chip which specify input and output conditions at very high clock input rates. Basically, dynamic testers determine how well the IC works.

Static parameters include the Vcc and Icc required and the fundamental ability of the logic chip outputs to correctly follow the various logic inputs for that chip, according to its respective truth table, i.e., "Does the chip work?"

Since ICs are manufactured to stated minimum and maximum specifications (reflected in their dynamic and static operation parameters), there is a very good degree of assurance that the IC will function properly in a circuit, if its static operation is correct and it is not required to operate above its minimum guraranteed operating frequency or otherwise outside its specified dynamic operating characteristics. The great majority of applications for digital TTL or CMOS ICs do not require near-maximum performance of the chip (except, perhaps, for VHF or UHF prescalers and gigabit logic applications, where ECL or other newer technologies are used). Therefore, a static tester will be quite adequate since, if the chip does work correctly, it can be expected to meet at least its minimum dynamic operating

Table 6-3. Static and Dynamic Testing Comparison.

	Static testing	Dynamic testing
Type of testing	Functional 　Truth table compliance 　Icc (quiescent drive 　　current)	Full parameter 　Propagation delay 　Max toggle frequency 　Noise immunity 　Min. clock pulse width 　Fan-out/Fan-in
Clock frequency	Low to medium	Near maximum rated
Complexity	Simple—switches and 　logic indicators	Very high—normally 　special-purpose or 　custom-designed ROM or 　microprocessor-controlled
Commercial cost	$500—$1000	$1500 (benchtop) to over 　$50,000 (production line)

characteristics in a particular design application.

Since it is terribly frustrating and time-consuming to either debug a logic circuit prior to getting it working or troubleshoot the circuit to find out why it has stopped working, a few simple quality-control procedures taken before installing the IC can save many headaches later. The easy-to-perform burn-in process, described later, followed by static testing with the tester detailed, will give you a very high measure of confidence that the ICs you use are good when you install them and that they will remain in working condition for many, many years,   if not subsequently damaged electrically or physically.

Circuit Description. The function block diagram for the IC tester is shown in Fig. 6-20. Each pin of the device under test (DUT) is interconnected by the switching matrix, as required for the particular IC tested, to +Vcc, GND, up to six logic input and six logic output lines, as well as four other signal lines. One side of the matrix has a line for each of the 16 pins and the other matrix side connects to the input and output circuits and controls.

Many switch matrix alternatives were evaluated in terms of cost, physical size and switch capacity required. The 20 × 20 pin-plug matrix board was discarded as both too bulky and costly. A programmable solid state switching array was found to be possible, but not practical, in terms of components and supply current required. A miniature jack and jumper pin matrix suffers the same problems as the plug-pin matrix and, additionally, introduces the requirement to store many jumper wires. Matrix cost and size were minimized by using an array of twelve 16-pin DIP switches, each containing eight SPST rocker switches. However, with only 96 switches available, they must be utilized to the very best advantage to interconnect the inputs and outputs of those ICs most likely to be tested.

The particular matrix organization shown in Fig. 6-21, was painstakingly selected to allow the maximum number of ICs to be tested with a convenient number of DIP switches. If desired, additional DIP switches can be added to increase the number of matrix crosspoints interconnected, consistent with the panel space available. Although it is apparent that many matrix crosspoints are not provided an interconnection by one of the switches, with only a very few exceptions, all input and output logic lines for the ICs listed in Table 6-2 are testable with this matrix arrangement. For those ICs having more than six input or output lines (e.g., an eight-bit multiplexer), the DIP switch matrix can easily be reprogrammed to test those logic lines not tested with a single DIP switch setup.

Input logic signals to the DUT are generated by depressing the correct combination of push-button switches, in conformance with the logic input requirements specified in the truth table—up=logic 1, down = logic 0. In addition, an internal clock can be supplied as a logic input signal for input #1, and fed through the switch matrix to the selected input pin of the DUT. Three clock speeds have been incorporated—a slow speed (1 Hz) for monitoring latching operatings, a medium speed (3 Hz) for toggling most gates and a high speed (10 Hz) for clocking multibit counters, shift registers, etc.

The +5 Vcc (Vcc and Vdd for CMOS), GND and other input signals lines are also connected to the proper pins of the DUT, via the switch

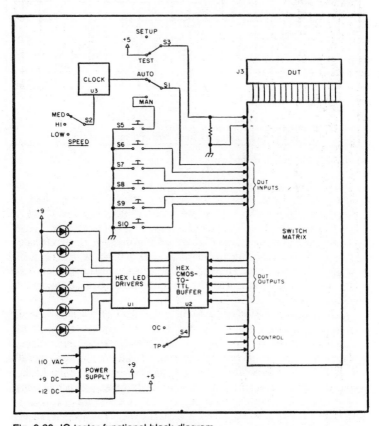

Fig. 6-20. IC tester functional block diagram.

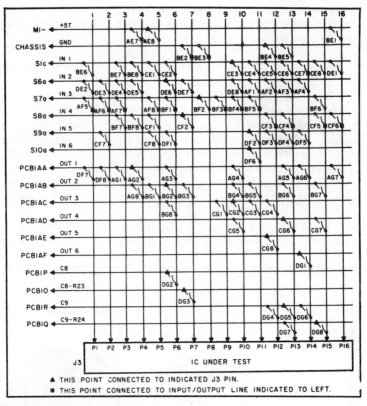

▲ THIS POINT CONNECTED TO INDICATED J3 PIN.
■ THIS POINT CONNECTED TO INPUT/OUTPUT LINE INDICATED TO LEFT.

Fig. 6-21. DIP switch matrix.

matrix. Monostable multivibrators (e.g., 74121, 84122, 74123) require timing resistors and capacitors for proper operation. These are R23, C8, R24 and C9 in Fig. 6-22, and have been selected to give approximately a ½-1 second low-high-low indication at the respective output monitor.

Output logic states of the DUT are led from the switch matrix through a hex CMOS-to-TTL buffer to individual LED drivers for display. A high output logic level (logic 1) from the DUT lights the LED, and a low output logic level (logic 0) is indicated by an unlit LED. Unused output LED indicators remain unlit.

During setup, S4 is placed into either the OC or TP position, depending on whether the DUT is of the open-collector or totem-pole output variety. This action connects or disconnects a pull-up resistor to each logic output pin of the DUT. The pull-up resistors are required only for open-collector ICs, since the output transistor collector is uncommited (Fig. 6-23) for high-voltage or high-current applications, whereas the totem-pole output transistor collector is connected internally to another transistor (Fig. 6-24) to provide direct TTL output levels without the requirement for an external pull-up resistor.

134

Power Supply. A conventional full-wave rectifier bridge (D6-D9) and a +5v regulator (U5) are included to power all logic circuitry, as well as the DUT. A separate +9v +12v DC input jack may be installed by those desiring complete portability. The external battery supply voltage is dropped to about +8 by D1-D4, to reduce the voltage drop across, and hence, the

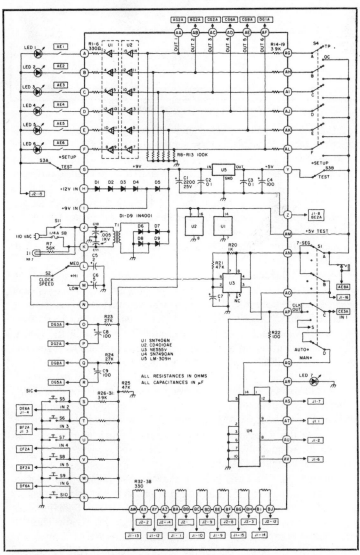

Fig. 6-22. Logic board (PCB-1) schematic diagram. Connection points inside rectangular boxes refer to external jacks (J1-6=pin 6 of jack J1) or PCB-2 switch matrix connection points (DF6A=pin A of DIP switch DF 6). N.B Insulate jack from chassis.

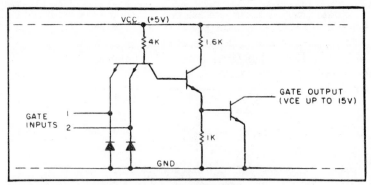

Fig. 6-23. Open-collector output gate (¼ SN7438N).

amount of heat dissipated by U5. The diodes also provide battery polarity protection. The +5v supply line powers the DUT and the tester circuitry only when S3 is in the TEST position. This prevents inadvertent application of input logic levels prior to application of Vcc to the DUT, since the input logic levels are themselves derived from the same +5v line. With S3 in the SETUP position, all input, output and Vcc lines to the DUT socket are returned to ground. An optional 0-100 milliammeter can be inserted in the Vcc line to the DUT to monitor the current required, or the Vcc line can be run to a closed-circuit three-way jack to permit external monitoring with your VOM.

7-Segment Decoder-Driver. Special provisions have been included on the logic printed circuit board to allow the popular SN7447A 7-segment decoder-driver to be tested. The output pulse stream from U3 is fed to an internal dedicated BCD counter, U4, and the four binary output lines are led directly to J1. Since this IC socket is dedicated for testing only the 7447A, the input and output lines required can be hardwired to J1 and J2. A common anode 7-segment display can be permanently installed at J2. When testing the 7447A, S1 is set to 7-SEG and J1 is used as the test socket instead of J3. U4, J1 and J2 are optional, but considered very useful to extend the tester's capabilities to the ubiquitous 7-segment decoder so popular in digital displays, counters, DVMs, etc. The internal BCD counter, which is hardwired to J1, allows the 7-segment display to cycle through all numerals 0-9 continuously. Additionally, the *all segments on* and *all segments off* tests can be performed on the 7447A under test.

Construction. The LMB 007-746 sloping panel cabinet was selected since it is both attractive and conforms to the physical layout requirements of the panel controls. The entire logic circuitry, shown in Fig. 6-22, including the power transformer, is implemented on a single PC board. The DIP switch matrix is on a second PC board. Thus, there are no components external to the PC boards except the required controls, LEDs jacks and switches. The switch matrix board (PCB-2) is wired (wire-wrap or point-to-point) prior to installation, to reduce interconnection time required between it, the main logic board (PCB-1) and the panel controls. The use of IC

136

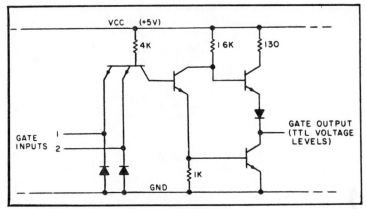

Fig. 6-24. Totem-pole output gate (¼SN7400N).

sockets on PCB-2 permits the DIP switches to be removed temporarily for other uses, if desired. PCB-1 is laid out to accept IC sockets for each IC, although the use of sockets is not mandatory. Ribbon cable is used to interconnect the PCBs and controls, although a combination of wire-wrap and point-to-point wiring can be used.

A significant measure of flexibility can be added to the tester by using an IC clip and ribbon cable which connects to J3. This permits very rapid testing of many ICs of the same type and also greatly extends the lifetime of J3. Either 14- or 16-pin Ics can be tested directly, since the DIP switch matrix programming accounts for pinout differences as long as pin 1 of the IC clip lead (or J3) is always connected to pin 1 of the 14-or 16-pin DUT when testing.

Operation of the IC Tester. Due to the wide variety of IC functional pinouts, it is necessary to set up the DIP switch matrix to interconnect the input/output signal lines to the proper IC pins for each different functional pinout, but not necessarily for each different IC. For example, Fig. 6-25 and Table 6-4 show a sample truth table page, which can be used for testing many IC types. There are four related parts to each page—the listing of ICs testable by the matrix settings on that page, the matrix settings, the truth table and, for easy reference, the pinout configuration for each IC listed on that page. For a complete parts list for this project, see Table 6-5 and Figs. 6-26 and 6-31.

Table 6-4. Truth Table Notation.

O =	Off
● =	On
❶ =	Off-on-off
↑ =	Up (Steady)
↓ =	Down (Steady)
X =	Up or Down (Don't care)
⌐ =	Release
⌐ =	Depress

137

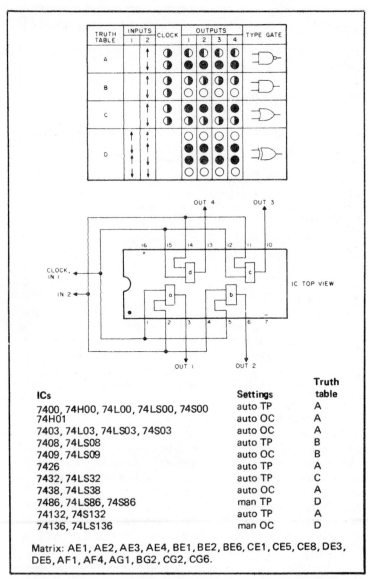

Fig. 6-25. Sample truth table. See Table 6-4 for explanation of notations.

To test an IC (e.g., SN74LS08N), the switch matrix and other required panel controls should be properly set before inserting the DUT into J3. S3 is first put into the SETUP position, S4 is set to either OC or TP and the DIP switch matrix is programmed. S3 is then placed to TEST and the operations called for in the input portion of the truth table are then performed. Each gate output is monitored separately on (in this case) output LED indicators

138

Table 6-5. Parts List.

C3	0.01μF/25 V disc
C2	0.1 μ F/25 V disc
C7	1 μF/10 V axial
C5	2 μF/10 V axial
C6	5 μF/10 V axial
C4, C8, C9	100 μF/10 V radial
C10, C11	.005 μF/1kV disc
C1	2200 μF/15V axial
D1-D9 (5*)	1N4001
I1	NE-2
J1-J3 (2*)	16 pin panel mount socket
J4*	RCA phono jack
M1*	0-100 mA meter
LED 1-6	Red LED
LED 7	Green LED
R22	100 Ohm, ¼W
R1-R6, R32-R38	330 Ohm, ¼W
R20	1k Ohm, ¼W
R14-R19, R26-R31	3.9k Ohm, ¼W
R23, R24	27k Ohm, ¼W
R21, R25	47k Ohm, ¼W
R7	56k Ohm, ¼W
R8-R13	100k Ohm, ¼W
S1	(Calectro E2-168 or equivalent) 4P3T rot. sh. sw.
S2	(Archer 275-325 or equivalent) SPDT on-off-on switch
S3	(Archer 275-1546 or equivalent) DPDT on-on switch
S4	(Calectro E2-169 or equivalent) 6PDT rot. nsh. sw.
S5-S10	(Archer 275-1547 or equivalent) SPsT MC NO PB sw.
S11	(Archer 275-611 or equivalent) SPST on-off rock sw.
T1	(Archer 273-1384)**6.3 V 300 mA transformer
U1	SN7406N
U2	CD4010AE
U3	NE555V
U4*	SN7490AN
U5	LM-309H (7805 is an acceptable substitution)

SPST DIP rocker, 16 pin (12 required) (Grayhill 76B08 or equivalent)
16-pin wire-wrap IC sockets (13 required) (1*)
16-pin low profile IC socket (1*)
14-pin low profile IC sockets (2*)
8-pin low profile IC socket (1*)
16-pin DIP clip (Pomona 3916 or equivalent)
7-segment display (Opcoa SLA-1 or equiv.) (1*)
Miscellaneous:
LMB 007-746 sloping panel cabinet
Ribbon cable (8 or 16 conductor)
AC line cord
Knobs
Hardware
Term. strip cinch CJ2005
Min. 3½ mm 3-way closed-circuit jack (Calectro F2-844 or equivalent)*
*optional
**PCB laid out to accept physical size of transformer listed

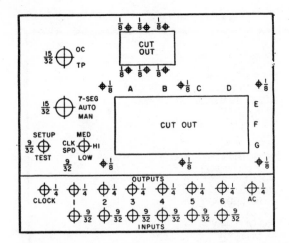

Fig. 6-26.
Front panel.

1-4. If three outputs agree with the truth tables and the fourth does not, the pinout diagram will tell you how that gate is being tested by which inputs. You may want to remove the pins to and from the defective gate and use the three functional gates later in another application.

To prevent the possibility of damage to the IC under test, return S3 to the SETUP position prior to removing the IC from the test socket or IC clip. No other precautions are necessary to use the tester for CMOS or TTL ICs, except for proper IC handling procedures and ensuring that pin 1 of the IC is always connected to pin 1 of J3.

A very flexible property of this IC tester is that, if equipped with the ribbon cable and IC clip, in-circuit monitoring of up to six logic lines of an operating IC can be done. The only precaution necessary for this type of operation are to ensure a common ground between the IC tester and monitored IC, input lines from the tester (by opening the appropriate DIP switches in the matrix) and to place S4 in the TP position. The six output LED monitors, however, can be connected, as desired, by the DIP switch matrix to monitor either input or output lines on the IC. The IC tester uses a CMOS-to-TTL hex buffer in the output monitoring circuit; therefore, loading of the monitored IC, whether TTL or CMOS, is not a problem since the typical buffer gate input current of 10 pA is low enough to not alter the fan-out of the monitored IC.

IN-QC (Quality Control). The procedures just outlined will test an IC to determine if it is working now, but cannot guarantee that the IC will not fail 10 minutes from now. Digital ICs are fairly well standardized and spot sample tested prior to shipment by most manufacturers. However, even in *prime quality* lots, a certain percentage may be found which are initially defective or which are prone to failure during the infant mortality period of use (first 48 hours under power). The percentage defective may vary widely between IC types of the same family; for instance, combined initial and infant mortality has been found to be as low as 1 in 200 for the SN7400N and as high as 1 in 15 for the SN7473N for the prime quality lots tested.

140

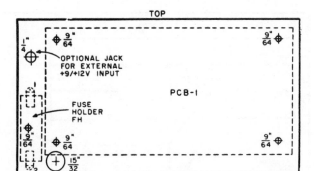

Fig. 6-27.
Rear panel.

In order to ensure that the ICs used in your projects are good, and will remain good for many years, the easy bake-in and testing sequences which follow are strongly recommended, particularly if IC sockets are not used in your projects. These procedures, while not absolutely fool proof, can identify up to 95 per cent of the faulty or faultprone ICs for 90 per cent of the common failure modes. ICs passing these tests have a very high probability of a useful lifetime in excess of 20 years, if not subsequently subjected to undue electrical or physical stress.

The procedure is accomplished by temperature stressing the IC and then testing its static operation in the tester described. Place the ICs to be tested on a cookie sheet and bake at 250°F (120°C). Remove and place in

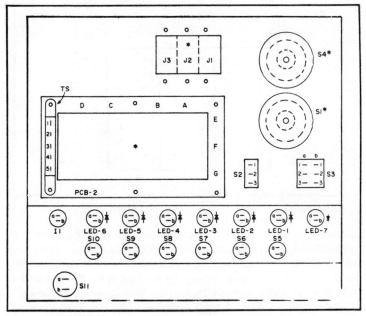

Fig. 6-28. Front panel interior parts plaecment. *See Fig. 10.

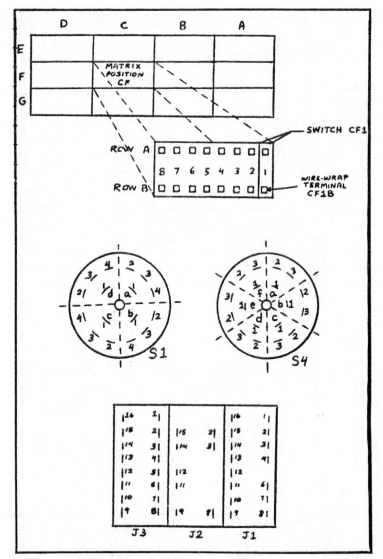

Fig. 6-29. Expanded views of matrix, S1, S4, and J1-J3.

the coldest freezer you have for 15 minutes. Rebake and refreeze a second, third and fourth time. Finally, bake the ICs at 250°F for 48 hours. After they cool, visually inspect each for casing cracks and test them on the tester. Experience has shown that only about 1-2 per cent of the ICs that tested good before the baking process failed the post-bake-in testing. These were the failure-prone ICs, which probably would have failed later in the operating circuit and required troubleshooting of the circuit.

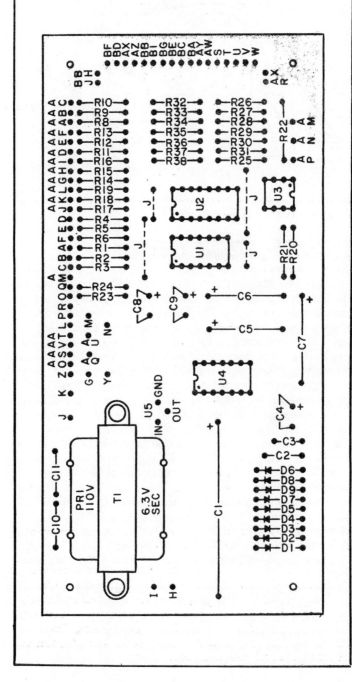

Fig. 6-30. PCB-1 parts placement.

143

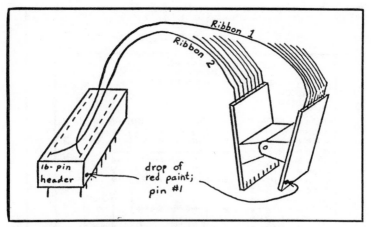

Fig. 6-31. DIP clip ribbon cable connections.

How to Find a Forgetful Computer Memory

Simple memory diagnostic programs for 8080-based systems have been described and illustrated for its use in troubleshooting 4K memory boards. While those methods are effective for localizing gross defects, such as chips which have one or more bits permanently set, more subtle defects may evade detection. For example, after discovering four bad chips in 8K of memory an attempt was made to load and run SCELBAL—but to no avail. After many hours of checking and reloading scattered addresses with the correct contents, it dawned on the programmer that when the contents of some addresses were corrected, the contents of other addresses were sometimes changing. What obviously was needed was a more thorough memory *debugging* approach. This project brings us one step closer to that end and SCELBAL is now running fine.

Two basic techniques are employed. One technique takes the value of an address and loads it as a data word into that same address. This is done for all of the 256 addresses on a page. Next, the values of the addresses are compared with their respective contents. If they are dissimilar, an error printout is produced. For example, if the eighth data bit (bit 7) at address 023 is always set to 1 due to a chip defect, the actual contents of that address will be 223, which would generate an error. However, if bit 0 of address 023 was always set to 1, no error would be detected. Needless to say, this would cause problems in an actual program which required that bit 0 at address 023 be set to 0.

One way that this problem can be overcome is to use a slightly different approach. Typically, this second technique loads a random number into all ¼K addresses of a page and then sequentially examines the contents of the 256 addresses to see if they still contain the pattern which was loaded into them. If enough random patterns are tried, a bit at some given address which is permanently set will sooner or later be discovered. However, a not uncommon problem is for two or more addresses to interact. If addresses 023 and 025 were interacting in such a manner that whatever was loaded into bit 7 at address 025 changed the corresponding bit in address 023, then this second technique would not detect the error by virtue of the fact that the whole page was loaded with the identical data word.

As if this isn't trouble enough, programs which check only one page of 256 addresses at a time are unable to detect interactions between two or more of the four blocks of 256 bits which actually reside on a single 2101 type memory chip.

This program overcomes the shortcomings just listed. When fully implemented, it provides a potent tool for detecting memory problems.

About the Program. In this system, the program, which is listed in Table 6-6 resides on page 043, address 200-320 (octal). Relocation should provide little difficulty once the mechanics of the program are understood.

The program loads a Test Word (TW), either 377 or 000 (octal), into one address in a 1K segment of the board. It then fills the remaining addresses of the four pages with a Field Word (FW), either 000 or 377 (the complement of the TW). When the four pages have been filled with the FW, the program retrieves the TW which was previously stored in memory and examines it to see if it was altered. If the TW has been altered, an error printout will be produced. If no error has occurred at that address, the next sequential address is loaded with the TW while all other addresses are loaded with the FW. Then the stored TW is once again tested. This process is repeated until each address in the four page block has been tested.

The program repeats the above procedure eight times (your option). The reason for repeatedly testing a 4 page block is that the repeated accessing of a particular area in memory will generate additional heat, consequently increasing the chance that marginal chips with thermal defects will be detected.

The programmer then sets up the program to check the next block of 4 pages on a board until the whole board has been checked. In order to keep the program simple for beginners, it was opted to have the programmer initialize each successive 4 page block. Obviously the computer could be made to do it, albeit at the cost of making the program more complex and harder to follow. After checking out the whole board, the programmer should change the TW and FW to the values given in brackets in the program, and then test the whole board again. You may encounter a situation where the 377 test word won't be altered by an interaction with a 000 field word, but a 000 TW will be altered by a 377 FW.

Table 6-6. A Typical Printout from DEBUG.

020	167	375	020 134 010
020	222	376	020 135 020
021	026	177	023 163 100
022	140	337	023 272 004

Table 6-7. Improved Memory Diagnostic Program.

Symbolic Address	Location	Machine Code	Mnemonic	Comments
DEBUG	043-200	021	LXI D	Set TW pointer
TWP	201	xxx		to start of block
	202	yyy		to be tested
	203	006	MVI B	Load reg B
TW	204	377 (000)		with test word
	205	353	XCHG	Switch regs D-E & H-L
NEXT	206	160	MOV M,B	Load reg M with TW
	207	353	XCHG	Switch regs D-E & H-L
	210	041	LXI H	Set FW pointer
FWP	211	xxx		to start of block
	212	yyy		to be tested
LOAD	213	175	MOV A,L	Test to make sure
	214	273	CMP E	FW isn't loaded into
	215	302	JNZ	same addr as TW
	216	225	OK	
	217	043		
	220	174	MOV A,H	Still testing to
	221	272	CMP D	make sure FW isn't
	222	312	JZ	loaded into same
	223	227	SKIP	addr as TW
	224	043		
OK	225	066	MVI M	Load reg. M
FW	226	000 (377)		with FW
SKIP	227	043	INX H	Advance FW pointer
	230	174	MOV A,H	Test to see if FW pointer
	231	376	CPI	is at start of next block
NB	232	aaa		Page number of next block
	233	302	JNZ	No, continue loading FW
	234	213	LOAD	
	235	043		
TEST	236	353	XCHG	Switch regs D-E & H-L
	237	176	MOV A,M	Retrieve stored TW
	240	270	CMP B	Has stored TW changed?
	241	304	CNZ	Yes, call ERROR routine
	242	277	ERROR	
	243	043		
	244	043	INX H	Advance TW pointer
	245	174	MOV A,H	Test to see if TW pointer
	246	376	CPI	is at start of next block
NB	247	aaa		Page number of next block
	250	302	JNZ	No, go back and load TW
	251	206	NEXT	into next addr in this block
	252	043		
LOOP	253	076	MVI A	Fetch loop counter value
LCV	254	010		from addr 254
	255	075	DCR A	Decrement LCV
	200-256	312	JZ	LCV=0?
	257	267	DONE	Yes - have tested whole block
	260	043		"LCV" times, so jump to DONE
	261	062	STA	Store the loop counter value
	262	254		in LCV
	263	043		
	264	303	JMP	Make next pass through
	265	200	DEBUG	the program for the
	266	043		present block under test
DONE	267	076	MVI A	Reset the loop counter
	270	010		value
	271	062	STA	Store the loop counter
LCV	272	254		in LCV
	273	043		
	274	303	JMP	Jump back to
	275	bbb	MONITOR	your MONITOR
	276	ccc		
ERROR	277	305	PUSH B	Save the TW on the stack
	300	114	MOV C,H	Move page number to reg C
	301	315	CALL	Output the page
	302	ddd	OCTOUT	via your octal conversion
	303	eee		and print routine
	304	115	MOV C,L	Move addr to reg C
	305	315	CALL	Output the address
	306	ddd	OCTOUT	
	307	eee		
	310	116	MOV C,M	Move the errant TW to reg C
	311	315	CALL	Output the errant TW
	312	ddd	OCTOUT	
	313	eee		
	314	315	CALL	Output A
	315	fff	CRLF	carriage return and
	316	ggg		line feed
	317	301	POP B	Retrieve TW from the stack
	320	311	RET	Return to main program

Table 6-7 illustrates a typical printout. Note that the page, address and incorrect data are outputted. As in the previous program, this program assumes the existence of an octal conversion and print routine ("OCT-OUT") which it can call.

A Few Tips. When you are making the test, do it once or twice with the case off the computer when it is relatively cool. Then put the case on and run through the procedure with the computer thoroughly warmed up. It would be possible to find four defective chips with the program when testing cool and four more when testing hot. Two of the cool defects may show up only sporadically under the hot condition. Also, this is a time when effective use can be made of the spray-type component coolers. You might also wish to sandwich the memory board(s) between other boards in order to entrap the heat.

Build the IC Experimenter To Get Started with TTL and CMOS

For both the novice experimenter and the advanced digital circuit designer, the two most essential tools of the trade are a basic power supply and a suitable signal source. Most modern digital circuit families (DTL, TTL, CMOS and NMOS) either require, or will satisfactorily operate from, a regulated 5v DC source. A one-ampere supply will suffice for most medium-size TTL projects and will more than exceed the needs of even the most ambitious CMOS undertaking. The signal source should provide squarewave (true and complement are helpful) and pulse outputs over a reasonable range of frequencies. A bounce-free variable one-shot output is also an absolute must.

The instrument about to be described will meet all of the above requirements; it is immune to accidental shorts which often occur in the course of trying a new circuit and will, therefore, be a useful addition to any laboratory.

Condensed Specifications. *Power supply*: 5 ± 0.25v DC at 1 amp, short-circuitproof; automatic thermal shutdown. *Astable oscillator*: 10 Hz to 100 kHz, providing simultaneous true and complementary square waves and 1 μs pulses, TTL-compatible, short-circuit protected; manual and/or remote gating. *Monostable oscillator*: Simultaneous true and complementary pulses variable from 5 μs to 50 ms. TTL-compatible, short-circuit protected; manual or remote triggering. The complete parts list is found in Table 6-8.

Power Supply. The power supply is shown schematically in the lower portion of Fig. 6-32. It is a basic full-wave rectifier, capacitor input supply, followed by an IC regulator capable of delivering in excess of 1 ampere at 5v, while being fully protected against sustained shorts and overloads. The power transformer secondary is rated at 16v c-t at 1.5 amps. These are the minimum parameters to guarantee proper operation under worst case line and load. The IC regulator (7805) has internal thermal protection and must be heat sinked to at least 15 square inches of 1/16″ aluminum sheet metal.

Astable oscillator. The heart of the oscillator circuit is U1, a CD 4047 IC. Its operating mode, i.e., astable or monostable, is controlled by the logic

Table 6-8. Parts List.

C1	4000 μF, 16 V, electrolytic capacitor
C2	0.22 μF
C3	0.022 μF matched set, polycarbonate
C4	0.0018 μF capacitors
C5, C8	120 pF
C6	10 μF electrolytic capacitor
C7	1000 pF, disc capacitor
C9	0.1 μF disc capacitor
CR1, CR2	1N4001 rectifier diodes
CR3, CR4	1N914 switching diodes
F1	0.25 Amp slow-blow fuse
Q1	2N2907 transistor
R1	1000-Ohm wire wound potentiometer
R2	1000-Ohm trimmer potentiometer
R3, R4, R5, R10	5600-Ohm, ¼-Watt resistor
R6	3300-Ohm, ¼-Watt resistor
R7	100-Ohm, ¼-Watt resistor
R8, R9, R11	1000-Ohm, ¼-Watt resistor
R12	3400-Ohm 1%, ⅛-Watt, metal film resistor
R13	4020-Ohm 1%, ⅛-Watt, metal film resistor
R14	Trim-on-test, ¼-Watt resistor, approximately 1000 Ohms
S1	SPST slide switch
S2	DPDT slide switch
S3	1-pole, 4-position rotary switch
S4	NO momentary push-button switch
T1	Power transformer, 115 V 60 Hz primary; 16 V c-t at 1.5 Amps secondary
U1	CD 4047AE oscillator
U2	74L00 low-power quad two-input NAND gate
U3	7403 open-collector quad two-input NAND gate
VR1	UA7805 IC voltage regulator
Misc.	Perforated circuit board, suitable enclosure, line cord, 5-way binding posts, control knobs, fuse clips, hookup wire, solder, hardware.

state at terminals 4 and 8. As shown in the schematic, with pin 4 grounded and pin 8 high, the astable mode is established. The oscillator may be inhibited by raising the level on pin 4; this is achieved by grounding either of the two inputs of gate U2A. The frequency of oscillation is determined by the charging time of the frequency range capacitors, C2-C5. When used in accordance with the IC manufacturer's recommendations, the frequency would be continuously varied by varying the resistance between pins 2 and 3. Unfortunately, this yields a frequency calibration proportional to 1/R and a highly nonlinear dial for a linear taper potentiometer. This problem has been overcome in this circuit by replacing the variable resistor with a variable current source, consisting of Q1 and R2 + R13. The base voltage of Q1, adjusted by linear potentiometer R1, linearly varies the current through R2, and hence inversely varies the effective resistance between pins 2 and 3. A perfectly linear frequency calibration is thus achieved. CR4 and R14 act as a temperature-compensating network for the current source.

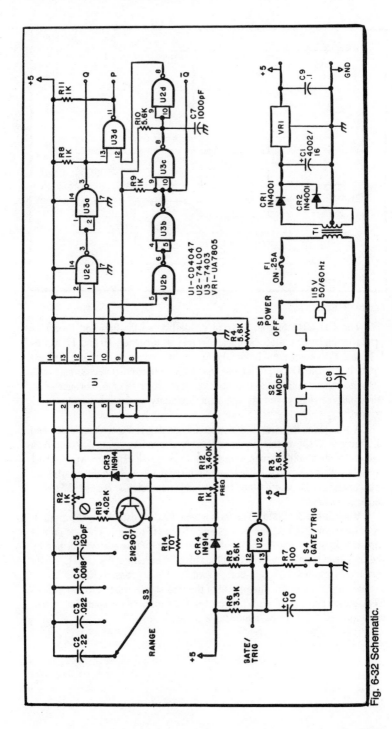

Fig. 6-32 Schematic.

149

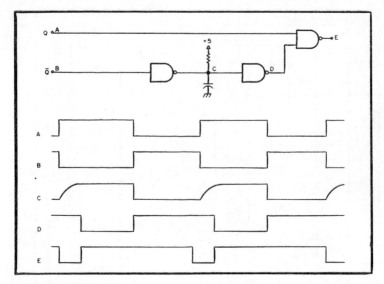

Fig. 6-33. Schematic.

The CD 4047 has an internal divide-by-two flip-flop and therefore produces highly symmetrical square waves at terminals 10 and 11. Gates U2C and U3A couple and buffer the Q output, and gates U2B and U3B act similarly for the Q̄ output of the oscillator. An open-collector gate was chosen as the output device, in spite of its slower rise time, because of its wired-OR capability and its inherent immunity to accidental shorts to ground.

A 1 μs pulse is generated by combining a delayed Q̄ with an undelayed Q. Figure 6-33 shows this circuit and its corresponding timing diagram. As a rule of thumb, a 6000-ohm pull-up resistor and a 1μF capacitor will yield a 1 ms delay. This relation is quite linear with capacitance. If, in your particular application, a pulse width other than 1 ms is needed, just change C7 to the required value.

Monostable Oscillator. The monostable mode is established by grounding U1-8 (through U2A) and by applying a logic "1" to pin 4. The one-shot is triggered by momentarily raising the level at pin 8. This is accomplished by the output of gate U2A, which will go high when either the GATE/TRIG terminal is grounded or push-button switch S4 is depressed. R7 and C6 act as a debouncing circuit. The period of the one-shot is related to the astable frequency, such that $T \approx 1/f$. A correction capacitor C8 shunts all range capacitors (C2-C5) in the monostable mode.

Chapter 7
Digital Voltmeters

With all the interest in digital electronics, and with prices of the related parts dropping, it seems natural to that much of the equipment in labs could be *"digitized."* The result would be that equipment would be much easier to read (no analog meters to interpret) and perhaps the readings would be more accurate. This is part of the justification for this project. It is the design of a digital voltmeter that is cheap and easy to build. It should be included with your power supplies, communications gear, etc. It would also be nice to have a digital multimeter, too.

You may be aware of the fact that many semiconductor houses offer *digital voltmeter* chip sets. These consist of an analog to digital converter chip and a digital counter array chip. All you have to add, at the most, are a power supply, a system clock, a reference source and a display system. That can easily mean six chips and up! This means high cost and lots of construction. But there is a better way.

Motorola has introduced a chip set that requires only a power supply and a display system. It's easy to work with and easy to calibrate. And the best part is that it's available and cheap (under $18 for the set), too! Accuracy is rated at 0.05 percent and that is probably worst case! Your voltmeter accuracy will depend upon the accuracy of the standard you use. This project you are about to build is the system the "Cheap and Dirty DVM" is based on.

Some of the disadvantages of the DVM should be mentioned. The first one is that it won't measure negative voltages. You have to reverse the input leads for that, like on an analog VOM. The second disadvantage is that the chip set has an input impedance of 4k ohms, but this is changed to 10 megohms input with a $2.00 op amp in this unit. None of these problems are serious—the first requires a quad op amp and some parts. The second has already been taken care of. If you want to read further about the Motorola chip set, call or write them for the data sheets.

Construction isn't too difficult, but there are several areas to watch. Probably the biggest problem you will have is finding a box to house the finished unit. If you want a gereral purpose meter to be used around your workshop, a minibox will work fine. However, if you want to build a digital panel meter to be installed permanently in a piece of equipment, you are in a sticky situation because digital panel meter cases are very hard to get. Use a small minibox if you're stuck. Cut a rectangular hole in the front of the box and put a red filter behind it. Then cut and bend aluminum stock into two L brackets large enough to fit on the ends of the box. You could then cut a hole large enough to accommodate the box front in your panel and use the L brackets to secure it. Possibly a better way would be to make up an aluminum plate larger than the front of the box and use it to secure the box to the panel. You would have to duplicate the display cutout in the plate, of course, but the result could look pretty neat!

If you have never built a digital voltmeter before, you are in for some interesting construction! If you have, these construction hints will sound

Table 7-1. Parts List.

C1—4.7 μF, 10 volt electrolytic cap

C101—100 pF disc cap

C2 —47 μF, 10 volt tantalum or electrolytic cap

C3 —0.01 μF disc cap

C4 —220 μF, 10 volt electrolytic cap

C5 —0.0033 μF disc cap

˜6 —2200 μF, 25 volt electrolytic cap

C7 —100 μF 25 volt electrolytic cap

D1 —1N4148 diode

D2, D3 —1N4002 diode

D4 —1N751 zener diode, 5.1 volts

IC1—LM308 Radio Shack

IC2 —MC1405L Circuit Specialists

IC3 —MC14435VP Circuit Specialists

IC4 —MC14511P Circuit Specialists

IC5 —LM340-5 Circuit Specialists

Q1 —2N3904 transistor

Q2-Q4 —2N3904 transistor

R1 —5k trimmer

R2 —20k trimmer

R3—10k resistor

R4 —10k resistor

R5 —470 Ohm resistor

R6 —2.2 k resistor

R7-R13 —150 Ohm resistor

R14 —27 Ohm, 2 Watt resistor

R15 —820 Ohm, ½ Watt resistor
All resistors ¼ Watt unless noted

DIS 1 —HP 5082-7730 Common
Anode Display

DIS 2 to DIS 4 —HP 5082-7740
Common Cathode Displays
DIS 5 Man 5024 LED
Misc: Case for meter, perfboard, assorted wire, etc.

familiar. The layout isn't too critical, and the parts may be placed nearly anywhere. For a complete parts list, see Table 7-1. The heat generating power supply should be located as far away as possible from the MC-1405 chip. In fact, try to get it in a separate module. Why the caution? Heat may cause drift of calibration if it is great enough. By the way, this circuit draws around 100 mA from the +5 volt line. Also, when you wire up the LM-308 and the MC-1405 circuitry, the grounds are critical. Improper grounding causes calibration errors, drift and other problems. You'll notice how all grounds come together at one point on the schematic in Fig. 7-1. This is the way to wire your unit. Remember that this voltmeter can be a laboratory quality instrument if you follow these simple precautions.

Six different units should be built. They should all be pretty much alike. First, assemble the display board. Three HP 5082-7740 LED readouts are used for the multiplexed part and an HP 5082-7730 readout is used for the 1 on the display. Note that the 7740 readouts are *common cathode* and the 7730 is *common anode*. You could use all 7740s if you wish. Or even Data Lit 704s. You name it in common cathode! A separate LED is used for the decimal point. This makes for easier reading at a distance. The three 7740s are wired up for multiplex operation. This means that all A segments are wired together, all Bs and so on. Homemade L brackets were used to attach the completed display to the rest of the electronics board.

The next part of this project is to get the electronics built and running. Figure 7-1 shows the schematic of the basic meter. This is the circuitry that is recommended by Motorola for their chip set, so you'll find it well described in their ap notes and bulletins. The schematic should help you get your unit built and on the air, pronto!

You could build your first units on scraps of copperclad perfboard, known as *groundplane* board. This method of construction works fairly well, because the grounds are very easy to make, but extra time is required to drill out the copper from holes where parts are going to be mounted. Regular perfboard works fine though, and you are welcome to use this method. It's also cheaper and you don't have to worry about shorting IC pins to a copper ground plane. You could also start building your units by installing the IC sockets and wiring up the grounds. On the non-ground plane boards, bring the grounds from all ICs to one point—pin 8 on IC2. Number 24 bare wire can be used for all connections. Then with the hard part over, wire up the rest of the unit. Note that the pin numbers on IC1 are for the TO-5 can. If you use the DIP version of the LM-308, you have to look up the new numbers. The driver transistors, Q1 through Q4, are not critical. Almost any silicon NPN unit with a beta of 100 or better will work for Q1. And any silicon PNP units with betas of 100 and up will work for Q2 through Q4. Consistently good results have been obtained with 2N3904 and 2N3906. If you can't easily buy the ICs, try Circuit Specialists, PO Box 3047, Scottsdale, AZ 85257. They can help. The MC1401 is $8.95 and the MC14435VP is only $7.95. The LM308 is available from Radio Shack.

As you finish up the electronics board, don't try to squeeze by through subbing single turn pots. They will be very hard to adjust and you probably

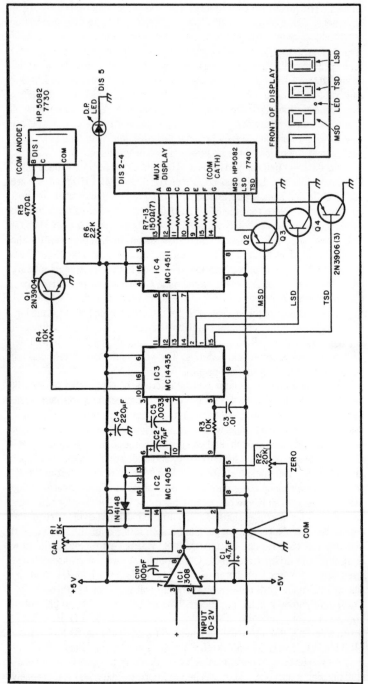

Fig. 7-1. Basic meter.

154

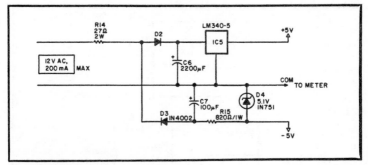

Fig. 7-2. Power supply.

won't be able to calibrate. Always use 10 turn wire wound units or better yet 20 turn units if you have the space. You can find them for 50 cents each at Poly Paks. The exact resistance values aren't especially critical, so you can sub pots fairly easily. Another thing you can do with the pots is to "remote" them by mounting them off the electronics board and attaching them with wires. Be safe and bring out separate ground wires for each pot. As before with the other wiring, terminate them at pin 8 of IC2. Finish up by attaching the display board to the electronics board with homemade L brackets on the units. Use resistors R7 through R13 to attach the display segments to IC4 and save yourself some wire.

You may need a power supply. You'll need plus 5 volts at about 100 mA, and minus 5 volts at about 10 mA. You might be able to borrow these voltages from other equipment if they are regulated. If not, build the optional power supply shown in Fig. 7-2, and power your voltmeter from a 12 volt filament transformer. One of the power supply's internal transformers had an extra 12 volt, 250 mA winding, so it was used. You could also use one of those line plug transformers used for calculators if you open the case and remove the diodes that are usually there.

You might want to add a range switch to the input of your new meter, or perhaps just change ranges. As built, it measures 0 to 2 volts. Figure 7-3 shows some ideas for attenuators, both simple and deluxe. The simple one was built into the power supply voltmeter. A 0 to 20 volt meter was needed, so a × divider was made. RB was 10k, 1 per cent and RA was 100k, 1 per cent. Ideally, these resistors should add up to 10 meg, the standard digital voltmeter input resistance, but in a power supply, these resistances aren't critical (no loading problem!). You can use whatever precision resistors you have for ×10, ×100 or even ×1000 dividers. If you don't need a divider at all, connect a 10 megohm resistor across the voltmeter input to cut zero drift when the test leads are open. The deluxe divider shown is what you would use in a digital multimeter. It features 10 megohms input resistance and overload protection. For best results use 0.1 per cent resistors throughout. You will also want some kind of decimal point switching on the display. That will mean another deck on the range switch and two more LED/dropping resistor combinations.

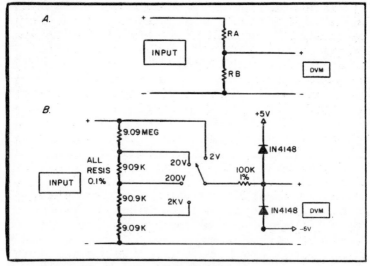

Fig. 7-3. Input attenuators, simple (a) and deluxe (b).

Calibration is quick and easy once you are set up. Obtain a DC voltmeter calibrator that is at least 0.01 per cent accurate. Or, lacking that, you can build the calibrator shown in Fig. 7-4. It is designed around the National LH0070IH 0.01 per cent 10 volt reference, which sells for about $5.00. Circuit Specialists might be able to get one for you. Power is supplied by two 9 volt batteries. Lacking this calibrator, you can either calibrate your voltmeter against another DVM, OR (shudder) with a 1.34 volt mercury battery. The battery is a last resort, because you might be able to get only about 1 per cent accuracy or so, depending upon the condition of the battery. It should be fresh and unused.

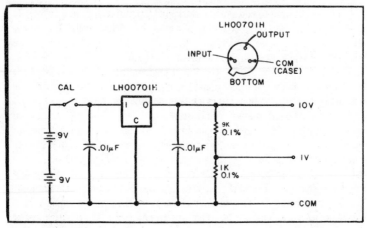

Fig. 7-4. Calibrator.

With that out of the way, connect up the voltmeter and apply power. Let it settle down for about 20 minutes or so and then short the input leads together. Adjust the zero pot, R2, for a 000 reading. You'll see the meter read something like 087-093-097-098-099-000-001-002 if all's well.

Once the zero is adjusted, apply either ×2.000 volts from the commercial calibrator (preferred) or +1.000 volts from this calibrator. This is assuming you have the basic 0 to 2 volt unit; for other units increase the input voltage by ×10, ×100 or whatever. Then tweak R1, the calibration pot, for a 1.999 to .000 reading (or 1.000 with a 1 volt calibrator). Go back and short the input leads to check the zero; if it is necessary to adjust zero, recheck the calibration, too. These adjustments interact somewhat.

I hope you like your new digital voltmeter. It really adds a touch of class to have equipment in the workshop with digital readouts. The price is right, too. You probably spent about one-quarter as much for your meter as you would have for a commercial unit!

Build a 3½ Digit DVM

If you have been looking for something to update your meter, but the price of a new one and the price of a portable digital voltmeter has scared you away, you'll love this project.

One of Motorola's new CMOS devices—the MC14433 —is an analog-to-digital converter with a 3½ digit display which can be set up for either a 200 mV or 2v full scale reading. The MC14433 is a high performance, low power 3½ digit A/D converter combining both linear CMOS and digital CMOS circuits on a single monolithic IC. It is designed to minimize the use of external components, and with two external resistors and two external capacitors, the system forms a dual slope A/D converter with automatic zero correction and automatic polarity selection. For ease of use with batteries, the MC14433 may operate over a wide range of power supply voltages.

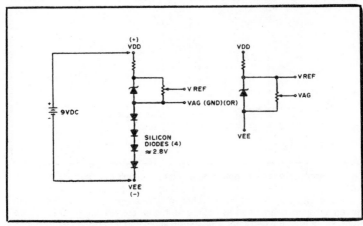

Fig. 7-5. This circuit is not recommended for use with LED displays.

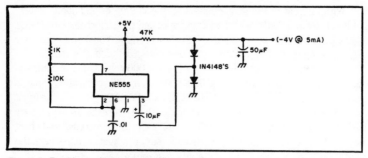

Fig. 7-6. Positive voltage to negative supply.

When you get one, make sure that it is on its own little piece of conductive foam, and don't take it off until you are ready to put it on the circuit perfboard.

This project is really a bare-bones layout with a minimum amount of functions, but all that is required to upgrade to a full function DVM is a few more resistors and switches. Since the MC14433 requires both a positive and a negative supply, it necessitates either the use of two batteries, or some other way to generate a negative voltage from a positive source. This is really quite easy and there are a couple of different methods to obtain it. One easy way to get it is by the method shown in Fig. 7-5. In this example, a 9 volt supply can be used, with 3v between V_{ag} and V_{ee}, leaving 6v for V_{dd} to V_{ag}. This sytem leaves a comfortable margin or battery degeneration (end of life). Note that due to the current requirements of the LEDs, this method is recommended for use only with LCDs. Another method is shown in Fig. 7-6. This method uses the old reliable NE555. Since this thing generates a square wave, why not use only the negative cycle? Looking at the circuit in Fig. 7-7, we see that the 555 is connected in a regular oscillator fashion. The

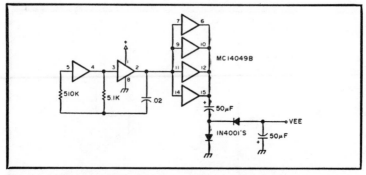

Fig. 7-7. Another method of obtaining a negative supply form a positive supply. When only + volts is available, a negative supply voltage can be generated with this circuit. Two inverters from CMOS hex inverter are used as an osillator, with the remaining inverters used as buffers for higher current output. This square wave output from the oscillator is level-translated into a negative-going signal. This signal is rectified and filtered. A voltage of +5 V will result in a —4.3 V output.

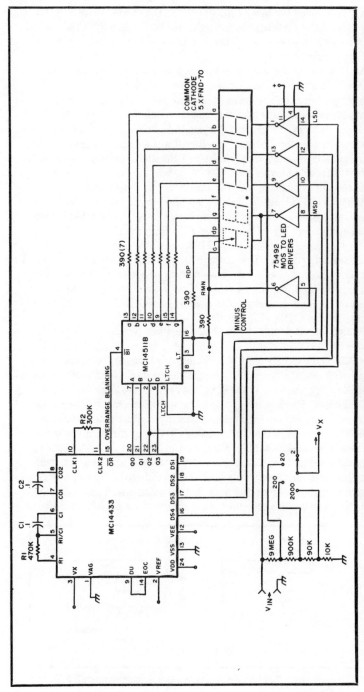

Fig. 7-8. For discrete LEDs, parallel all segments, but connect only segments B and C to MD (1). When using 5 display units, tie the cathode of MSD and the minus display unit together. For 2 V full scale: R1 = 470K, Vref = 2 V. For 200 mV full scale: R1 = 270, Vref-200 mV.

output, pin 3, is fed through a capacitor to the junction of two diodes. D1 allows the negative cycle to pass through it, and D2 allows the positive cycle to go through it to ground. After filtering, this negative wave is amazingly transformed into DC!

An idea for saving money is to use potentiometers in the area of precision resistors on the input circuitry. Sure, precision resistors would be the way to go, but as long as we are being cheap about this thing, let's go all the way. There are definite values of resistors required for proper operation of the input circuitry, but instead of trying to find the closest thing in your junk box and hoping for the best, we will start with a value less than what is required and supplement it with a miniature 10 turn pot, which on the surplus market is relatively inexpensive. That way, you will have an even more precise resistor combination than you could get by ordering it.

Although a DVM can be made on a printed circuit board, a perfboard with sockets will do just fine, as parts layout is not really critical. One precaution, though: Try to keep wires away from the clock resistor and wave-forming circuits. Unlike with TTL devices, one cannot leave unconnected leads unconnected. Due to the extremely high impedance of these CMOS devices, you must tie the unused leads to a high or to a low. This circuit can be used with LEDs or LCDs with some changes, but in the interest of the local economy (your wallet), go with the popular FND-70 common cathode LEDs.

Printed Circuit Boards. More should be about printed circuit boards. Double-sided PC boards with plated-through holes are available, and the price will be in the vicinity of $4.00 to $6.00. This board has provisions for a few more frills and the price of a kit using that board sells for $39.95. Write for details to Dacron, Inc., 12609 Blackfoot Trail, Round Rock, TX 78664.

Calibration. The first thing to do in the way of calibration is to set the 200 mV reference voltage (or 2 volts, depending on which option you take). Do this with any accurate meter or another DVM, as the accuracy of the DVM depends upon this setting. Next, with an ohmmeter, set the value of your resistor strings to equal the required resistance, e.g., 5 megohms with a 5 meg pot for the required 9 megohms. Do this with all the resistors. When you have adjusted these to their approximate value, insert them into the circuit. Now you can fine tune the pots to the exact value. Note that when

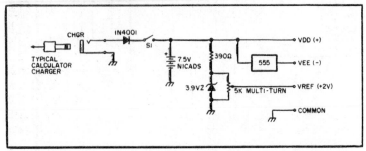

Fig. 7-9. Voltage chart. Total power requirements are approximately 60-70 mA.

you adjust one pot, it will affect the values of the other ranges. This may take a while, but when it is complete, you will have a very accurate voltmeter.

Operation. This is the easiest of all. All that is required is that you feed it the voltage commensurate with the range it is in. While other common voltmeters can take a few "prangs" with the meter movement, do not try to measure 150 volts with the switch in the 2 volt range, for if you do, you will find yourself ordering another MC14433. The schematic and voltage chart for this project are found in Figs. 7-8 and 7-9.

Super DVM with LCD

If you are not only interested in replacing your old reliable Simpson 260, but are also interested in knowing what makes those funny liquid crystal displays work, your next voltmeter should contain only devices which won't suck up all the juice out of the batteries. Before we get into the actual circuit construction, let's start from the top with design theory.

The Display. The operation of a field effect liquid crystal display depends on changing the optical properties of a liquid crystal by applying an electric field. The best short description of a liquid crystal is that it is an ordered fluid. Crystals of this type (liquid) which are used in displays belong to a class called *nematic*. Fluids of this type consist of cigar-shaped organic molecules with the long axis of each molecule pointing in the same direction. There are three main types of chemicals which are used in displays. These are *Schiff-bases, esters* and *biphenyls*. At the present time, the Schiff-bases are the best choice for displays, taking into consideration switching times, reasonable threshold voltages, lifetimes, good temperature ranges and expense.

The Motorola MLC 400 is constructed from two pieces of glass coated with transparent indium oxide conductors. These conductors are shaped to form the segments of a numeric display. The glass surfaces are also specially treated to align the liquid crystal molecules in a particular direction. Alignment is parallel to the plane of the glass, with the alignment direction of the top rotated 90° relative to the alignment of the bottom plate. This causes the liquid crystal molecules in the cell to assume a twisted orientation, when viewed from top to bottom. The plane of polarization of plane-polarized light will follow this twist and emerge from the cell rotated 90°. Thus, if the cell is placed between crossed polarizers, the polarizers will transmit light. Where an electric field is applied, the liquid crystal will align parallel to the field, twist will be destroyed and that portion of the cell will appear dark between crossed polarizers.

The MC14433. The MC14433 is a high performance, low power, 3½ digit A/D converter combining both linear CMOS and digital CMOS circuits on a single monolithic IC. The chip is designed to minimize use of external resistors and two external resistors and two external capacitors, the system forms a dual slope A/D converter with automatic zero correction and automatic polarity. The MC14433 is ratio-metric, and, by itself, may be used over a full scale range from 199.9 millivolts to 1.999 volts. Systems using the MC14433 may operate over a wide range of power supply voltages for ease of use with batteries. In addition to DVM/DPM applications, the MC14433

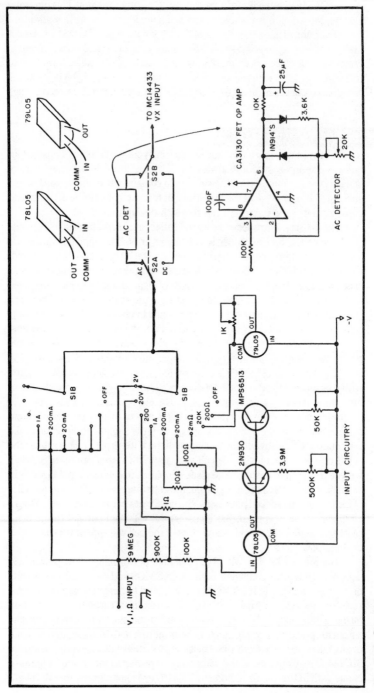

Fig. 7-10. Circuitry.

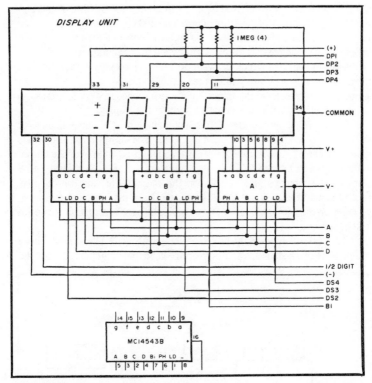

Fig. 7-11. Display unit.

finds use in digital thermometers, digital scales, remote A/D, A/D control systems and MPU systems, and has an input impedance of greater than 1000 megohms!

This A/D system performs a ratio-metric A/D conversion. That is, the unknown input voltage, Vx, is measured as a ratio of the reference voltage, Vref. Therefore, a full scale voltage of 1.999v requires a reference voltage of 2.000v, while a full scale voltage of 199.9 mV requires a reference voltage of 200 mV. Both the Vx and Vref are high impedance inputs.

The circuit in Fig. 7-10 performs parameter-to-voltage conversions and scaling and function switching. The AC/DC DPST switch changes the input path and the signal is then fed into the MC14433. A 10 megohm voltage divider consisting of three precision resistors provides 2, 20 and 200 volt ranges. Three precision shunt resistors are connected directly from the input to ground, providing 1 A, 200 mA and 20 mA scales.

The resistance scales are established with calibrated current sources using the MC78L05 and MC79L05 voltage regulators. A stable 5 volts above the minus supply is produced by the 78L05 positive regulator. The current sources are simple base emitter biased transistors. A 2N930 with a guaranteed beta at 1 microamp is used for the 2 megohm scale, and an MPS6513 is used as a .1 mA source for the 20k ohm scale. Each is adjusted by a single 10-turn pot. The 200 ohm scale current source uses a 79L05 negative regulator. Its input is connected to the negative supply and a scaling resistor is placed between the common and output pins. When not in use, this circuit draws only a few microamps of bias current, even though it sinks 10 mA when measuring a connected load. All current sources are biased from the minus supply to increase battery life; thus, all resistance scales produce a negative sign on the display.

When breadboarding the circuit, you could take the minus supply for the input circuitry from the common -6.2v bus. The only problem with this is that the 78 and 79L05 regulators require about 2 to 3 volts over the output voltage to work properly. Thus, the regulators will require at least 7 to 8 volts on the input to give 5 volts on the output. For this reason, it is recommended that you use 9 volt batteries. Note that absolute maximum voltages on the MC14433 are + and − 8v, so be sure to drop that portion of it somehow.

The real substance of the project is shown in Figs. 7-11 and 7-12. Three MC14543N LCD latch/decoder drivers are used to demultiplex, decode the

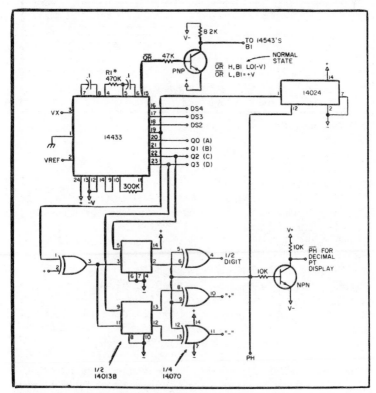

Fig. 7-12. For V full scale = 199.9 mV, set Vref = 200 mV and R1 to 27k. For V full scale = 1.999 V, set Vref = 2 V and R1 to 470k.

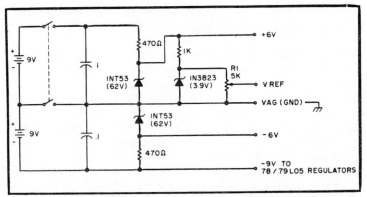

Fig. 7-13. Power distribution. Set R1 for an output voltage of 2 V. It must be accurate and with load connected. Any zenor diode of equivalent value may be used.

three digits and drive the LCD. The half digit and polarity are demultiplexed with the MC14013B dual D flip-flop. Since the LCD requires an AC signal across it, the low frequency square wave drive for the LCD is derived from the MC14024 binary counter, which divides the digit select output from the A/D. Although this is a convenient way to obtain the required square wave, it is not necessary to take it from here. The frequency should be about 4 kHz, as this will provide for the best contrast on the display unit. This low frequency square wave is connected to the backplane of the LCD and to the individual segments through the combination of the output circuitry of the 543B and the exclusive OR gates at the outputs of the 013B. All of the decimal points are tied to PH through a 1 megohm resistor, and, to display a particular DP, it is switched to PH.

The overrange pin (15) goes low when Vx exceeds Vref. It is normally high. The 543Bs require a ground on pins 7 (blanking) to display. In our case, the ground is actually the most negative supply (to get the maximum amount of voltage swing on the output). This normally high OR pin on the 433 is tied to a PNP transistor, which is tied between −V and +V. In its normal state, the transistor is not conducting, thus allowing the −V to be on the collector. When the input goes low, indicating an overrange condition, the transistor conducts and places a high (+) on the blanking input of the 543s, thus blanking them. Note that the first digit has no provision for blanking.

You can wire-wrap the LCD socket and the 543Bs. One reason for this is because the pins of the socket of the LCD are very close together.

Calibration. The first thing to do in the way of calibration is to set the reference voltage. Note on the schematic that it can be set up for 2 volts or 200 mV full scale (Figs. 7-13, 7-14 and 7-15). It should be set for 2 volts due to possible noise problems, but, even with the 2 volt scale, it can be read to .001 volt. Be sure to be accurate with this reference voltage, as the accuracy of the entire instrument depends upon it. A short word should be said about the quality of components used in the frequency determining resistors and capacitors, especially in the capacitors. These .1 uF caps should be of the high quality polyester or mylar. Using cheap caps here can lead to inaccurate readings.

If you used precision resistors for the voltage divider network, the next step is to calibrate the ohms scales against known resistance values. Just tweak the control until the correct value is displayed. If you used pots in place of the precision resistors (in the interest of saving money), you must set these up. The way to do this is to get a calibrated voltage source. Set the voltmeter to the 200 volt range, and adjust the 100k pot for a correct reading on the meter with about 90 to 120 volts applied to the input. Watch out, as the 433 doesn't really like all that voltage, especially if it is applied directly to the 2v scale. It might blank out forever! After you have calibrated the 200 volt scale, don't touch that pot again. Apply about 16 volts to the input and adjust the 900k pot for a correct reading. Next, do the same thing for the 2 volt scale, using the appropriate voltages. After this, you will probably have to touch each of the pots up, as each will interact with the other. With a little time, you can save yourself a little money, and probably have a more accurate meter than if you were to buy precision resistors!

There are a few things which you should be on guard for and those are in the area of the AC detector circuit. Since this is essentially an amplifier/detector circuit, anything that is placed on the input will show up as a DC potential on the output. The Ac detector circuit takes more time to refine than does the rest of the DVOM. On the switch assembly, don't run the PH and PH lines with unshielded wire because the input is a very high impedance, and therefore doesn't take very much to drive it. So, a little shielding and careful placement should take care of it.

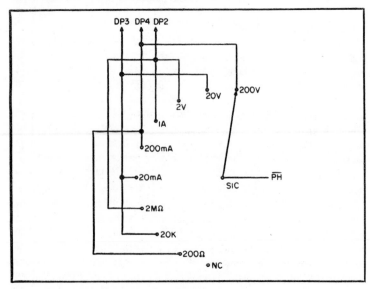

Fig. 7-14. Decimal point switching. To energize DP, tie line to PH.

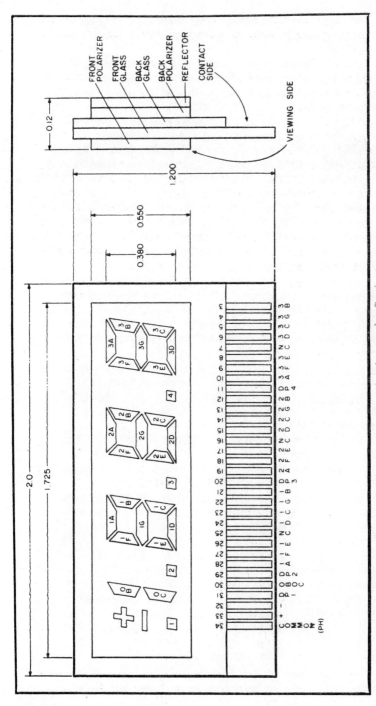

Fig. 7-15. Diagram of MLC400 liquid crystal display. Use connector Teledyne Kinetics S202U.

167

To calibrate the AC scale, the easiest way is to set up the voltmeter to the 200 volt scale and measure the line voltage of your house. If it does not read correctly, turn the pot until it does.

Another little hint—be sure not to hook up the analog ground to chassis ground as if you are measuring current in a high voltage circuit. This will place a high potential on the bare metal chassis. The chip is available from Tri-Tek for $18.95.

PC boards for a basic 2v DVM using LEDs can be purchased from Dactron, Inc. 12609 Blackfoot Trail, Round Rock, TX 78664.

Ecstasy In Multimeterland with an Autoranging Marvel

The most frequently used and trusted piece of test gear is the VOM. New ICs have changed the appearance of the VOM by taking a giant step in providing numeric readouts in the place of the sometimes difficult-to-read meter movement. Even if the greater accuracy was not considered, the ability to match voltage currents and resistors was greatly improved. However, not until recently have the IC manufacturers been able to provide the function we have all been waiting for—autoranging!

The operation, construction and calibration of a complete 4-3/4-digit autoranging, autozero, autopolarity digital multimeter capable of measuring DC or AC volts (.1 millivolt to 2.9999kv), DC or AC current (1 uA to 2.9999 amps) and ohms (.1 ohm to 29.999 megohms). The autoranging feature will automatically provide the proper decimal location to give the most significant digits possible. A single 5v supply is all that is needed for operation of the meter.

Circuit Description. The block diagram shows how simple the circuitry has become with the introduction of Intersil's 8052 analog signal conditioner and General Instrument's AY-3-3550 4-3/4-digit DMM integrated circuit. (Figs. 7-16 through 7-19).

A voltage of unknown amplitude is applied to the variable gain amplifier. Ideally, the output of the amplifier will be between +2v and −2v. The output will be converted to DC if the AC/DC switch is in the AC mode. This unknown voltage will be applied to the input of the 8052 (IC11) only when the AY-3-3550 (IC3) is ready to make a measurement. When ready, the sample switch will be enabled for 10,000 counts, the sample switch is turned off, the comparator output is sensed for polarity by IC3 and the polarity data is used to force the dual slope integrator to integrate in the opposite direction. IC3 will store the count required for the integrator to cross zero. This count will be directly related to the amplitude of the unknown voltage. If the integrator does not reach 0v by 20,000 counts, the unknown voltage is too large. In this case, the variable gain amplifier is set to reduce the unknown voltage by a factor of 10 and start the sampling over again. If the count is less than 1800, then the variable gain amplifier has insufficient gain. In this case, the gain will be increased by 10, and the sampling will repeat. Note that as the gain is changed, the decimal point is shifted.

Between voltage samples, the autozero circuit of IC3 will recalibrate specific capacitors so that the new measurement will start from zero.

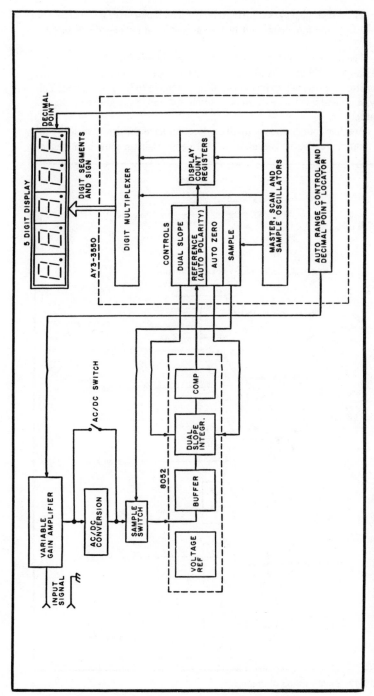

Fig. 7-16. Autoranging DMM block diagram.

169

The 5-digit display is controlled by the digit multiplexer internal to IC3. An LED to the immediate left of the display is on for negative measurements.

The circuitry is complicated by the requirement of an accurate positive and negative voltage reference for dual-slope integration. A "flying capacitor" technique is used to bias all input voltage to IC11 to +1 volt. This way, a ground level appears to be −1 volt, and only a positive reference is required.

Another complication is that IC3 does not have the drive capability for the numeric readouts. Therefore, a 7447 decoder driver must be added for

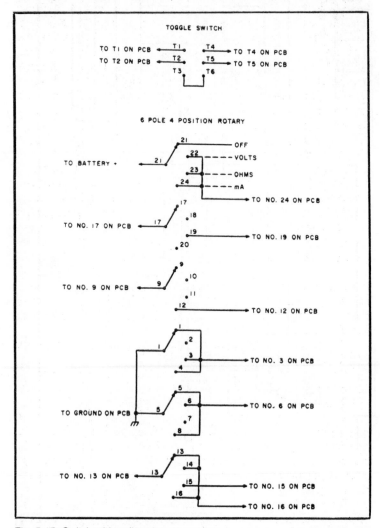

Fig. 7-17. Switch wiring diagram.

segment and discrete transistors must be used for digit drivers. Resistors are required to limit base currents and additional resistors are required to limit segment current.

In order to control the variable gain amplifier, sample switch and autozero circuit, analog switches must be used. The three analog switch packages are powered from a ±8.2v supply. The control input to these switches must also be ±8.2v. Therefore, a level converter is necessary to convert the +5 ground level of IC3 to ±8.2v. The three LM339 ICs were added for this purpose. Leftover portions of the LM339s were used for gating functions.

IC3 has inputs which provide the capability of changing the upper and lower limits of the autorange circuitry. Since all three functions use different limits, a 4052 (IC4) multiplexer was required. IC4 senses the mode of operation and applies the desired range limits to IC3. For example, the upper range for volts in nnnn.n and the upper range for ohms is nn.nnn.

The ±8.2volt supply was developed experimentally. A 555 (IC7) oscillator drives the primacy of a transformer. Each side of the centertapped secondary forms a half-wave recitifier circuit with one being positive and the other being negative. Each voltage is regulated at 8.2 volts to limit the maximum output voltage.

Since IC3 is a digital multimeter, not a digital voltmeter, use of existing circuitry to measure resistance and current is simplified. In addition to providing two inputs to IC3 for the function desired, slight modification to the variable gain amplifier is required. When reading ohms, the unknown resistor becomes the feedback resistor for IC12, and the feedback resistors become the input resistors for IC12. Current measurement is made possible by measuring the voltage drops across a 1-ohm resistor.

Construction. The use of a double-sided plated-through printed circuit board simplified construction. For a complete parts list, see Table 7-2. The most important decision, prior to inserting parts as shown on the assembly drawing (Fig. 7-20), is to consider the mounting of the PC board. The front panel is ¾" from the board. It was with this in mind that most capacitors were mounted on the bottom side. The transformer also presents a problem and is mounted on the bottom.

The four 18-turn pots can be mounted on the bottom. This will permit calibration without removing the front panel.

The rotary switch must be insulated from the board using nonconducting washers.

Do not remove the nut on the switch; doing so can change the stop pin for the switch. Wiring from the switch to the PCB is simple, since the points on the board contain the same number as the points on the switch. The common lugs are slightly recessed. When wiring to a common lug, go to the number specified and use the recessed lug (the one closest to the shaft). Switch lug-to-switch lug wiring should be done according to the switch wiring diagram (Fig. 7-17).

The AC/DC toggle switch should be wired as shown in Fig. 7-17. The PCB is marked for easy switch-to-PCB wiring. Be careful when installing this switch in the panel as it is easy to put it in upside down.

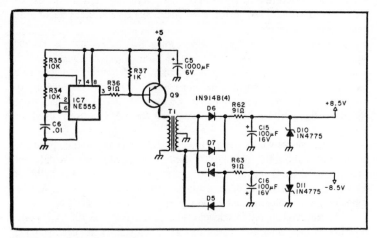

Fig. 7-18. +8.2-volt supply.

Use caution when inserting the ICs. Note that pin 1 is located away from the NRO, except for IC4 and IC5.

T1 wiring is aided by PCB marking. The white, black and yellow leads are closest to the NRO. Only the yellow and white leads are used (marked Y and W). The white, blue and red leads are connected to the W, B and R points marked on the PCB.

Calibration. The most critical adjustment is R40—the 1.0000-volt reference. This is the one calibration point where it is desirable to borrow an extremely accurate voltmeter. An easy point to pick up this line is switch 2 lug 15. Simply adjust R40 until a 1.0000-volt reading is measured. With the DMM in the volts mode; short the voltage input to ground and adjust R48 for a zero reading on the DMM display.

Connect a known-value AC voltage to the voltage input, and adjust R50 for a proper reading on the DMM display. Make sure the AC/DC switch is in the AC position. Insert a 1k to 1.799k precision resistor across the ohms input (AC/DC switch in DC position), and adjust R39 so that the DMM display gives the correct value for the resistor.

If an accurate voltmeter cannot be obtained, then calibration can be accomplished by the following procedure:

- Short out the voltage input leads, set the meter to DC volts and adjust R48 for a zero reading.
- Obtain a precision resistor above 1800 ohms (10k to 17.9k desirable), set the meter to the ohms mode and connect the resistor to the meter leads. Adjust R40 until the DMM gives the proper value.
- Obtain a precision resistor below 1800 ohms (1k to 1.79k desirable), set the meter to the ohms mode and connect the resistor to the meter leads. Adjust R3 until the DMM gives the proper value.
- Set the DMM to read AC volts. Connect the voltage probe to SW2 pin 15 and adjust R50 until the DMM reads 1.0000 volts.

172

Table 7-2. Parts List.

R1-R6, R15, R16, R35, R37	1k, ¼W, 5%
R17, R18, R32-R34,	10k, ¼W, 5%
R-7-R14, R36, R62, R63	91 Ohm ¼ W, 5%
R19*	10 meg, 1%, 1 Watt, 3500 volt (TRW CGH-1 or equiv.)
R20-R28, R38, R46, 1R47,	
R52-R53,	100k, ¼W, 5%
R29	270k, ¼W, 5%
R30	24k, ¼W, 5%
R31	22 meg, ¼W, 5%
R39	500 Ohm pot
R40	1k pot
R41	680 Ohm, ¼W, 5%
R42	100k, 1%, RN55C
R43	10 meg, 1%, RN55C
R44, R56	10k, 1%, RN55C
R45	1 meg, 1%, RN55C
R48, R50	7.5k, ¼W, 5%
R49, R51	10k pot
R54	100 Ohm, 1%, RN55C
R55	22.1k, 1%, RN55C
R57, R60, R61	20k, 1%, RN55C
R58	6.2k, ¼W, 5%
R59	15k, ¼W, 5%
R64	1 Ohm, 1%, 3W
R65	1k, 1%, RN55C
C1, C7, C8, C4	.1 μF disc ceramic
C6, C9, C11	.01 μF disc ceramic
C2	.001 μF disc ceramic
C3	100 pF disc ceramic
C5	1000 μF, 6 volts
C10	100 μF, 6 volts
C12	2.2 μF, 6.3 volts solid tantalum
C13	330 pF disc ceramic
C14*	.22 μF polypropylene
C15, C16	100 μF, 16 V
C17	10 pF disc ceramic
C18, C19	1 μF mylar
IC1, 2, 6	LM339
IC3	AY-3-3550 General Instruments
IC4	CD4052
IC5	SN7447
IC7	NE555
IC8, 9, 10	CD4066B
IC11	ICL8052ACPD Intersil
IC12	LF355
IC13	747C
D1-D9, D12, D13	1N914B
D10, D11	1N756A, 8.2 V zener, 5%
LED 1 and 2	FLV150
NR01-5	FND507
Q1	TIS 92
Q2-Q8	TIS 93
Q9	2N2905
T1*	Archer 273-1381 transformer
SW1	DPDT subminiature toggle switch
SW2	RCL 16-ECB-4J 6-pole 4-position rotary switch

Miscellaneous
Printed circuit board
Spacers, screws, banana jacks, pointer knob, case, front panel
*Substitutions not recommended.
The following items are available from : SOA Products, P.O. Box EG0256, Melbourne, FL 32935:
1 double-sided plated-through hole PCB-$15.00
Kit #1: 1 PCB (see above), 1 General Instruments AY-3-3550, 1 Intersil ICL8052ACPD: total price-$44.95
Kit #2: Complete kit of all components itemized above including case and front panel: total price—$99.95
All orders add $2.50 postage and handling. Florida residents add sales tax. Master Charge and BankAmericard accepted.

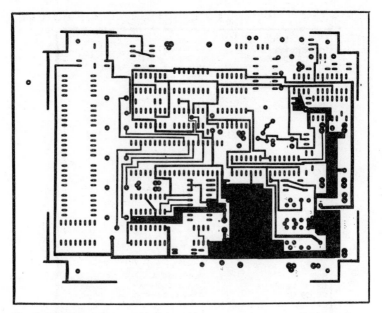

Fig. 7-19. PC board.

Operation. The actual operation of the DMM is relatively easy. To measure voltage, move the function switch to VOLTS, and insert the meter leads in COM and VOLTS. The reading you get will always be in volts. To

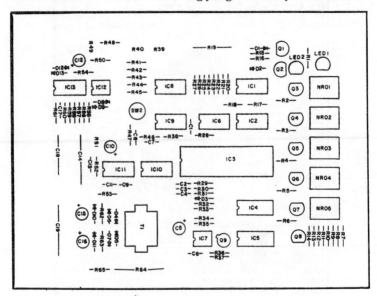

Fig. 7-20. Component layout.

measure current, select the mA function and insert the meter leads in MA and COM. All measurements will be in mA. To measure ohms, select the ohms function and insert the meter leads in OHMS-C and OHMS. All measurements made without the MOHMS indicator on will be in kilohms. If the MOHMS indicator is on the reading is in megohms.

Chapter 8
Calibrators and Frequency Standards

Inexpensive HF-VHF Frequency Standard

There have been many frequency standard projects published in books in recent years using advanced technology integrated circuits, however, there is a need for a foolproof circuit that is both inexpensive and easy to build in a modest workshop. Most projects neglect the need for 30 kHz and 300 kHz markers in VHF FM operations, where 30 kHz is the standard channel spacing. Also, most 2m FM frequency standards neglect the needed 100 kHz and 10 kHz markers are used by the HF man. This prompted an interest for a simple design frequency standard which could develop both 30 kHz and 10 kHz markers as well as a host of other frequencies.

This standard uses readily available TTL integrated circuits. The integrated circuits should not cost more than $4.50 total. You may have a crystal in your junk box, but an International EX Crystal should work as well. It costs $4.95.

The block diagram of the frequency standard is shown in Fig. 8-1. Note that a 5 volt regulator was used to furnish the 5 volts needed by the TTL circuits. By using a regulator, an automobile 12 volt battery can be used to power the frequency standard when tuning up the mobile rig in the car. Also, a lantern battery may be used to power the unit for portable use as on field day. The 9 Mhz oscillator is composed of a hex inverter operating in the linear mode. The 9 MHz output from the oscillator is divided by three to generate the 3 MHz signal used by the rest of the circuit. Two divide-by-ten circuits are then used to generate the 300 kHz and 30 kHz marker outputs. When 100 kHz outputs are required, another divide-by-three circuit is inserted between the 3 MHz signal and the divide-by-tens. Since there are two different outputs on each BNC connector, a small LED is used to signal when the second divide-by-three is in the circuit. The LED monitors the output of the divide-by-three and when it is outputting, the LED is illuminated corresponding to 1 MHz, 100 kHz or 10 kHz selections. When the LED is not illuminated, then the outputs are 3 MHz, 300 kHz or 30 kHz.

The circuit diagram of the frequency standard is shown in Fig. 8-2. A 7404 TTL hex inverted is used as the crystal oscillator. A small 2-8 pF trimmer capacitor is used to permit zeroing the crystal with WWV. A 7476 TTL dual J-K flip-flop is connected in a standard divide-by-three configuration. This same circuit is used later as the divide-by-three in order to

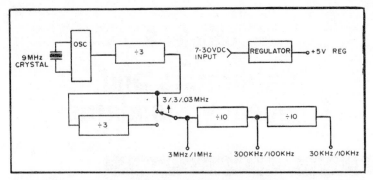

Fig. 8-1. Block diagram of the frequency standard. Selection of the 1 MHz/.1 MHz/.1 MHz is made by the toggle switch.

produce the 100 kHz markers. A toggle switch selects either the second divide-by-three or bypasses it depending whether 100 kHz markers are needed. Two 7490 TTL decade dividers are then used to furnish the lower frequency markers.

The TTL circuit layout is not critical. The standard, however, should be built close together to keep down stray capacitances. To help those who wish to duplicate the circuit, circuit boards are being made available. Alternatively, the circuit may be built on a small vector board. Sockets may be used to help in the point-to-point wiring.

The board was mounted in a small aluminum box and BNC connectors were used for the outputs. The frequency standard is quite readable at 150 MHz and probably higher. Because of the square wave output, the harmonic content is very high. To calibrate the oscillator, the 100 kHz mode is chosen. The 100 kHz is then compared with WWV at 5, 10 or 15 MHz. The small trimmer capacitor is then adjusted until a zero beat is noted. Using this method, the harmonics at 146 MHz should be no more than a few Hz off.

The World's Cheapest Calibrator Even Works on 2M!

Having trouble netting your transmit crystals on your 2 meter handie-talkie or transceiver? Can't buy or conveniently borrow a frequency counter to do the job?

With today's proliferation of sensitive, narrow band repeaters, it is increasingly important to be right on frequency to obtain maximum range and good audio quality from your own gear when "making the system." Even when operating simplex the tolerant wide band receivers are rapidly disappearing. Most would agree equipment should be within about ±1 kHz of the nominal channel frequency for optimum results.

You may be tired of borrowing a frequency counter every time you put new rocks in your equipment or change the channel locations of some of them and can't see putting out $100 or more for a frequency counter. How about adding a 2 meter converter to your low band receiver with its stable 1 kHz calibration? So what if the low band receiver is for SSB-CW and your 2 meter gear is FM. The frequency of the unmodulated FM carrier would be just as recognizable as a continuous CW signal.

In looking around for a 2 meter converter for your receiver if it should happen to be a Heathkit SB-301 low band receiver, you'll discover Heath Company no longer makes the Model SBA-300-4 2 meter converter. However, you can pick up a used one in a flea market for about $10.! The Heath Model SBA-300-4 2 meter converter output is a 28-30 MHz signal that is inputted into the four SB-301 half megahertz 10 meter band increments. With the proper crystal in the converter, any 2 MHz segment of 2 meters is tunable with 1 kHz resolution on the SB-301. Other 2 meter converters are available or you can build one yourself.

Netting 2 meter crystals is a snap now. First, check the calibration of your receiver and converter. You can verify it by comparing it to the output frequency of one or more of the accurately calibrated repeater stations when keyed by an unmodulated FM carrier with the receiver in the CW position. Use a minimal antenna or back off the rf gain if necessary, so your receiver is not overloaded and the S-meter reaches a maximum without pinning the needle. Reset your receiver dial hairline if required for accurate calibration.

The actual netting of your 2 meter crystals is simple. Feed your transceiver, preferably into a dummy load. Turn your low band receiver sudio gain down to minimum to prevent audio feedback. Use little or no receiver antenna or reduce rf gain if necessary to avoid overloading. With the low band converted receiver set to the desired 2 meter frequency, adjust the netting capacitor for maximum S-meter reading by alternately keying and adjusting the 2 meter gear. Double check your work by varying the receiver frequency slightly while keying your unmodulated 2 meter

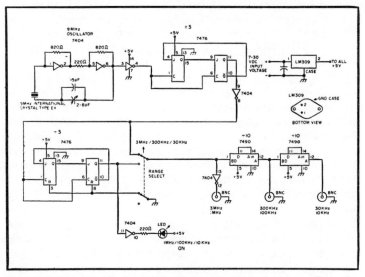

Fig. 8-2. Schematic of the inexpensive frequency standard using readily available parts.

179

transceiver. The S-meter reading should peak sharply at the desired frequency, especially with the receiver in the CW mode.

You can get to within about ±1 kHz using this—and for $10 that's a good bargain! As a bonus, a low band receiver-converter can be used to receive FM (somewhat poorly though) in an emergency by detecting it off the bandpass slope in the low band receiver.

All Band Frequency Marker

Crystal controlled marker generators are useful adjuncts in any frequency determining situation requiring high accuracy, such as locating band edges, sub-bands and calibrating receivers. If you've been entertaining thoughts about construction of one, a version is described here which uses the new C/MOS integrated circuits powered by a 9 volt transistor radio battery. And instead of the usual rotary harmonic selector switch, a multipin IC connector strip and three test plugs serve as a miniature patch panel to enable various divisions of the reference crystal, with a maximum countdown of 256. Rocks from 100 kHz to 4 MHz oscillate readily in this circuit. In this model an FT241 xtal set to 400,000 Hz has been chosen for control and has usable receiver calibration divisions down to 2.5 kHz. The harmonic spectrum extends to at least 160 MHz, the tuning limit of a transistor superregen used in testing. When used in densely occupied HF bands, an AM beeper can be switched on as an identification aid.

Referring to Fig. 8-3, one third of a hex inverter makes up a crystal controlled oscillator and buffer, another third is a slow rate pulser and the two remaining units function in the dividing section. These are all standard circuits described in RCA's COS/MOS Data Book #SSD-203. An emitter follower minimizes loading on the IC outputs, speeds up rise time to increase harmonic content and provides a low impedance output. The AM beeper is a simple clamp that gates rf on or off to following stages. A complete parts list can be found in Table 8-1.

All components are mounted on Vector P pattern perfboard that fits inside a Bud mini box. Sleeving ⅜" (10mm) long is slipped over the wire trap terminals of the contact strip to space it up from the board. A DPDT center-off miniature toggle switch acts as one board to panel spacer. Diagonally across from it, a 4-40 threaded rod conducts emitter follower output up through the front panel via a ½" (13mm) insulating spacer and plastic shoulder washers. Two regular 4-40 screws and spacers complete the four corner mounting. This spacing allows the contact strip to project partly through a panel cutout so that it is mechanically secure without fastening.

The completed assembly has a stick-on label with patching connection callouts for various division ratios. If only one crystal is employed, labeling could indicate most used frequencies instead. A typical frequency vs. division listing for this model is shown in Table 8-2. You can easily make up a complete table of all possible ratios, remembering that each CD4015 shift register divides by even numbers *only*, starting at 2 and ending at 16.

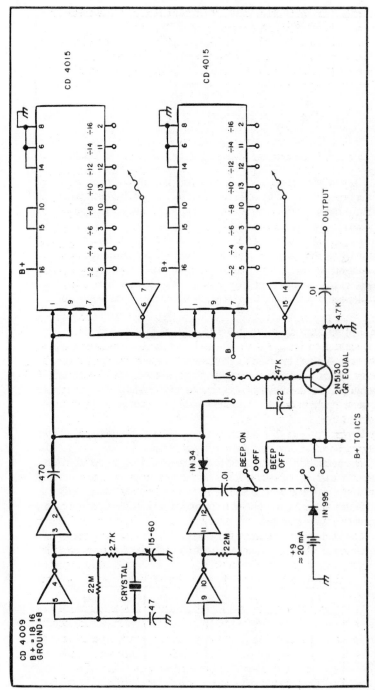

Fig. 8-3. Schematic.

181

Table 8-1. Parts List

1	CD4009	1	15/60pF trimmer
2	CD4015	1	.01 μF
1	1N34	1	xtal; see text
1	1N995	1	xtal socket
1	2N5130 or equiv.	1	Alco #MST205P switch
1	2.7k	1	Bud #CU-2115HG minibox
1	4.7k	1	Vector #44P29-062 perfboard
1	47k	1	#216 battery
2	22M	1	Battery connector
1	22pF	3	16 pin IC sockets
1	47 pF	3	Augat patch pins or equiv.
1	100pF	1	20 pin contact strip
1	470 pF		SAE # Series 7000

Uses to which a marker generator may be put have been described before: i-f alignment, BFO, scope linearity, etc. A type that divides down to the audio range like this one is especially useful in checking superhets. A very broad and flat spectrum of overlapping signals is generated and an audio tone will be heard no matter where the set is tuned. If its tracking and sensitivity are top-notch, the S meter will hold steady over the tuning range. Tracking adjustment amounts to tweaking for maximum meter reading or loudest audio tone. Then patch for 100 KHz markers and check calibration. It's a lot faster and easier than using a conventional signal generator.

A Synthesized IC Frequency Standard

Synthesized receivers are gradually appearing in the marketplace. Their numbers will increase as the cost of complex ICs drops. They all operate basically the same, using the phase lock loop (PLL) technique of generating L. O (local oscillator) frequencies. One crystal is used to generate the whole range of L.O. frequencies—each frequency as accurate as the crystal.

Here's a project that will provide you with several precise (.005 per cent) frequency standards and also get you familiar with phase lock loops.

The heart of this unit is the NE562 phase lock loop. Internally it has a phase comparator, adjustable low pass filter, a VCO (voltage controlled oscillator) and two outputs. It also has provisions for inserting a programmable divider between the phrase comparator and VCO, thus permitting frequencies to be changed.

The frequency range of this device is 1 to 10 MHz in 1 MHz steps, with an accuracy of .005 per cent at each frequency (greater accuracy is possible

Table 8-2. Frequency vs. Division Listing.

Divide by	Output, kHz	Divide by	Output, kHz
1	400		
2	200	20	20
4	100	40	10
8	50	80	5
10	40	100	4
16	25	160	2.5

with a closer tolerance xtal). The output is a symmetrical square wave at each frequency that is useful as a scope timebase calibrator, checking bandwidth and frequency response of HF amplifiers, calibrating communication receivers with the 100 kHz output or as a "clock" for future IC projects.

All of the ICs (including the NE562) are readily available from electronic mail order houses. The remaining few parts you can get from your junkbox.

How It Works. Before we get into the details of construction, a little basic operation of phase lock loops may be helpful. A basic phase lock loop consists of a phase comparator, a VCO and a low pass filter, as shown in Fig. 8-4.

The phase comparator (or phase detector) has two inputs—the reference frequency and the output of the VCO. The output of the phase comparator is a DC control voltage proportional to the phase difference between the two input frequencies. The control voltage equation is $V_c = K$ (Θ ref-Θvco) where K is a constant in volts/radian and (Θref-Θvco) is the phase difference in radians. Now the VCO frequency is controlled by a DC voltage. As the frequency of the VCO tends to drift, a phase error develops and is fed back to the comparator which compares it with the reference signal and produces a DC voltage that changes the frequency of the VCO in the direction that will reduce the phase error. Since the VCO is controlled by a DC voltage, any ripple on the voltage will FM (frequency modulate) the VCO. That's the purpose of the low pass filter between the phase comparator and VCO, to reduce this ripple. It also helps to set the capture range of the loop.

One drawback of Fig. 8-4 is that it will only work at one frequency. The output frequency is the same as the reference and there is no way to change the frequency. If we put a variable divider ($\div$N) between the VCO and the phase comparator feedback path, we can vary the frequency of the VCO, then divide it down so that the phase comparator inputs are always the same. This allows us to change the VCO yet maintain the same inputs to the phase comparator (Fig. 8-5).

The control voltage equation now is:

$$V_c = K \left[\Theta ref - \frac{\Theta vco}{N} \right]$$

where N is the number programmed into the variable divider. The operation of this system is the same as the basic loop just described.

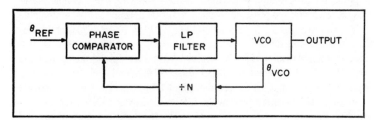

Fig. 8-4. Basic phase lock loop.

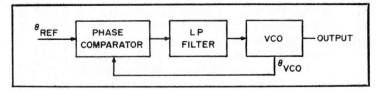

Fig. 8-5. Variable divider added to basic phase lock loop.

The Reference Circuitry. The circuitry (Figs. 8-6 and 8-7) of Q1 is a 10 MHz .005 per cent crystal controlled oscillator. R2 and R3 is the bias network. C12 is the feedback capacitor and is part of the resonant tank along with C13 and L1. If you don't have a .7 uH inductor for L1, you can make your own by winding 19 turns of #30 magnet wire (closewound) on a ½ watt resistor (100k or larger). C13 is 2 680 pF capacitors in parallel. X1 is a 10 MHz .005 per cent (series resonant) crystal.

C25 is a coupling capacitor to the following IC. Normally, a shaping circuit would be included here. But the rise and fall times at 10 MHz are adequate to trigger the IC. We will need a diode though (CR1) to clip the negative going transistions and prevent V_{eb} breakdown in the IC. IC1, IC2 and IC3 are wired to divide by 10. The output of IC3 is at 10 kHz. This will be the reference signal for the phase comparator. You'll note that IC3 is wired a little differently than IC1 and IC2. This is done so that a symmetrical 10 kHz square wave is fed to the comparator. R5 reduces the drive level into the comparator.

The Programmable Divider Circuitry. Pin 4 on IC4 is one of the VCO outputs. R10, R11 and C20 couple the signal out while still maintaining a DC path for the emitter in IC4. This signal is fed into IC9 which is ½ of an SN7413. The SN7413 is a dual Schmitt trigger. A Schmitt trigger improves the rise and fall transitions over a wide range of input frequencies. Since the output frequency changes from 1 to 10 MHz, suitable transitions are necessary over the full range of frequencies to insure positive triggering for the following IC. IC7 is a prescaler. It divides the output of the VCO by 10. This was done so that the programmable divider following (IC6) will operate at a lower range of frequencies, from 100 kHz to 1 MHz. Propagation delays and race conditions can cause false triggering at high frequencies. IC6 is the programmable divider. This is where all the switching of the system takes place. Basically, we want to divide all frequencies out of the VCO so that the output of the programmable divider is always 100 kHz. For example, if the VCO is at 5 MHz, the prescaler reduces it to 500 kHz and the programmable divider is switched to ÷5 and the output is 100 kHz. If the VCO is at 9 MHz, the programmable divider is switched to ÷ 9 and the output is 100 kHz. IC6 can be switched from ÷2 to ÷9. Now at 1 MHz the output of the prescaler is 100 kHz. We don't need the programmable divider here so it is simply bypassed. The output of IC6 is further divided by 10 in IC5, wired to provide a symmetrical 10 kHz square wave out. R12 is the same as R5, to reduce the drive to the phase comparator. R15, CR2 and CR3 are part of the ÷7 decoding. CR2 and CR3 can be any general purpose *germanium* diodes. IC8

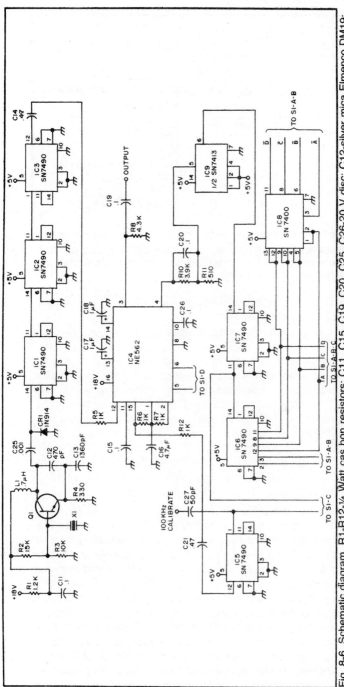

Fig. 8-6. Schematic diagram. R1-R12-¼ Watt cas bon resistors; C11, C15, C19, C20, C25, C26-20 V disc; C13-silver mica (2-680 pF in paralled) DM 19; C14, C21-35 V Sprague type 196D, 105XOO35HAi; IC1, IC2, IC3,IC5, IC6, IC7-SN7490 decade counter; IC4-NE562 phase lock loop;IC8-SN7400 NAND gate; IC9-SN7413 NAND Schmitt trigger; Q1-2N5172;X1-10MHz .005% crystal (series reasonant) from Jan Crystals, 2400 Crystal Dr., Ft. Myers FL 33901;L1-.7 microhenry inductor; S1-rotary switch 4 pole 10 position, Centralab PA-1015.

185

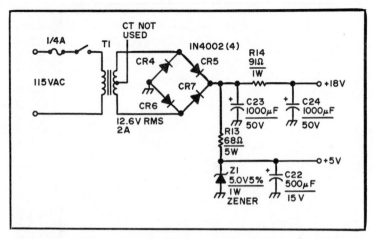

Fig. 8-7. Power supply. R13, R14-carbon resistors; C22-15V; C23, C24-50 V; T1-12.6 V rms power xfmr Allied 6K36HF; Z1-+5% zener diode 1N4733A; CR4-CR7-1N4002 (PRV 100 V, 1A).

inverts the A, B, C, D outputs of IC6 to provide the A, B, C, D, outputs to decode the programmable divider.

So now we have described the development of the two input signals to the phase comparator, one signal a fixed reference (Θref) and the other containing the phase error information (Θvco) from the VCO.

Pins 5 and 6 on the NE562 (IC4) are provided for capacity tuning the VCO. As each frequency is switched, the capacitor across pins 5 and 6 has to be switched. The free running frequency of the VCO has to be tuned within the capture range of the loop. As mentioned earlier, the low pass filter determines the capture range of the loop. C17 and C18 on pins 13 and 14, respectively, on IC4 are the low pass filter capacitors. When the free running VCO is tuned within approximately 300 kHz of center frequency, the loop will lock. Table 8-3 gives the approximate tuning capacitors for F_c at each frequency. They should all be silver mica (Fig. 8-8).

Construction. Except for the power supply and divider-frequency switch, all the parts are mounted on the 3-7/16″ × 5″ etched PC board (Fig. 8-9). Holes are provided on the board for all the wires going to S1. Sockets can be used for all the ICs to simplify troubleshooting, if necessary, or they can be hard wired to the board. Molex terminals are available at electronic stores and work well here.

They're inexpensive and are satisfactory sockets. Holes are also provided to strap all the IC power connections. Use magnet wire. Several circuit jumpers are also needed as shown on the parts layout of Fig. 8-10. C1-C10 should be mounted right on the switch (S1), with all the common points brought together in the center. Run a wire to the circuit board.

Final Checkout. Before applying power, check that all the jumpers are in and the ICs are mounted in the correct positions. Set the frequency switch to the 1 MHz position. After you're sure everything is fine, apply power and check that the supply voltages are present. +18 volts should be at the junction of R14 and C24. R14 can be changed to make it +18 volts. +5 volts should be at the zener diode. A counter would come in handy here for the following check but a scope will do. A triggered scope is even better. At the junction of C14 and R5, there should be a 10 kHz square wave approximately 3 volts peak to peak. This one point checks that the oscillator is working and also the divider chain. The same wave shape should be at the junction of C21, R12. If it is, the loop is locked at 1 MHz .005 per cent. Repeat this for each position of the frequency switch. If you reach a position where both wave shapes are not the same, check the switch wiring of the programmable divider. If it's alright, you'll have to adjust the tuning capacitor for that particular channel. Calibrate the scope using the reference 10 kHz signal. Display one square wave for 10 cm on the scope (junction C14, R5). Move the probe to the junction of C21, R12 and adjust the tuning capacitor until the same wave shape appears.

If the output is to be used as a "clock" to drive TTL circuitry, the circuit in Fig. 8-11 is suggested.

Use low capacity cabling on the output if you're going to run it any distance to preserve the wave shape. With the front panel switch in the 2 MHz position, a symmetrical 100 kHz signal is available on the front panel for calibrating communication receivers; simply couple the output (loosely) to the antenna input. A miscellaneous parts list for this project is found in Table 8-4.

Ultra-Flexible Crystal Calibrator

A very unique crystal calibrator manufactured and sold by Rainbow Industries (P.O. Box 2366, Indianapolis, IN 46206) is the Rainbow FS-200 calibrator. What makes it so different from the usual circuit providing calibration markers down to every 1000, 100 and sometimes 25 or 10 kHz is that it does all this and goes much further. It provides strong switch-selected, square wave markers every 1 MHz, 500, 250, 100, 50, 25, 10, 5, 2.5 and 1 kHz, as well as every 500, 250, 100, 50 and 25 Hertz! Thus, it can be used for a number of additional applications, such as an oscilloscope timebase calibrator, audio generator, frequency counter timebase, accurate signal source for analog and digital projects, hi-fi response testing and even code practice. It features CMOS (complementary metal oxide semiconductor) integrated circuitry, constructed on a 3¼" × 2½" board, and has a trimmer capacitor for precise frequency adjustment. It will operate on any power source from 6.5v DC to 15v DC at about 15 mA. A front-panel BNC

Table 8-3. Tuning Capicitors.

C1	300 pF	C6	43 pF
C2	150 pF	C7	39 pF
C3	100 pF	C8	33 pF
C4	68 pF	C9	30 pF
C5	50 pF	C10	27 pF

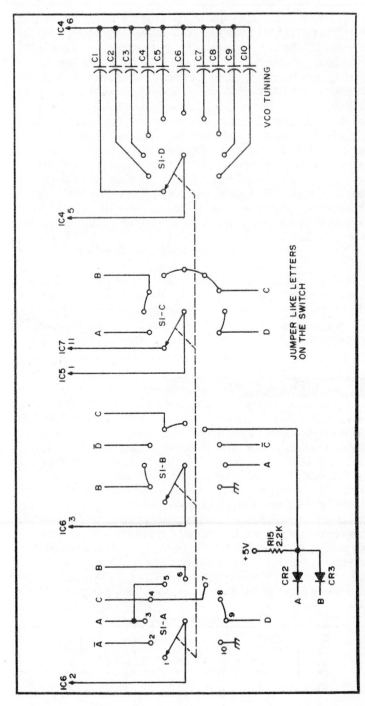

Fig. 8-8. Programmable counter switching and VCO tuning diagram. C1-C10-Elmenco DM10 or DM15 silver micas; CR2, CR3-1N118 (germanium).

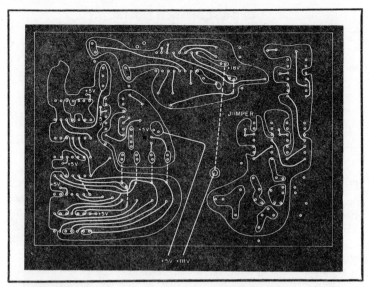

Fig. 8-9. PC board.

connector is provided for connecting to the output stage directly or for using a short piece of wire as an antenna. Desired output frequency is switch-selectable from the front-panel *range* and *multiplier* switches.

The calibrator is available either fully wired and tested (Model FS-200C for $34.95) in a minibox enclosure or as a prewired printed circuit board

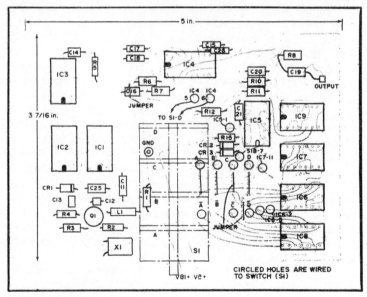

Fig. 8-10. Parts layout.

Table 8-4. Miscellaneous Parts List.

3½″ × 5″ copper clad board
Aluminum box
Line cord
On-off switch
Connector
Misc. hardware
ICs available from: B&F Enterprises, 119 Foster St., Peabody MA 01960; or Solid State Systems, P.O. Box 773, Columbia MO 65201.

(Model FS-200B for $19.95) less controls, switch, connectors and cabinet. Buy one of the fully-assembled units, hook it up to a 9-volt battery and you'll be pleasantly surprised to discover that it works just great. It provides strong markers at practically any desired internal frequency well into the VHF region (over 100 MHz for some units). The unit we used easily calibrated against WWV by adjustment of a trimmer capacitor on the PC board. You could also calibrate and cross-check against a frequency counter of known accuracy. If you calibrate by zero-beating against WWV, it's best to use the highest frequency receivable so that you would be using a high-order harmonic of the basic oscillator in the calibrator, for highest accuracy. One point to remember is that, as you approach zero-beat, the audio *beat note* becomes lower and lower, so it may be difficult to detect true zero-beat. Try using the receiver's S-meter. It will usually try to track the swing through zero-beat and can generally be relied on to get down to within a few cycles accuracy.

The fact that the FS-200 can produce markers down to 25 Hz suggests other uses for it—perhaps as an oscilloscope calibrator/timebase and audio signal generator. By interrupting its supply voltage, it could even be used as a code practice oscillator.

As fine an instrument as the FS-200 is, it does present a minor problem. It is another little convenience gadget for the workshop that needs another 9-volt transistor radio battery. Though they're inexpensive and easily available at Radio Shack, each time one poops out, you have to open up the cabinet and replace the battery. This is a slightly inconvenient task when done repeatedly. The addition of the FS-200 brings into focus the need for a small, cheap power supply, providing a source of multiple low-voltage outputs. The FS-200 could provide the *piggyback* vehicle for it, along with some simple interface circuitry for the calibrator. The only modifications required on the calibrator are mounting two rear-apron miniature phone jacks (external power and key), running leads to the battery clips and mounting an RCA phono jack on the rear apron for rf output (wired in parallel with the front-panel BNC jack). Be sure to remove the PC board while drilling to prevent possible damage to the components.

The circuit in Fig. 8-12 is what evolved when this was tried. Mounted in a 3-¼″ × 2-3/16″ × 4″ minibox (Radio Shack #270-251 or equivalent) and connected to the FS-200 by a short length of cable, it provides interface

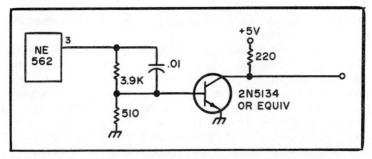

Fig. 8-11. Output circuit to drive logic. (Note: Circuit from Signetics Handbook.)

circuitry for the calibrator and a means of low-voltage power distribution. The circuit provides direct receiver rf output coupling for the FS-200, a small antenna or auxiliary rf output and adjustable rf and audio outputs. Also provided are nonregulated 6- and 12-volt outputs, as well as regulated 9-volt outputs to power various and sundry gadgets. Regulated power for the FS-200 is simply routed back to it through a short length of cable. Using a closed-circuit jack on the FS-200 rear apron allows internal 9-volt battery operation when external power is removed. An LED provides power-on

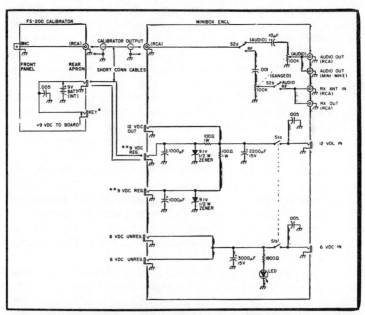

Fig. 8-12. Interface diagram. Additional 6-, 9-, and 12-volt outputs may be added as desired. All values are nominal. You may have to experiment with zener and LED dropping resistors for best performance. *Insulated from the chassis (mounted on the plastic rear apron of the FS-200). **Low current drain (20-30 mA). Use high-wattage zeners and dropping resistors for higher current capacity. Use separate zener circuitry for each regulated device.

indication for the interface/distribution box. Wiring follows conventional point-to-point practice.

The power source used for the unit is simple. Locate two junk box AC adapters (battery eliminators). One provides about 12v DC (no load) at 300 mA and the other supplies about 6.2v DC at 300 mA. The 12-, 9- and 6-volt circuits are kept separate to keep interaction down and regulation simple. Zener diodes in the 9-volt circuit provide a small degree of voltage regulation. Though the output of the battery eliminator (AC adapter) doesn't have enough real current reserve for good regulation under heavy loads, it can easily power most of the small low-current accessory devices which require four penlight cells or 9-volt transistor radio batteries as power sources. A 9-volt adapter can be used if a 12-volt unit can't be located, with the dropping resistor values changed accordingly. Voltage regulation would, of course, be minimal. No attempt was made to regulate the 6-volt circuit, though this could be done if necessary for your purposes.

In any case, ensure that the adapters used are not cheapies having a direct connection to the AC line, which would present shock and grounding-out problems. Also, most adapters are not adequately filtered to provide good DC output. High-value electrolytics (1000-3000 uF) across the supply lines will do much to filter out any trace of hum or ripple.

Various types of connectors can be used on the piggyback box for audio, rf and power connections, whichever meet your needs. RCA-type phono jacks can be used for rf (routing the receiver coax antenna lead through the unit for signal injection), RCA-type phono jacks and miniature Amphenol-type mike connectors can be used for audio and miniature phone jacks for dc power input and output.

Note that if you use the calibrator with a transceiver, you will have to provide some means of switching it out of the circuit on transmit to prevent transmitter output from damaging the unit. If your transceiver has an *external receiving antenna* input, you can usually route the output there— check your transceiver's schematic first!

Suggested application notes and circuitry are furnished by the manufacturer.

For code practice, it's easiest to simply run the rf output directly into the receiver. The FS-200 is acting, in effect, as a mini CW oscillator. The receiver doesn't even need a bfo. By playing around with the *range* and *multiplier* switches, you can come up with MCW (modulated CW). Try it for code practice—it really works.

Don't try to use the piggyback supply for IC circuits requiring good regulation under heavy load; it isn't designed for such use. But, for the multitude of simple projects and gadgets (audio filters, compressors, preamps, signal generators, etc.) requiring but a nominal source of low-voltage DC, it fills a real need at a minimum investment, not tying up an expensive, heavy-duty regulated DC supply. If you don't own an FS-200, try building the power distribution box anyway. You'll find it will interface well with other less versatile calibrators, signal and marker generators, and audio generators and the power supply will indeed come in very handy when the battery in your speech compressor or keyer fails!

Super Standard Goes Right Down to 1 Hz

With the proliferation of subbands and interference-obscured net frequencies, the use of an accurate secondary frequency standard is both good operating procedure and also helps fill legal requirements for frequency measurement capability under FCC regulations. In addition, such a standard can be used as a timebase or signal injector to test digital logic circuits.

The WA7VVC frequency standard was designed to be a low cost answer to the need for a good secondary frequency standard. It generates marker signals of 1000, 500, 100, 50, 25, 10, 5 and 1 kHz and 100, 10 and 1 Hz. With harmonics usable well beyond 30 MHz, markers are available to denote subband edges, align receiver dials, find net frequencies and measure the frequency of unknown signals. Should your latest logic circuit not perk properly, two TTL level outputs are available as substitute clocks or signal injectors.

Short term accuracy is approximately 1 part in 10^6. The unit is easily aligned to WWV with a short wave receiver. An attenuator is included to permit matching signal strengths with the received signal so that a zero beat can be easily and accurately identified. Only one trimmer need be adjusted to align the standard.

With the standard still attached to the receiver and providing an audible measure of its accuracy, a counter or other TTL-compatible device may be attached to either TTL output and that device's accuracy checked.

A frequency burst mode is provided to allow identification of the standard's markers in a crowded receiver passband. Enabling the burst turns the output on and off 10 times per second, resulting in an easily recognized "beep-beep-beep." This burst is also available on one TTL output.

An external DC input makes operation in the field possible. Nine-15 volts at 250 milliamps is all that you will need to have an accurate standard available for Field Day. This is cheap insurance against FCC out-of-band citations.

Circuit Description. The active devices in the frequency generation chain are 7400 series TTL. They are readily available, easy to use, inexpensive and capable of the fast rise times necessary for high level high frequency harmonics.

The oscillator shown in the schematic of Fig. 8-13 uses a 7404 hex inverter, A1, with a 200 kHz crystal as the feedback element for frequency stability. This circuit provides clean output. The circuit will work with either a 2000 kHz crystal or a 1000 kHz crystal. In fact, you can eliminate one 7474 package by using a 1000 kHz oscillator.

Frequency division is accomplished by 7474 dual type D flip-flops A2, A9 and A10, and 7490 decade counters A3-A6, A11 and A12 wired for division by 10. The 7474s are used as toggle flip-flops by connecting Q to the data input. Preset and clear are not used and are tied high through 1.8k resistors. The 7490s are ripple counters and are prone to spikes and level changes in their output. Proper bypassing of all ICs is necessary to prevent these devices from putting spikes on the power buses. The .1 uF bypass capacitors are not superfluous—use one at each IC.

The various frequency outputs are selected by a rotary switch and fed to A8c, one section of a 7400 quad gate. The selected frequency can either be passed without charge or gated with the 10 Hz output of A7, an NE555 astable oscillator, producing an easily identified frequency burst at the output terminals.

The NE555 timer A14 is used as an astable oscillator whose output is a train of 24 millisecond low-going pulses with a period of .81 seconds. A15a inverts this to a train of positive-going pulses. A15b and A16a gate the pulse train under control of the STEP push-button switch.

On each low-going pulse edge at pin 14, the 7493 binary counter A17 increments by one. A16b forces a reset on a counter output of 1011, permitting outputs from 0000 through 1010 to select the 11 frequencies of the standard.

The binary output of A17 causes the 74150 multiplexer A19 to select a signal from its inputs to be fed to the standard's output buffers. A18, a 74154 binary to one of 16 decoder, enables the corresponding LED.

To change frequencies on the standard, depress the STEP push-button. The standard will step through its 11 outputs, one every .8 seconds until the button is released. When the desired frequency is reached, as indicated by its LED, you have three-quarters of a second in which to release the button before the standard steps again.

Because of the additional current required by the electronic switch, the 2200 uF filter capacitor should be changed to 4700 uF if this circuit is added.

The remaining gates of the 7400 are used as output buffers. The two used for TTL outputs will drive 10 TTL loads apiece. One of these gates buffers the 1000 kHz output of A2a and makes it available at a BNC jack on the rear panel. The other takes the output of A8c, which is controlled by the frequency selector and the BURST switch, and makes it available at a BNC jack.

The output to a receiver is from A8b through a 100 pF capacitor and a 500 ohm pot used as a signal level attenuator. Connection of the pot as shown on the schematic prevents the receiver sensitivity from being affected by the attenuator setting.

The power supply uses the ubiquitous LM309K +5 volt regulator. Since the circuit draws only 250 milliamperes, the project case can be used as the heat sink. Dissipation of the 309 is only 0.7 watts.

Substitution for the surplus 7 volt power transformer used is easy. Use a 12.6 volt center-tapped filament transformer in the configuration of Fig. 8-14.

In order to use the unit in the field, provision is made for an external DC input. The .01 uF capacitor removes stray rf from the power lines and the series diodes prevent damage from polarity reversal. Two diodes were used

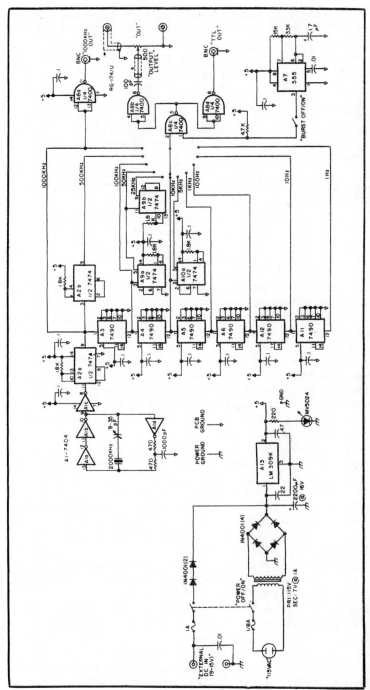

Fig. 8-13. Schematic.

195

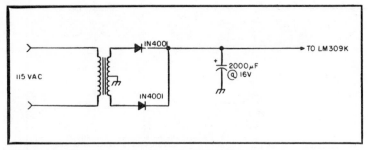

Fig. 8-14. Configuration for a filament transformer.

because the 14.5 volts of a car battery under charge come perilously close to the filter capacitor's 16 volt rating. With the two diodes shown, applicable DC input voltage is 9-15 volts. With the appropriate filter capacitor voltage rating and a single diode, voltages in the range of 8 to 25 can be used.

Do not leave out the .22 uF bypass capacitor on the 309 input. It prevents oscillation of the device should a remote battery be used as a power source.

Construction. Because of the high speed switching characteristics of TTL and the high frequency harmonic content of the output waveforms, each IC is a transmitter and each interconnecting wire is an antenna. A thoughtful layout and careful construction are important to minimize unwanted radiation.

Switching transients appearing on the Vcc and ground lines can add unwanted noise to the output. Use a prototyping board with Vcc and ground planes to minimize glitches. A printed circuit board would be even better. If you wish to build the circuit with wire wrap sockets on vectorboard, use bus wire for the power leads to the sockets. An effective technique is to interleave Vcc and ground bus wires for each row of IC sockets, with a row not containing more than six ICs.

Bus wire should also be used to connect the board directly to the 309. Do not invite problems by grounding the board to the chassis. Connect the chassis to the 309 case, the board to the 309 case and the rectifier ground to the 309 case. This prevents the board ground from rising above the power ground and developing noise problems.

Asynchronous TTL devices such as the 7490 generate plenty of switching transients. Put a .1 uF bypass capacitor between Vcc and ground at each device socket and another at the power input to the board. Bypass the 309 input and output as shown in the schematic to prevent spikes and oscillation.

The leads connecting the frequency divider ICs and the selector switch make excellent antennas, so shield the output from the board to the attenuator pot and from the pot to the output terminals with RG-174 coax to reduce unwanted pickup. Long unshielded leads will reduce the effective control range of the pot.

Mechanical stability will be reflected in electrical stability. Rigidly mount the crystal and trimmer capacitor close to the 7404 oscillator.

Whenever a frequency standard is brought up, usually the discussion turns to the National Bureau of Standard's shortwave broadcasts on station WWV, but using these transmissions at any distance from Fort Collins, Colorado, where the transmitter is located, is difficult when reception is poor.

If you service CB radio equipment, commercial transmitters or radio gear, an accurate means of measuring frequency is essential. If you are an experimenter who has a frequency counter, perhaps you have either not calibrated it or let its calibration slip because you did not realize that an atomic frequency standard was as close as your color television receiver.

For some time now, NBS has been pushing the color TV burst frequency as as method of frequency dissemination, and once you use the system it will become apparent that it is both very accurate and inexpensive to implement.

The basic idea is that all three major TV networks use atomic oscillators to produce the 3.5795455454 MHz color subcarrier frequency which is used to code and decode the color information in the video signal. This subcarrier frequency, or more exactly a piece of it, is broadcast with the video which is then used by the color receiver to regenerate a continuous carrier in the set for decoding purposes. The continuous frequency generated by the set is locked in frequency and phase to the transmitted piece of carrier or *color burst* as it is called, which thus produces an exact replica of the output of the network's atomic oscillator. While there may be minor phase shifts due to path length changes from network switching, the frequency stability is basically that of the generating source.

Thus, your access to an atomic frequency standard involves two simple steps:

- Bringing the color burst frequency out of a TV and into your counter.
- Waiting for a network program to come on. Local stations use crystal oscillators which, while good, can be a couple of Hertz off frequency.

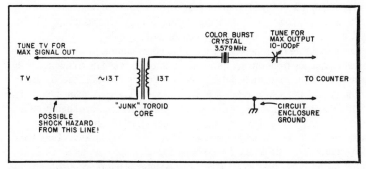

Fig. 8-15. Color burst isolator and filter.

The tapping of the color burst frequency in your set may take a little imagination, since each set is a little different. Receiver subcarrier regeneration systems fall into two broad categories: phase locked oscillators and ringing tuned circuits. To get into the right territory, look for the color burst crystal on your set's schematic diagram. Examine the circuit around the crystal to determine which type of system you have. If the crystal is an oscillator, then you have the *phased locked type* and it is the output of that oscillator that you want to bring out of the set. If the signal seems to pass through the crystal as a filter, then you have the *ringing* system and it is the output of the amplifier stage following the crystal that you should tap. Do not try to take an output from the crystal itself, as this can lower its Q and could stop oscillations altogether.

In either case, look for a good low impedance output point. Careful probing about with a scope should turn up a nice clean waveform you can use. Some set schematics have waveform pictures that can often be helpful clues. A word of caution is needed to remind you that many portable receivers have no power transformer, so the chassis could be 115 volts hot!

For this reason, use the circuit of Fig. 8-15 on your portable set. The toroid transformer isolates the set from the line. Include an additional color burst crystal to peak the waveform into nice clean sine wave and filter noise pulses. This circuit easily drives your counter. You may find that the tint, color or fine tuning controls on the set, as well as the trimmer on the isolator circuit, may need to be used to peak your output waveform.

The tap-off circuitry tends to affect color reception, so you may want to have a jack to remove your output circuit when you are not using the set for calibration purposes. An alternative is to only use the set as a frequency standard, especially if it lost one of its colors in the picture tube and was deemed not worth fixing. This is a very common occurrence. Since you really need only one color to tell if a program is network (you could probably get by with just sound), so long as the color burst section is working, you have your atomic standard.

Your home-built counter can have a timebase that can be set as long as 100 seconds. This means that the counter, when adjusted, can read color burst frequencies to .01 Hz (or has an accuracy of .1 Hz at CB frequencies or .4 Hz at 2 meters). If your counter reads with a 1 second timebase, you can only set it to 1 Hz at color burst frequencies, giving you an accuracy of 10 Hz at CB and 40 Hz at 2 meters when calibrated.

To calibrate the timebase oscillator on your counter, just feed the color burst frequency from the set into your counter and read its frequency with your maximum resolution setting. Now adjust your timebase oscillator trimmer until the display reads 3,579,545.35 Hz. The reason for the .35 Hz rather than the ideal .45 Hz is that all network frequencies are slightly offset due to a change in the international frequency standard after the oscillators were installed. You should measure with a good clear stretch of network programming and look out for commercials that may be local. Greatest timebase accuracy is attained when your counter is left on continuously. In any event you should remember to allow a good warm-up period before calibration.

While the network frequencies have been 3,579,545.35 Hz for years, if you really want to stay on top of this situation you can subscribe to the NBS monthly time and frequency services bulletin, which lists the exact offsets for each network on a week basis. It is free on request, the only cost being a questionnaire they ask you to fill out about once a year asking which NBS service you use (WWV, TV, data, etc.), how often and why (ham radio, CB service, hobby, etc.). With the bulletin, great accuracy can easily be achieved. Those people with clock and frequency standard hobbies have found the TV system to be a tremendous boon to obtaining synchronization of secondary standards with little effort or expense. For extreme precision, a beat-frequency method can be used (but it is more complicated than the direct count method described here).

If you choose to leave your counter running continuously, you may wish to consider dividing your timebase to 50 Hz and building one of the numerous clock kits available today. Set your digital clock chip to run from 50 Hz and feed the timebase signal into the circuit where it previously was connected to the power transformer. (The 60 Hz line is still used for power.)

Instant Counter Calibration Using Your TV Set

It's rather unfortunate that the development of affordable frequency counters paralleled the recent rapid growth of two meter FM, because the serious FMer should have (or have access to) a reasonably accurate frequency counter for netting the crystals in his or her rig. The luxury of being *talked-in* on the larger systems has passed. For the moment, let's suppose that you have your own frequency counter. How does one go about accurately calibrating (or at least verifying the calibration of) the little gem? As already mentioned, you are probably a little surprised at what a handy little frequency reference a TV set can be when tuned to a network station.

All four TV networks, NBC, CBS, ABC and PBS, presently use rubidium frequency standards to generate their color burst, horizontal sync pulses and vertical sync pulses. Some local stations also use rubidium standards locked to the network with which they are affiliated, but unless you are sure, you can't count on it. These rubidium standards are traceable to NBS (National Bureau of Standards) of WWV fame, inasmuch as the networks are monitored by NBS and offsets are published periodically.

At this point, it should be explained exactly what a rubidium frequency standard is. Rubidium 87 is a metallic element whose automic resonance is 6834.6826 MHz. This natural atomic resonance of rubidium 87 is very stable and not easily upset by external factors (particularly when properly shielded and operating in a temperature-controlled cavity). The rubidium unit influences control on a crystal oscillator, which then adopts the same order of stability as the rubidium reference. The crystal oscillator feeds a frequency systhesizer that contains outputs of 5.0 MHz and 3.579554 MHz (color burst frequency). The longterm (one year) stability of commercially available rubidium standards is $\pm 5 + 10^{-11}$, and the short-term (one second) stability is $\pm 1 + 10^{-11}$. These figures improve even more when correlated to NBS offsets.

As stated earlier, the TV station uses the 3,579,545.4 Hz color burst signal as a reference for its sync generators, which then derive the familiar horizontal and vertical sync signals used to lock the sweep oscillators in the home TV receiver.

Here's how you can take advantage of those highly accurate signals that are beaming around us, all of 18 plus hours a day. The easiest way is to pick up the horizontal sync signal, which is in abundance inside the cabinet of a TV receiver due to its use for the derivation of the high voltage that's needed to run the picture tube. This high voltage spike can be found anywhere around the flyback transformer or the picture tube yoke (which is a safer area to work in). If the set has a wooden or plastic cabinet and your counter is sensitive enough, you may find enough signal to lock on even outside of the cabinet (try the picture tube screen for openers). Note that there is no need to connect either lead from the counter directly to the TV receiver. In fact, unless the TV set is equipped with a power transformer, it would be disastrous to do so (especially if the counter is grounded, which it should be)! Some of the newer portable TV sets have no power line isolation transformers and use bridge rectifier circuits for AC to DC conversion, with the result that even reversing the AC line cord won't place the DC common at ground potential (it's always hot). This situation makes it impossible to hook up externally grounded test equipment to these sets unless an external isolation transformer is used. The preceding was mentioned only to protect the innocent. During a network color broadcast, the horizontal scanning frequency will be 15,734.265 Hz, which can be read on the Heath IB-1103 (if you happen to have one) by simply placing a well insulated lead from the counter over the deflection yoke on the neck of the picture tube. You'll notice, if your counter will read below 1 Hz, that the "265" will vary between counter sampling periods because you're not phase locked to the signal.

It should hit ".265" occasionally. On most counters, if you're within a couple of cycles of 15,734 Hz, you're in good shape. Incidentally, you can use a black and white TV instead of a color set because its horizontal oscillator is locked to the color standards (H and V sync pulses) as well. Just be sure you are tuned to a network color program. Most network color mobile units now have rubidium standards on board, with the possible exception of the small news mini-cam units. Even in-the-field sporting events will provide accurate sync signals. One more thing: Don't be in a hurry. Give yourself enough time to average the reading over a 10 or 15 minute period. This will give the clocking oscillator in the counter time to stabilize after adjustment, and also will permit you to observe the medium term stability of your counter. Whether or not you actually watch the program on the TV screen is strictly up to you.

There are some inherent errors in the system just described, such as distortion in the microwave relays used for cross-country TV signals, transmission and multi-path distortions within the local *ether* and distortions that take place within the TV receiver itself, but these are *phase* and not

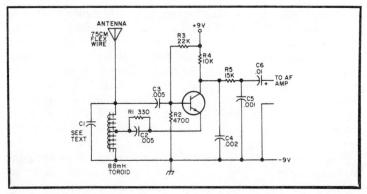

Fig. 8-16. Simple receiver.

frequency distortions. As long as you're not trying to read to three places below one cycle (Hertz), don't worry about them.

There you have it—no digging into the circuitry, handy at most times of the day and night, available throughout the country and very accurate.

A 15.75 kHz Oscillator Simple TV Test Unit

This is to describe a device of considerable value to the many experimenters that service TVs either in the shop or at home. It is simply a regenerative receiver set on 15.75 kHz, the TV horizontal oscillator frequency.

With it the horizontal oscillator frequency can be set correctly without a signal and the adjustment make in seconds without any doubt as to whether to increase or decrease the oscillator frequency. Many of us have spent valuable time in the shop blindly turning the slug in and out without the slighest idea where it should be for the correct frequency. If the oscillator is not working, there will be no signal regardless of adjustment. Adjustment of the horizontal oscillator in the usual manner only brings the frequency near enough that the sync pulse can lock it in step and thus is no assurance that in the free-running state it is on frequency.

This receiver may be built into any small case such as that from a defunct transistor radio. Its variable condenser, audio amplifier and speaker may also be used if good. It may be made small enough to carry in a shirt pocket on house calls.

It is a simple project (Fig. 8-16). To conserve space and avoid hand capacity effect, an 88 mH toroid coil was used rather than a regular horizontal oscillator coil which would have required some shielding.

This circuit using the collector at ground potential for rf was chosen to simplify the audio takeoff. R5C5 provides additional filtering to keep rf out of the audio output. Use the audio amplifier in the original case, if convenient. If you must build your own, a small IC is suggested for compactness.

A regenerative receiver is most sensitive when not oscillating at full strength, so it is a good idea to use variable resistors to determine the best values for reliable but not excessive feedback, replacing them with the nearest fixed small resistors. It is safe to assume that at least 100 types, requiring different bias resistors, etc., may do well here.

C1 is made up of one or more fixed mica condensers in parallel with a small variable condenser or mica compression trimmer; the latter is definitely second choice. The total capacity required should be around .0018. Marked values are seldom correct. Temperature sensitive condensers are to be avoided for tuning. Silver micas are preferred. The temporary use of an external variable condenser of considerable capacity will expedite finding the proper value and frequency.

With this receiver's antenna near the horizontal area of an operating TV, listen for the 15.75 kHz signal when you are sure the receiver is oscillating. When zero-beat is obtained, with final tuning condensers in place in the final assembly, no further adjustment is required and it is ready for use.

With no signal or antenna on the TV to be serviced and with front panel control, if any, set at midrange, adjust the horizontal slug for zero-beat with the receiver and you are finished with that part of the job—no guesswork! If the sync doesn't take control, then that is a different problem, and there is no need to twiddle with the horizontal oscillator.

Chapter 9
Signal Generators

For certain aspects of equipment testing and alignment, as well as for circuit experimentation, a good rf signal is a must. Yet many experimenters do not own such a generator because of the cost of one and because it is not as frequently needed an instrument as; perhaps, a scope or a VTVM. This project is a simple rf signal generator that covers from 100 kHz to about 75 MHz in six bands. It is useful for checks of low frequency i-f circuits on up to VHF circuits (using harmonics up to 220 MHz).

The cost of the unit can be as low as $5, if one has an exceptionally well-equipped junk box, or it can cost up to about $20-25, in the average case. Still, it is relatively inexpensive for a wide range transistorized and portable signal generator.

The basic generator puts out a CW signal only over its operating range. Various optional circuits can be added to provide tone modulation and a sweep frequency capability. But we would suggest that these circuits be added later, once the basic generator is functioning properly.

One of the first things to notice about the generator is that there is no frequency readout dial scale. There are several reasons for omitting it. Usually, the frequency readout scale on most inexpensive generators is more fiction than truth. To do any meaningful alignment work today, one needs a counter to set any signal generator correctly—even some very expensive commercial units. One doesn't need a calibrated scale just to sweep past a 455 kHz i-f or to determine that the front end of an HF or VHF receiver is basically functioning. Lastly, the lack of the not-so-useful scale allows the unit to be far more economically and compactly constructed.

The electrical circuit of the generator is shown in Fig. 9-1. One FET is used as the basic oscillator in a Hartley-type circuit. The second FET is lightly coupled through the 5 pF capacitor in its gate lead to the oscillator. This stage functions as source follower isolation stage. The last stage, the 2N3866, is designed to boost the signal level up to about 1 volt output on most bands. This level is far more than what is required for most receiver-type work, but the increased level comes in handy when doing transmitter exciter work, where the generator might substitute for a vfo. The output of the 2N3866 stage can be regulated by the 500 ohm carbon potentiometer, which will provide about a 30 dB variation in output level. A 47 ohm resistor can also be switched in across the output, so a true nominal 50 ohm generator source impedance can be simulated for tests such as receiver sensitivity. The switching in of this resistor also serves as a high-low output level selector for the generator.

Although the generator is not complicated electrically, its true potential will not be achieved unless it is carefully constructed. Fortunately, no elaborate construction work is required, but attention should be paid to the few details mentioned here. All of the circuitry is mounted on a single-side copperclad board measuring about 2½" by 2½". The board is wired point-to-point, using the isolated pad technique, starting with the oscillator stage toward the back panel of the enclosure and progressing forward to the 2N3866 stage towards the front panel. There is nothing critical about the wiring, whatever technique is used, but the circuitry should just be stretched out to provide maximum separation between the oscillator and 2N3866 output stage.

The heart of the signal generator lies in the band-switched oscillator coil assembly and the variable tuning capacitor. The tuning capacitor is a readily available broadcast receiver type, which contains a single section AM section of about 300 pF and a single section FM section of about 25 pF. Such capacitors can often be found with built-in tuning shaft drive reductions of 3:1 to 6:1. The Burstein-Applebee catalog is one source for such a capacitor, although various similar types should be available from Radio Shack, Lafayette and the mail order suppliers. A simple alternative to the AM/FM type is the even more readily available standard dual section AM type, where one section, designed for local oscillator usage, has fewer plates. Remove more plates from the oscillator section, so it is left with four stator plates and three rotor plates.

The coils for the six bands can either be purchased or constructed from a mixture of home brew and commercial coils. As a completely purchased set, one can use the Conar CO-69 through CO-74 series, at a total cost of $4. These are replacement coils for an old-fashioned National Radio Institute tube-type signal generator, but they work just fine in this FET oscillator. The coils are available from Conar, National Radio Institute, Washington, D.C. 20016.

Another alternative is to just purchase the coils for the lower frequency bands, which would be almost impossible to home brew, and wind the other coils. In this case, for the first three frequency ranges one can use prewound J.C. Miller-type coils, which have the necessary tapped windings. The types are 9015, 9013 and 9013. For the highest three frequency ranges, one can selfwind the necessary coils on ⅜" diameter slug-tuned forms. The windings necessary and the tap points for each of the three coils are as follows: 15 turns tapped at four turns from the ground end; seven turns tapped at three turns; and, four and one-half turns tapped at two turns. The latter coil is wound using #18 wire while the other two coils use #24 wire.

The coils can be mounted directly on the 3P6T rotary bandswitch. The coils are secured to the bandswitch with epoxy cement and wired in place. In order to ensure a good ground connection for the coils, a piece of sheet copper was cut out to resemble a six-legged starfish and placed over the bandswitch shaft, so one of the "legs" could be soldered to each coil. This

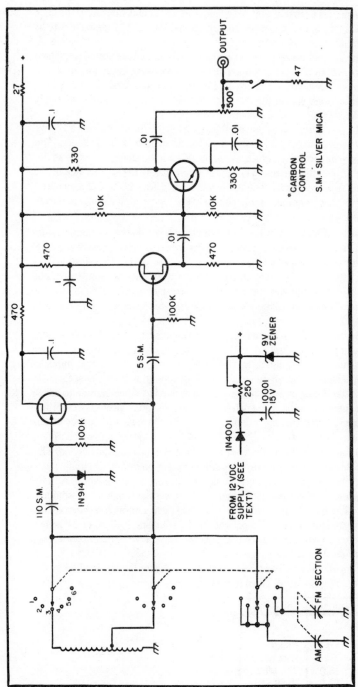

Fig. 9-1. Schematic of the generator. Only one of the six oscillator coils is shown for clarity. The FETs are HEP802 or MPF102 types. The output transistor is a 2N706. Other details are covered in the text.

arrangement is probably a bit overdone. Ground connections from the coils to several ground lugs, equally spaced around the shaft of the bandswitch, should suffice just as well.

The generator is assembled in a standard commercial enclosure, which measures about 5″ on each side. The dimensions were based on the size of the bandswitched coil assembly, tuning capacitor and circuit board. With a bit of effort, one should be able to fit the generator into the more readily available 4″ × 5″ × 6″ aluminum enclosure.

The generator can be powered either from the AC line or from a 12-volt battery source, making it ideal for both fixed and portable applications. The power supply for the generator is *not* included in the same enclosure as the generator, and this seems to contribute significantly to the total lack of AC hum on the output signal. The AC power supply is an AC wall plugmounted 12-volt DC battery replacement supply of the type commonly sold to power transistor radios.

Within the generator enclosure, and as shown on the schematic in Fig. 9-1, there is only a diode to protect against reverse voltage polarity, a 1000 mF filter capacitor and a 9v zener regulator. For portable application, 12v from a battery pack, or even 9v from a transistor radio-type battery, can be used. The 250 ohm variable resistor before the zener is adjusted for the maximum resistance value that still allows the zener to maintain a constant 9v output.

Because of the lack of a frequency readout scale, there is not the usual need to try to adjust the low and high frequency range excursions on each band. However, they should be checked with a counter to see that sufficient overlap exists between ranges. The slug tuning of the coils suffices to correct the tuning on each range. Although one has some latitude to adjust the frequency coverage on each band to suit individual preferences, Table 9-1 shows a typical arrangement, starting with the lowest frequency band.

The stability of the oscillator proved to be good enough on all ranges except the highest, so that temperature compensation was not needed. This is probably due to the low power operating requirements of the circuit and to the fact that the power supply is mounted externally to the generator. Since the highest band was not used extensively, it was not temperature compensated. But, by selection of small value NPO capacitors placed across the small section of the tuning capacitor and by watching the output frequency change on a counter, it should be possible to achieve excellent frequency stability on the highest range also.

Tone modulation can be added to the generator by the circuit shown in Fig. 9-2. The circuit provides a single frequency tone modulation, which is

Band	Low end	High End
1	100	570 kHz
2	400	1400 kHz
3	1.2	4.5 MHz
4	4.1	17.0 MHz
5	15.0	39.0 MHz
6	25.0	75-80 MHz

Table 9-1. Typical Frequency Coverage.

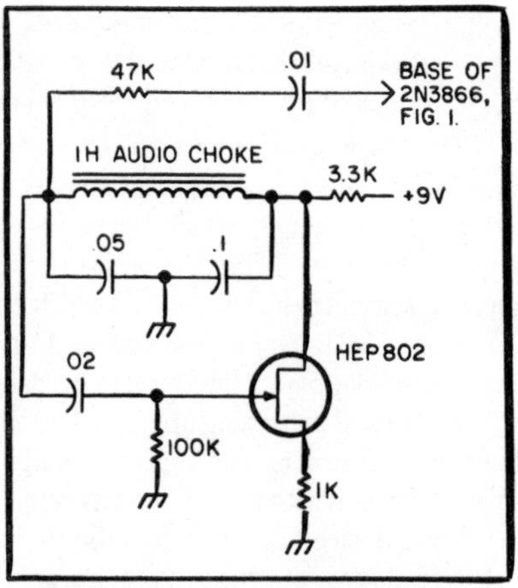

Fig. 9-2. Audio oscillator which can be added to the basic signal generator to aid in identification of the rf signal.

useful to identify the presence of the rf signal when working with a receiver having only an envelope (AM) detector. By placing the output of the audio oscillator on the gate of the second FET (instead of on the base of the 2N3866 as Fig. 9-2 indicates), a slight FM modulation of the oscillator will occur. So, the signal generator can be utilized with SSB/CW, AM or FM receivers.

A sweep frequency capability can also be added to the generator, by means of a varactor diode connected across the gate terminal of the oscillator FET to ground, and driven by a suitable sawtooth of triangular waveform. The five for $1 varactor diode selections available from Poly Paks are very suitable for this purpose.

You now have the basic construction information needed to put together a very good wide range rf signal generator.

Updated Universal Frequency Generator

Basically, this generator is a collection of IC oscillators and dividers that can generate square waves from the HF range all the way down to the sub-audible AF range and can generate markers all the way into the VHF range. There is nothing sophisticated about the generator, and some of its functions can be performed today by more advanced ICs with even greater versatility. But, the generator is hard to beat as a simple straight-forward device that can be built at a very low cost (none of the ICs costs more than $1.00, except for one optional $3.00 type). It makes an excellent little project for those who still haven't started to experiment with digit ICs.

The uses for the generator are about as versatile as those of a grid-dip oscillator. Also, like such an oscillator which is not a very advanced type of instrument today, one nonetheless always keeps on discovering new and handy uses for the instrument. The board uses still include:

- frequency marker generation
- generation of crystal-stability RF or AF square wave signals
- crystal activity checker

207

- range extension for present AF or RF generators
- divider chain to allow HF oscillator stability and calibration to be checked by a low frequency receiver for those who lack a counter

The block diagram of this updated generator is shown in Fig. 9-3. It consists of a selectable oscillator section, a fixed string of divide by 10 stages and two divider stages which can be switched in at various points along the divide by 10 stages. The first oscillator is a 1 MHz crystal controlled stage. Improved circuitry has been used which ensures more stable oscillation with any good 1 MHz frequency standard crystal. An LED indicates that the stage is oscillating. The second oscillator stage can be used with any external crystal extending up to the low VHF range. An LED again indicates that the oscillator is working. The stage can be used as a simple crystal activity checker with the LED, or crystal frequencies can be used which will give some desired output when the oscillator is fed into the divider chain. The third stage is not really an oscillator stage in itself, but an oscillator input stage. It will accept any external sine or square wave input, condition it and then apply it to the divider chain. The fourth oscillator stage really increases the versatility of the generator. It is a square wave generator whose frequency can be controlled by an external capacitor only. By proper selection of this capacitor, frequencies from several Hz to several MHz can be generated. Thus, in combination with the divider chain, any desired frequency or marker can be generated. The fifth oscillator stage, which is optional, is similar to the fourth oscillator except that it is intended mainly for the entire HF range up to about 25 MHz. Its frequency of oscillation is also controlled by an external capacitor, but its main feature is that it is a voltage controlled oscillator. By varying the voltage with a potentiometer to one section of this stage, a variable or sweep frequency output can be obtained.

The divider chain is a simple series of four divide by 10 stages. The outputs from a 1 MHz input will be at 100 kHz, 10 kHz, 1 kHz and 100 Hz. The stages are similar to those found in the timebase of any frequency counter or many crystal calibrators. The two divide by two and divide by five stages can be switched in along the divide by ten string. Using the 1 MHz oscillator input example, the outputs of the divide by two stage will be 500 kHz, 50 kHz, 5 kHz, 500 Hz and 50 Hz. The outputs of the divide by five stage will be 200 kHz, 20 kHz, 2 kHz, 200 Hz and 20 Hz. This example may appear very obvious, but when dealing with inputs other than a simple 1 MHz one, it is important to list all of the various output frequency possibilities to avoid confusion. Of course, looking the other way, that is towards VHF marker frequencies, the listed frequencies represent the intervals at which marker frequencies would appear since the output of the digital stages is a square wave with a very rich harmonic content.

The actual circuit of the generator is shown in Fig. 9-4. Use is made, except for the optional fifth oscillator stage, of only simple SN7400 family ICs. The SN7400 1 MHz oscillator stage makes use of two of the gates for the oscillator itself, one for an output buffer and one to drive the LED oscillator activity indicator. The oscillator can be beat against WWV for accurate calibration using the 25 pF trimmer. Do this by connecting the

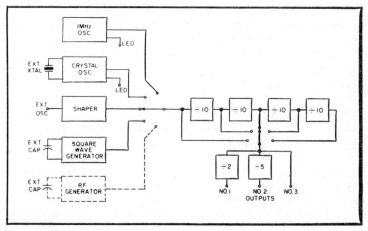

Fig. 9-3. Overall block diagram of the generator.

output of the oscillator through a small coupling capacitor if necessary, to the antenna input of a receiver tuned to WWV on 5, 10, 15 MHz. During the 10 second tone pause on WWV, adjust the trimmer for zero beat. The second crystal oscillator will operate over a very wide range. The only component that may have to be changed with frequency to ensure stable oscillation is the 150 pF capacitor from one side of the crystal socket to ground. The 150 pF nominal value will operate satisfactorily with most HF crystals. The oscillator range can be extended higher or lower by making this capacitor have a value (in pF) equal approximately to 500 divided by the crystal frequency in MHz. Overtone as well as fundamental mode crystals will work.

The third oscillator, a SN7405, is a hex inverter which uses a single external feedback capacitor to produce a square wave output ranging from a few Hz to several MHz. The capacitor values needed will range from 300 uF to 30pF. The output waveform is not exactly a square wave in the truest sense since the on and off times for each cycle are not exactly equal. Normally, this will not make any difference, but the situation can be corrected by connecting a 22k resistor from pin 4 to the supply voltage.

The fourth oscillator input stage has been found to function well. About a 1.5 to 2 volt input is required from any external sine or square wave generator to activate the multivibrator. The operation of the stage will be very obvious and even the output from a low voltage filament transformer through a current limiting resistor can be used to check that it is operative.

The fifth oscillator stage, a Motorola 4024, is a very interesting voltage controlled oscillator. Its circuit simplicity belies its very versatile usage. The operating frequency *range* of this oscillator (multivibrator) is controlled by the value of a single external capacitor connected between pins 3 and 4. The tuning or output *frequency* of the oscillator *within* the range established by the fixed external capacitor is determined by a variable 1 to 5 volt DC voltage applied to pin 12. For example, with a 430 pF capacitor between pins 3 and 4, the output frequency range as the voltage on pin 12 is varied between 1 and 5 volts is approximately 200 kHz to 1100 kHz. For a 100 pF capacitor, the range is 5 MHz to 25 MHz. The IC will actually operate up to 30 MHz with a bit of care as to lead dress, etc. The leads to the external frequency range determining capacitor should be kept as short as possible. Also, the ground leads and bypassing to the supply voltage pin should be short or the full frequency range of the oscillator may not be realized. Unlike the SN7400 series ICs which are widely available, one may have to look twice for a source of the MC4024. One source (at $3.00) is Circuit Specialists, P.O. Box 3047, Scottsdale, AZ 85257.

The divide by 10 chain uses SN7490s in their conventional arrangement. Once the layout of a single stage has been determined, the rest need simply be duplicated. They can easily be wired on perforated stock as can the other ICs in the generator.

The divide by two and divide by five stages are all actually contained in one SN7490 which sort of does double duty as a frequency divider.

The construction of the generator depends on individual taste. It can be constructed as an AC powered unit or as a portable unit. As an AC powered unit, a simple power supply using an LM309K regulator is recommended. My portable unit was constructed in an approximately 3″×4″×1¾″ metal enclosure. The perforated board was just slightly smaller than 3″ × 4″ to fit the enclosure. Portable power can be supplied by 4½ volt battery (Burgess 532 or equivalent) or 3 type C or D cells connected in series. There are two main bypassing details, however, which must be observed for good performance and which are not shown in Fig. 9-4 for sake of simplicity. Each IC must have a .1 uF disc capacitor going from its supply voltage terminal to ground. Also, a 100 to 500 uF/10v electrolytic must be connected across the battery terminals for a portable unit.

The many uses of the generator have been considerably expanded by these features. It can function as a highly accurate square wave frequency generator (down to fractions of a Hz) at any dividable frequency down from a crystal frequency reference. It can function as a highly accurate marker frequency generator in the HF or VHF range, depending on the crystal frequency reference used. It can produce harmonic markers at any frequency up to VHF by means of its internal oscillator which is variable in frequency. It can serve as a crystal activity checker for almost all fundamental and overtone type crystals. The combination of having a crystal reference frequency oscillator available in combination with a variable AF or RF oscillator allows comparison to be made so one does not fall too far out of range within any variable frequency range.

A final word might be said about the output waveform of the generator. The output is a square waveform and hence quite rich in harmonic output. It cannot function as a pure, fundamental frequency sine wave generator and was never intended to do so. Reasonably good sine waves can be produced, however, at any output frequency by suitable RC filtering. The situation is very similar to that of filtering a sharply *clipped* audio signal before it is applied to a modulator stage in a transmitter. Fig. 9-5 provides the details of

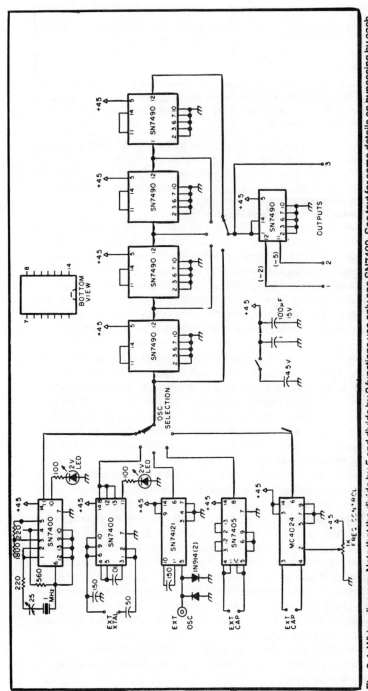

Fig. 9-4. Wiring diagram. Note that the divide by 5 and divide by 2 functions use only one SN7490. See text for some details on bypassing by each IC.

211

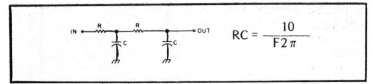

Fig. 9-5. Simple RC filter which will help round off square wave output so a quasi sine wave output can be obtained. Component values should be optimized using an oscilloscope if possible.

a two stage RC low pass filter which can be built for any AF to HF frequency to provide a reasonable sine wave output. Of course, LC filters will be better to get a pure sine wave, especially multi-section filters. The formulas for such LC low pass filters can be found in many reference texts on electronics.

Is It the Band or My Receiver?

Probably one of the most frustrating aspects of operation for the experimenter who does not use his station daily for several hours is to know when his station is functioning properly. Whether or not a transmitter is functioning properly can be verified by an in-line wattmeter for power and with a monitor receiver for modulation, at least to see that the rig is basically functioning. Answer check is a good basic check for an antenna system. But, what about the station receiver? An operator who uses his equipment frequently can pretty well judge band conditions on HF and knows that a band is *out*. But, the infrequent or weekend operator often checks a band and finds signal levels low and then starts to wonder if it is the band or the receiver. Obviously, various checks can be made to see what the real situation is like. If the receiver can tune outside the amateur bands, stations like WWV or selected shortwave broadcast stations can be used as a quick reference to conditions. If a second receiver of equal quality is available, it can be used for verification of the main receiver's condition. But, many experimenter's entire stations consist of a transceiver confined to tuning only the amateur bands.

In this case, some other means of checking receiver performance is necessary. A signal generator providing a calibrated output down to the microvolt would be ideal, but few hobbyists have such test equipment.

This is a different approach in terms of the *microvolter*. Basically it is a simple gadget—just an oscillator and a carefully made attenuator network to generate a signal of approximately one microvolt. It is compact, battery powered and uses a single transistor. But, it can almost instantly remove all doubt and confustion as to whether a *receiver is out* or the *band is out*. It was intended to be used by plugging it into the receiver's antenna terminals to make a check. However, it could be placed remotely and connected to its own antenna to see if an entire receiving system were functioning on any given band. Admittedly, there are pitfalls to this latter aproach, since when a signal source is close to an antenna system, good signal pickup may occur even though something has happened to the antenna system to change its directional properties.

The oscillator circuits presented here are mainly for the 20 through 2 meter bands since activity is usually great enough on the lower frequency bands that it is readily obvious if a receiver is functioning. However, if desired, the gadget may be designed to work on any amateur band.

As was mentioned, the microvolter consists of an oscillator and an attenuator network. Many circuits are possible which will work. However, the oscillator must be stable in regard to battery voltage, and construction of the attenuator network, while not critical, must be carefully done to avoid leakage signal paths.

Figure 9-6 shows a typical oscillator circuit which can be used and, by proper adjustment of the components, will work from 20 to 6 meters. It is crystal-controlled for simplicity, and the typical component values for 10 meters are shown. Note that the crystal does not have to have any particular frequency, as long as it falls within the band of interest, so advantage can be taken of the various surplus or odd-frequency crystals available at low prices from such outlets as Jan Crystals. The output is about ½ volt across 500 ohms, although this will depend on crystal activity and circuit tuning. The resistor attenuator network shown will bring the output down to about 1 microvolt across 50 ohms. The output of the oscillator has to be verified in some manner. The best is, of course, to measure it directly if good instrumentation is available. An alternative is to construct the generator and see what response the entire unit provides on a receiver that is known to be in good shape. After all, the main purpose of the instrument is to provide a quick, relative indication that a receiver has not lost sensitivity. If it can be accurately calibrated as to output it could be used to make direct sensitivity measurements on a receiver in conjunction with a VOM to check audio level changes, in dB, with and without the test signal being applied. A simple LED circuit is included both to indicate that the generator is on and to approximately indicate a low battery voltage condition. The pot in series with the

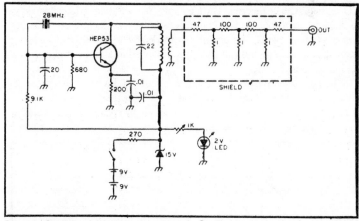

Fig. 9-6. Oscillator and attenuator network. Oscillator may be used on any band down to 6 meters by proper choice of LC circuit components. Coil: CTC LS5 form, 15 turns #22 enam., with 2 turn link.

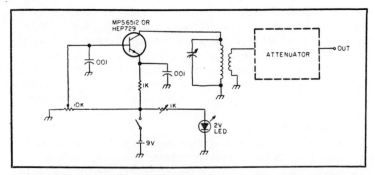

Fig. 9-7. Extremely simple VFO will work down to 2 meters by choice of proper tank circuit to resonate in desired band.

LED is adjusted so the lens of the LED is just barely fully illuminated. The difference between this point and no noticeable illumination of the LED is .15 to .2 volts for a typical LED. So, it can serve as a battery voltage indicator better than even a cheap meter.

Other crystal oscillator circuits can be used. K1CLL is a very good oscillator which, among other things, features a simple variable ouput level scheme by means of varying the emitter bias. For those who do not prefer to make their own oscillator, the International Crystal OX oscillator kits are very suitable. They cost only $3 with a tuned output circuit. The Lo kit takes crystals from 3 to 20 MHz and the Hi kit covers 20 to 60 MHz.

Simple VFOs can be used as the oscillator element but here, of course, the problem is that their output will vary with frequency. Nonetheless, if their output variations can be measured, there is no reason not to use them. For the purpose at hand, ultra frequency stability is not needed, so very simple circuits can be used. Figure 9-7, for instance, is about as simple a VFO circuit as can be desired. The frequency of oscillation is determined by the single LC combination. By choosing combinations which resonate in the various amateur bands, the circuit will work down to at least 2 meters. Leads should be kept and the oscillator enclosed in a metal enclosure to avoid hand capacity effects.

The attenuator network was shown in Fig. 9-6. Its construction is not critical, but it must be separately shielded even if the oscillator is enclosed in a metal enclosure. The easiest way to do this is by constructing it within a divided enclosure as shown in Fig. 9-8. The enclosure can be of brass sheet,

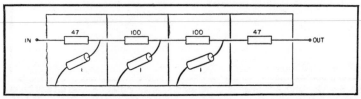

Fig. 9-8. Attenuator must be separately shielded. The output connector should be mounted on the attenuator shield.

available at many hobby shops, or of copper sheet, or even of sheeting salvaged from old tin cans. The only objective is to completely "button up" the attenuator network so signal leakage does not occur around it to the output jack of the oscillator.

This unit has been kept as simple as possible. There are various ways to embellish it, if desired. For instance, three output attenuator networks with different attenuation characteristics might be built: one as a microvolt output, one as an S9 output (usually taken as 30 microvolts) and one as a 40 dB over S9 output.

Tubeless Partability for the Heath Signal Generator

The Heath IG-102 radio frequency signal generator is a versatile general-purpose wide-range instrument of conventional vacuum-tube design which has been catalogued for several years. For those who are still looking for a good signal generator at a moderate price, this article provides a simple, economic and fast means of conversion for portability without sacrificing the original qualities. Modification of the IG-102 gives the instrument new capabilities. These include:

- Battery operation utilizing either dry cells or a nickel cadmium rechargeable bank capable of supplying 7 to 9 volts at a maximum current of 10 to 16 mA.
- With the use of rechargeable nickel cadmium cells of approximately 1.2 volts per cell, six to eight cells are sufficient to provide the required operating supply voltage. The basic radio frequency circuit will operate with a minimum of 4 volts throughout its frequency range. For charging purposes, a self-contained battery-charging circuit is used which employs the original components. These can be dispensed with if dry cells are installed. (Installation of the battery supply is simple, since there is ample chassis space available to mount battery brackets for accommodating the battery units.)
- All tube sections are replaced by four field effect transistors, such as those available from Radio Shack— #276-1623 or Calectro—#K 4-634 (Fig. 9-9). Some of these Radio Shack #276-1623 packages contain assortments of both rf and af field effect transistors. In testing them, the af types will not oscillate. One particular type which responds well at radio frequencies is the 2N5951. The white-black types are af and do not oscillate at radio frequencies; also, the metallic types are unsuitable.

Procedure for Modification. The following step-by-step instructions apply to changes and additions to the basic IG-102 radio frequency signal generator. Accordingly, the original instruction manual and drawings are used for reference to detail the modification procedure. This procedure involves simple changes, installation of new components and constructions. Before undertaking the modifications, spend some time studying and reviewing the basic circuit and assembly to become familiar with the original layout. Additionally, study the details of the field effect transistor connection lead designations. In the assortment provided in Radio Shack's #276-1623, three types are referred to in the designation of leads. With reasonable care, there should be no difficulty using any of the transistor types.

Construction.

- Refer to Fig. 9-10, a copy of the original circuit.
- Remove all tubes, if the set has already been constructed.
- Unsolder and disconnect power transformer connections (filament supply line and high-voltage rectifier-filter system), and reconnect the power supply components as shown in Fig. 9-11, using additional components as needed.
- Solder a 75-ohm, ½-watt carbon resistor across the R2 (33k, 2W) decoupling resistor used in the original circuit.
- Solder a 90-ohm, ½-watt carbon resistor across the R6 (4.9k, 2 W) decoupling resistor used in the original circuit.
- Spot solder one radio frequency field effect transistor across tube socket V1B, terminals 1-2-3, using the lead references given in the transistor package instruction sheet. However, if transistors are selected from the eight-transistor assortment given in Radio Shack's packet #276-1623, then lead orientation should be followed as per the drawing supplied in the packet.

To mount the transistor on the tube socket, do the following: Hold each lead with long-nose pliers (for a heat sink) as solder is applied. Use a 35- to 40-Watt pencil iron with a small blade tip ⅛″ wide and solder alloy 60-40. Apply a small drop of solder to the end of each lead. After thinning the leads, spread them to match the spacing of the lugs on the tube socket. Apply each lead to the required tube socket terminal, as designated. To facilitate connections, also apply fresh solder to the tube-socket terminals.

In spreading the transistor leads, a slight bend on the wire ends will help in making surface-to-surface contact for soldering. With the pliers, hold each lead against the surface of the required tubesocket terminal, and apply the iron to the opposite side of the terminal until solder flows well to form a good spot-solder joint. No mechanical connection is necessary to complicate the procedure. Once the first transistor lead is soldered, it will make the transistor self-supporting, and the remaining solder operation will be easily handled and completed. Remember to use the long-nose pliers as a heat sink for pushing the transistor leads against the tube socket terminals when soldering. Care in soldering and applying the transistor and applying the transistor leads will insure success and avoid thermal or mechanical damage to the components.

- Following the same technique for soldering as explained in the previous step, apply and connect another rf field effect transistor across the tube socket 1A, terminals 6, 7 and 8.
- Apply and connect the third rf field transistor to tube socket 2B, terminals 6, 8 and 9.
- Apply the fourth field effect transistor to tube socket 2A, terminals 1,2 and 3. If another type of FET transistor is used, make certain that the proper transistor leads are used to make connections.

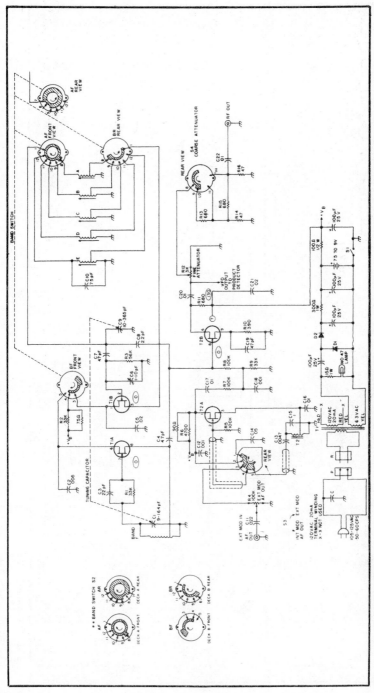

Fig. 9-9. Modified IG-102 circuit. T1A, T1B, T2A, and T2B-four rf FETs equivalent to Radio Shack package #276-1623 or Calectro #K 4-634.

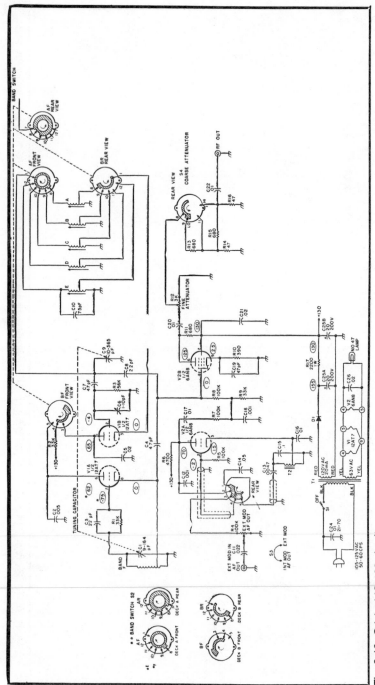

Fig. 9-10. Original IG-102 circuit.

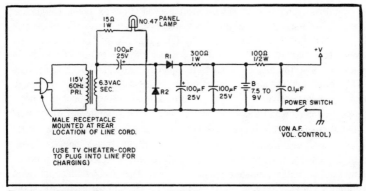

Fig. 9-11. Schematic.

Again it may be more convenient for making connections to turn the transistor over (round side up) and cross the D and G leads (use insulated sleeving).

This step completes the modification conversion of the basic IG-102 rf signal generator. Additionally, you can install rear tip jacks with the circuit connection leads to permit external testing of the battery supply (+) and (–).

Test, Operation and Adjustment.

- Before applying battery power to the circuit, check the positive-to-ground resistance to make certain that there is no short circuit or abnormally low resistance reading due to a defective component or wiring condition. Rotate the bandswitch while checking the resistance to ground (which should be several hundred ohms, at least).

- Insert a milliammeter in series with the battery and check the direct current on all bands, which should be 7 to 15 mA. Switching in the audio tone oscillator-modulator will increase it about 1 mA.

- Rf output. A diode detector rf voltmeter applied to the high end of the fine attenuator control (turned counterclockwise for minimum rf output) should indicate about 1 to 1.25 volts on band "A" (low-frequency end), and, on each successive band, it will drop off progressively. Nevertheless, rf output should be detectable on all bands.

  Vary the tuning on each band from low to high end—the rf voltage should vary smoothly (usually decreasing) without the sudden jumps or fall-off usually associated with parasitic absorption conditions.

- Audio tone oscillator modulator. The AC output of the audio tone oscillator, as measured across the audio output control, should be 1 to 2 volts. Check the tone frequency with a pair of headphones connected to a 0.1 uF coupling capacitor. Finally, use an allband receiver or grid-dip meter, if available, to check the rf output frequency. The frequency response in each band should be within a few per cent and not need any alignment or tuning adjustments.

Heterodyning with broadcast stations will show excellent frequency stability. On band "F," use an identifiable FM station to spot check the frequency calibration in the 88 to 108 MHz range. It may be necessary to *squeeze* the rf coil to about one-half its original length to get good frequency alignment. Use the long-nose pliers to squeeze turns. To check against an FM station, turn the audio modulation on, connect the rf cable and bring it near the FM receiver's vertical antenna. A good clean modulation note should be heard when the signal generator passes through an FM station.

Conclusion. This original circuit was modified by converting to the battery-operated solid state design, as described. In addition, a three-crystal oscillator frequency-spotting standard was installed with a product detector and audio amplifier speaker section. This combination provides frequency check intervals of 100 kHz, 1.0 MHz and 10 MHz, for spot-checking the internal six-band vfo or for external testing. Using the crystal standard, heterodyne testing showed excellent frequency tracking and calibration through the six bands. Since there is ample chassis space available, the three-crystal frequency spot circuit is left optional and is merely suggested to you. For this purpose, three crystals and oscillator design data can be obtained from Jan Crystals, 2400 Crystal Drive, Ft. Myers, FL 33901. Additionally, a dual-gate MOSFET product detector, a 250 mW integrated audio power amplifier, a miniature volume control and a 2½″ loudspeaker can be combined to provide the desired frequency-spotting function.

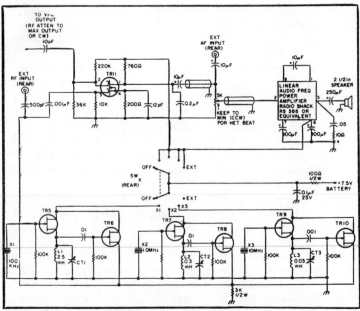

Fig. 9-12. TR4-TR10-Radio Shack #276-2039 or equivalent; TR11-dual-gate MOSFET; CT1-Arco 309 padder; CT2-Arco 306 padder; CT4-Arco 465 trimmer; L1 and L2-rfc; L3-20 turns #30 enamel wire on 2W, 100k resistor.

Figure 9-12 shows a circuit which incorporates the additional features just described—a three-crystal oscillator frequency section with a product detector and an audio power amplifier for monitoring the heterodyne reactions between crystal frequencies and the vfo spectrum. The entire unit can be easily and conveniently mounted in the rf signal generator chassis assembly or separately assembled externally as a sub-unit. Heterodyne activity can be detected up to the 50th harmonic. With additional wave-shaping amplifiers, the harmonic order could be extended considerably, but, unfortunately, that would involve high input current levels which would place excessive demands on the battery power supply.

VLF Generator

A major deficiency of the older or less expensive sine wave generator is that its output frequency doesn't go low enough for many vital purposes, the dial typically bottoming out at 20 Hz. A principal reason is the unwieldy values of resistance and capacitance required by the oscillator to tune sub-audio frequencies. The instrument about to be described here was designed specifically to complement the workshop sine wave generator by filling in at the low end of the spectrum. It circumvents the tuning problem by providing a number of discrete switch-selected output frequencies.

The frequencies chosen are 1 Hz and 2 to 20 Hz in 2 Hz steps (Fig. 9-13). Given different circuit values, however, the instrument is capable of a much wider range, as will be indicated. Maximum output amplitude is 3 volts rms or 8.5 volts peak-to-peak. A pot-and-switch attenuator allows the

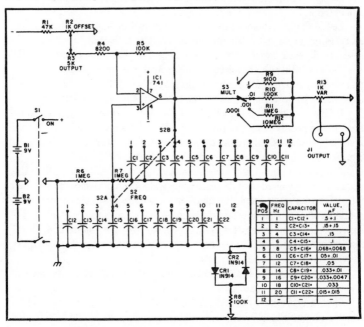

Fig. 9-13. LF sine wave generator, using Wien bridge oscillator.

POS	FREQ Hz	CAPACITOR	VALUE, μF
1	1	C1 • C12 •	.5 + .1
2	2	C2 • C13 •	.15 + .15
3	4	C3 • C14 •	.15
4	6	C4 • C15 •	.1
5	8	C5 • C16 •	.068 + .0068
6	10	C6 • C17 •	.05 + .01
7	12	C7 • C18 •	.05
8	14	C8 • C19 •	.033 + .01
9	16	C9 • C20 •	.033 + .0047
10	18	C10 • C21 •	.033
11	20	C11 • C22 •	.015 + .015
12	—	—	—

output level to be set with a fair degree of precision to any value within a range of five decades. The unit is powered by two 9 volt batteries of any type. Performance does not suffer until either battery potential drops below 6 volts.

How it Works. As shown in the circuit diagram of Fig. 9-13, the sine wave generator employs a common IC, the popular type 741, in a Wien bridge oscillator. This kind of circuit is characterized by a four-element r-c network connected in the feedback path of an amplifier. The four elements are R6 in parallel with the capacitor selected by S2A, and R7 in series with the capacitor selected by S2B. This network is connected in the positive feedback path of IC1, so oscillation occurs at the frequency where zero phase shift occurs. The two selected capacitors are always equal in value, to keep the reactance of the parallel combination the same as the series, so that the voltage-dividing action of the network gives the same amount of feedback at any frequency. The selection of these capacitors is discussed under the heading "Choice of Parts."

The gain of the amplifier, and hence the amount of positive feedback necessary to sustain oscillation, is determined mainly by the setting of output adjustment R3 in the negative feedback path. However, feedback in a sine wave oscillator is very critical; too little and the oscillator quits, too much and it saturates, generating a square instead of a sine wave. Nonlinear elements are usually introduced to deal with this problem. One system uses a lamp or thermistor where we have connected R3. Current sensitive elements such as these, however, introduce perturbations due to thermal inertia. Another method uses a voltage sensitive element like a zener diode, connected in the positive feedback path.

Zener diodes, though not suffering thermal undulations, have a deleterious effect on oscillator waveform, tending to flatten the peaks. We have introduced a variation of the zener idea. Ordinary silicon diodes CR1 and CR2 constitute the required nonlinear load, but waveform distortion is minimized by the mitigating effect of series resistor R8. This scheme yields a good sine wave without making the adjustment of R3 terribly critical.

A small DC offset results from the constraint offered by higher resistance in the positive input of the op amp than in its negative input. Hence offset adjustment R2 is provided to set the average output level of the op amp to zero, as described later under the heading "Adjustment."

Due to the shunting action of the network R8, CR1 and CR2, the formula normally used to find the frequency of a Wien bridge oscillator does not apply to our instrument. Furthermore, since this is a nonlinear network, frequency depends somewhat on the amplitude of the sine wave appearing across it. As a practical matter, with R3 adjusted to give a signal of 8.5 volts peak-to-peak at pin 6 of IC1, the oscillator frequency is $f = 0.6/C$, where C is the value selected by S2A or S2B, in uF.

It is important to keep in mind that the setting of R3 does not have an effect on frequency. Also, be advised that R2 and R3 interact slightly.

Choice of Parts. Frequency is determined by the two capacitances selected and by R6, R7 and R8. Precision resistors are recommended for

Table 9-2. Capacitance Values.

S2 Pos	Freq Hz	Capacitor	Value, μF
1	1	C1=C12=	.5+.1
2	2	C2=C13=	.15+.15
3	4	C3=C14=	.15
4	6	C4=C15=	.1
5	8	C5=C16=	.068+.0068
6	10	C6=C17=	.05+.01
7	12	C7=C18=	.05
8	14	C8=C19=	.033+.01
9	16	C9=C20=	.033+.0047
10	18	C10=C21=	.033
11	20	C11=C22=	.015+.015
12	—	—	—

use in these latter three circuit locations to give you the best chance of achieving the same output frequencies. Also, a precision resistor gives better long-term stability than an ordinary carbon composition type.

Likewise, stability is best served by using high quality film capacitors. Ceramics would be a dubious choice due to reports of frequent problems with stability, leakage, temperature sensitivity, tolerance, power factor and just plain open circuits (especially when capacitors of this type are obtained on the surplus market). Electrolytics or other types that polarize should not be considered.

If you want the output amplitude of the generator to be exactly the same for every setting of FREQ switch S2, the two capacitance values selected in each position must be matched exactly. Note that most of the values in Table 9-2 were obtained by connecting two capacitors in parallel. For example, C1 and C12 are each 0.6 uF, achieved by paralleling 0.5 uF and 0.1 uF. The two 0.5s should be matched and so should the two 0.1s. Use a capacitance bridge for this selection job. Of course, the closer the actual value of the capacitors to the figures given in the table, the more accurate will be the frequency settings.

If you don't have a capacitance bridge, you can match the capacitors in-circuit. Two capacitors are matched if the output level remains the same when they are interchanged. However, this cut-and-try method is tedious and tricky at best. Therefore, to keep the adjustment procedure reasonably short, we will proceed on the assumption that you have somehow managed to match the capacitors or else don't care if the output level varies slightly from one frequency setting to the next.

You may wish to extend the frequency range of your instrument. There is no lower limit, but the upper is dictated by the op amp itself, output amplitude suffering beyond about 5 kHz. Using a high frequency range to about 20 kHz. Also, it helps to connect a small capacitor from pin 2 of the op amp to common. The value is best determined experimentally but 250 pF would be typical.

Position 12 of S2 was left unused to permit the addition of jacks for external capacitors, if test signals outside the present frequency range of the device are required.

Construction. There are no special precautions to be observed in construction. The instrument can be built in a 5-¼" × 3" × 2-⅛" LMB box chassis No. 136. This box is admittedly a bit smaller to comfortably accommodate all those capacitors so you would be smart to use a bigger one.

The IC and a few other small parts can be wired onto a perfboard, which is then mounted to the chassis box by a spacer.

The batteries can be mounted inside the cover with double stick tape.

Adjustment. If you are using precision resistors for R6, R7 and R8, and have carefully selected the capacitors, the probability is very great that your output frequencies will be right on the money, and all you need for testing and adjustment is a VOM. If you wish to check output frequency, you will of course need suitable test equipment such as an oscilloscope with accurate timebase. To adjust the sine wave generator, proceed as follows:

- Set offset adjustment R2 to its minimum voltage, or ground, position, MULT switch S3 to the 1 position, VAR control R13 to its maximum output position and FREQ switch S2 to 20 Hz.
- Connect a VOM to output terminals J1. Set the VOM to measure AC voltage and adjust R3 to obtain 3 volts rms. (Don't use the "output" feature if your VOM has one, because its coupling capacitor may not be big enough to pass 20 Hz without some attenuation).
- Set the VOM to measure DC voltage on its lowest range and adjust R2 to obtain zero volts.
- Repeat the second and third steps until no further adjustment is necessary.

You may now calibrate the VAR control if desired. Your instrument can be calibrated in terms of peak-to-peak voltage, but you may prefer to use rms instead. An oscilloscope is more convenient to obtain values in the former units, a VOM in the latter.

If you have access to an oscilloscope, check frequency and amplitude of the sine wave at various switch positions. Variations in amplitude are caused by imperfect matching of the capacitor values. Errors in frequency at some switch positions but not others means inaccurately selected values.

If all frequencies should prove to be off by about the same percentage and in the same direction, the problem is easily solved by changing the value of R8 as required to make the output frequencies exactly correspond to the dial markings. Raising or lowering R8 by 20 per cent lowers or raises the frequency by about 8 per cent. However, varying R8 changes feedback, so you should repeat the adjustment procedure each time you try a different value.

There are no special tricks to operating the instrument, but remember that the output is DC coupled and a suitably large capacitor or some other means of DC translation is needed if you wish to drive any conductor on which DC is present.

A small DC offset appearing in the output as the minus battery runs down can be balanced out, if necessary, by touching up offset adjustment R2. For a complete parts list for this project, see Table 9-3.

Table 9-3. Parts List.

B1, B2	—9 volt battery
C1-C22	—Film capacitors (values given in table, see text for discussion of matching)
CR1,CR2	—1N914 diode
IC1	—Op amp, Fairchild Type FU5B7741393 or equivalent
J1	—Dual binding post, red and black. H H Smith Type 269RB
R1	—47,000 Ohm, ¼ Watt resistor, 10%
R2	—1000 Ohm miniature potentiometer, Mallory MTC-13-L.
R3	—5000 Ohm miniature potentiometer, Mallory MTC-53-L1
R4	—8200 Ohm, ¼ Watt resistor, 5%
R5, R10	—100,000 Ohm, ¼Watt precision resistor, 5%
R6, R7	—1 megohm, ¼ Watt precision resistor, 1%
R8	—100,000 Ohm, ¼Watt precision resistor, 1%
R9	—9100 Ohm, ¼ Watt resistor, 5%
R11	—1 megohm, ¼Watt resistor, 5%
R12	—10 megohm, ¼ Watt resistor, 5%
R13	—1000 Ohm linear potentiometer
S1	—Toggle switch, double pole single throw
S2	—Rotary switch, 2 pole, 12 postion
Misc. parts	—Battery connectors (2), perfboard, knobs with indices (3), suitable chassis, mounting hardware, etc.

Build an FM Signal Generator

The OSCAR amateur satellites have caused a renewed interest in VHF and 10 meter preamplifiers. Most of these preamps provide enough gain and noise improvement to allow them to be used with older general coverage receivers and surplus commercial "boat anchors." In order to be effective, however, the receiver used behind any preamp or converter must be properly aligned. Many available receivers employ a 455 kHz i-f section, which must be properly adjusted for optimum receiver performance. However, many hobbyists do not possess the signal generator necessary to adjust i-f stages, and unfortunately, many older kit type generators do not remain stable long enough to complete the tune-up.

This project is an easy to build FM fixed frequency signal generator for 455 kHz. It uses two readily available ICs, a 8038 function generator and a 741 op amp. A deviation control and output level control are provided, but the frequency is calibrated at just one frequency. It is possible to change the value to work with any i-f under 500 kHz by simply changing resistance or capacitor values. Because of the use of ICs, the other parts required are

225

relatively small. Even the power supply is optional if a good regulated 6 volt positive/negative supply is available.

The construction of a 2 meter receiver promoted the designing of this generator. The need for an FM signal generator had existed for several years, but this receiver construction forced the issue. A search of electronic catalogs showed that there really are no lower priced FM generators available, even in kit form. Next the government surplus list was reviewed with similar results. There are surplus units, but prices are very high, cover much wider spectrum than required and have many modes of modulation. At that point, the design specs were set up with cost and availablity of parts as prime considerations.

Circuit Description. Some information has been written on the 8038 function generator, but really not enough. It is a very useful device putting out sine, square and triangular waveforms of a quarter volt or more in amplitude (Fig. 9-14). Frequency range is from a small fraction of a Hz to about 1 MHz. Only a few external resistors and a capacitor are required to finalize the generator. Since an audio generator was already available for this project, it was decided to not make use of the full frequency capability of the 8038. Triangular output is available on pin 3 and the square wave by making the 10k resistor on pin 9 a pot.

The frequency of the generator is determined by the RC combination of C1, the 120 pF capacitor (mylar or polystyrene recommended), the series resistance of R3 (500 ohms) and the 680 ohm resistor. The variation of R3 shifts the generator frequency over a range of about 100,000 Hz. The 500 ohm pot across pins 4 and 5 sets the duty cycle or symmetry of the wave shape.

The generator output is taken from a voltage divider connected to pin 2, the sine wave output. An output of over 100,000 μV is available at this point. The output level can be modified by changing the ratio of the 10k and 4.7k ohm resistors.

FM modulation of the 8038 is obtained by applying a small varying signal between pins 7 and 8 and the positive voltage supply. This modulation voltage is obtained from a simple audio oscillator using just one IC, a mini-dip 741. Other op amps can be used, but experience has shown the 741 is one of the easiest to work with and the cost is very low.

This oscillator uses a notch network from the op amp output to the negative input. The result is a sine wave oscillation at about 1000 Hz. This signal is applied to the generator through the interstage transformer after being reduced in level by the resistor network.

The power supply is a simple bipolar (positive and negative) supply which is zener regulated and well filtered. The transformer can be two separate 12.6 volt units or a single 25.2 volt unit with a center tap. The four diodes could also be a rectifier bridge with at least 100 volt piv rating. This power supply could be eliminated if a good 6 volt bipolar supply is available. However, in this case, since it appeared much use would be made of the generator and since the power supply is quite inexpensive, it was built into the final unit.

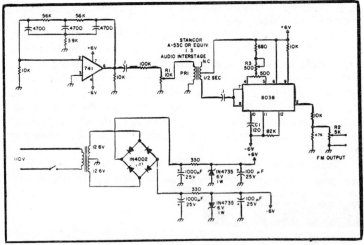

Fig. 9-14. One 25.2 V 300 mA CT or two 12.6 V 300 MA (primary in parallel, secondary in series). All fixed resistors ¼ Watt. Capacitors greater than 1 in pF, less than 1 in uF, unless otherwise marked.

Construction Details. In the original unit, the audio oscillator, frequency generator and power supply were each mounted on individual perforated boards. This was necessitated by several design changes made along the way. Packaging seems to work out best with the oscillator and generator on the same board, with the power supply separate.

Generally, construction is not critical, but the usual good construction practices should be followed. The entire device, including power supply, can be mounted in a 3″ × 4″ × 5″ minibox. The output level control, R2, the deviation level control, R1, and the output connector (BNC type) are mounted on the front panel. Since the output frequency is set at 455 kHz, this pot is mounted inside. The power cord is brought in the back, the transformers are mounted on brackets to the top and sides of the case.

Calibration. After checking all wiring several times, check it once more. Apply power to the generator and check to make sure that both plus and minus voltages are about 6 volts with correct polarity. With R1 set to the low end, look at pin 2 of the 8038 or the output of R2 with a scope. The correct setting for the 500 ohm pot between pins 4 and 5 on the 8038 is about the midpoint. If the output is badly distorted or is missing, adjust that pot to obtain a clean looking sine wave.

The next step is to adjust the frequency with R3. The best method of adjustment is with a frequency counter or meter. While continuing to monitor the waveform on the scope, adjust R3 to the proper frequency, 455 kHz. There will be interaction between the 500 ohm pot and R3. Continue to adjust back and forth until a good sine wave at 455 is obtained.

If a frequency counter or meter is not available, a receiver with known 455 kHz i-f could be used. By applying the generator output to an early stage (at low level) and measuring the output of the last amplifier stage, a peak in output would indicate the generator is set to the same frequency as the receiver i-f being used.

The only remaining item to check is the audio oscillator. Attach a scope or AC voltmeter across R1. There should be a good sine wave of several hundred millivolts present at about 1000 Hz. The generator is now fully operational and the box can be closed up. To actually look at an FM signal, connect a scope to the output of the generator. With R2 at full output, slowly advance the deviation control, R1. The trace should spread or smear horizontally as the control is advanced. Notice that there is no increase in amplitude since this is strictly frequency modulation. The FM generator is now ready for use.

Conclusions. The unit can be used to align the i-f amp and quadrature detector of an FM receiver under test. The unit is stable and provides more than ample deviation required for hobbyists work.

FM Alignment Oscillator

Surplus commercial transceivers still provide the most economical means of getting started on the amateur FM bands. Often, however, these rigs require considerable *tweaking* to bring them into our bands. An rf signal source to align the i-f and rf sections of the receiver is almost always required to accomplish tune-up.

This is a simple test oscillator built around the ever popular International Crystal "OX" oscillator board. Two oscillators provide a 10.7 or 21 MHz signal for i-f tune-up and a low level signal for rf section alignment. The crystal selected for rf alignment is in the range of 6 to 10 MHz—a harmonic provides output on the operating frequency. The 24th harmonic for two meter work was chosen here and has sucessfully tuned several 430 MHz rigs utilizing the 50th harmonic of a 9 MHz EX crystal. This allows an "OX-LOW" oscillator to be used in both sections of the test unit.

Construction. The unit consists of a 5″ × 2½″ × 2½″ minibox with a SO-239 coax jack and a SPDT switch mounted in opposite ends (Fig. 9-15). The two OX oscillators mount side-by-side with the supplied hardware. The rf output of one oscillator is directly connected to the output jack, with the second link coupled to the coil of the first with two turns of #18 insulated wire. This prevents the output of one oscillator being shorted by the other. The switch selects the oscillator to be powered by an external 9v battery.

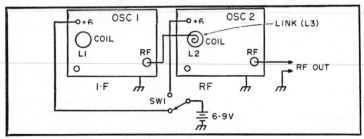

Fig. 9-15. Block diagram. Link consists of 2 turns #18 around L2; cold end of L3 is not grounded. SW1 is a center off SPDT.

Operation. The i-f crystal should be placed in the link coupled oscillator, as the strong fundamental output is utilized. When using the high frequency harmonic, directly couple the test unit output to the antenna of the rig. When the rf section is rough tuned, the oscillator is moved across the bench with a 10″ section of wire serving as an antenna. The noisy, not fully limiting, signal is then used for tuning of the receiver.

Although the EX crystal is of low accuracy (by FM standards), there is seldom a case where tweaking the receiver first oscillator trimmer did not tune the harmonic, even in the case of the 50th harmonic used for 430 MHz units.

The EX crystal frequency for the FM bands may be determined as follows:

$$144 \text{ MHz-Fx=Fo/24}$$
$$220 \text{ MHz-Fx=Fo/20}$$
$$430 \text{ MHz-Fx=Fo/50}$$

where Fx=frequency of EX Low crystal, and Fo=receiver frequency.

This unit is an invaluable aid in tuning a variety of surplus rigs. Once the rough tuning is complete, the rig may be *netted* by monitoring another ham or the output of your local repeater and zeroing the discriminator.

SSTV Test Generator

Probably the biggest obstacle facing hobbyists building their own SSTV monitor is finding a stable slow scan signal source to properly adjust the monitor circuitry. This is especially true if the reader decides to design his own monitor or deviates from an existing design. The monitor itself generally cannot be used as a test unit because all the circuitry has to be adjusted and operating properly before anything can be viewed on the monitor's CRT.

This is an SSTV pattern generator which produces a continuous, high quality, 4 × 4 checkerboard SSTV signal which can be used with a triggered sweep oscilloscope to follow the slow scan signal through the monitor circuitry. The generator can be used to improve the design of an existing SSTV monitor or as a diagnostic tool to repair one. This pattern generator is an adaptation of a circuit desgined by Bert Kelley K4EEU. NE555V timers are used as astable oscillators in place of the crystal oscillators, and in digital logic has been modified to produce a 4 × 4 checkerboard pattern.

You'll appreciate the need for such an instrument when you seriously begin designing and building your own monitors. Anyone undertaking such an adventure should have a similar instrument.

The front end of one hobbyists monitor contained a congolomerate of designs and circuitry. Breadboarding the circuitry was relatively easy for him, but finding a stable signal source proved to be the biggest problem. He was able to use an audio oscillator to initially adjust the limiter, pulse counting discriminator, video amplifier and sync separator, but fine tuning them for maximum performance was impossible using this method. A cassette recorder was used for a time, but was unsatisfactory from his

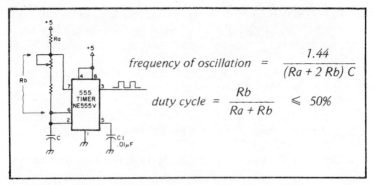

$$\text{frequency of oscillation} = \frac{1.44}{(Ra + 2Rb)\,C}$$

$$\text{duty cycle} = \frac{Rb}{Ra + Rb} \leqslant 50\%$$

Fig. 9-16. 555 timer connected as an astable oscillator. C1 is recommended by the manufacturer.

standpoint because the video content was constantly changing, as was the video quality.

1200, 1500 and 2400 Hz Reference Oscillators. The primary attraction of the K4EEU generator is its inherent stability, accuracy and lack of adjustments provided by three crystal controlled reference oscillators. This hobbyist felt, however, that the stability of 555 timers was adequate for most SSTV alignment applications and he designed his generator accordingly.

Several dozen Signetics NE555V timers manufactured over a two year period were tested over a temperature range of 25° C-55°C in an oil bath to determine their temperature coefficients (tempcos) and, therefore, their stability as reference oscillators when operated in the astable mode. A test fixture was constructed so that only the NE555V was placed in the oil bath and not the external timing components. Figure 9-16 shows a 555 timer connected as an astable oscillator. The tempcos of the NE555V timers measured .045 per cent to .07 per cent per degree centigrade (/°C). After the tempcos of the 555s were determined, a number of different timing components were included in the oil bath to check the overall oscillator tempcos. Best performance was obtained using precision wirewound resistors and metallized polycarbonate film capacitors with oscillator tempcos measuring .05 per cent to .08 per cent C. The worst performance was obtained using carbon resistors and disc ceramic capacitors with oscillator tempcos exceeding .3 per cent /°C.

As a compromise between performance and parts availability, he used RN55/60C (50 ppm) metal film resistors, cermet 15 turn trimpots and metallized polyester capacitors. Using these components, oscillator tempcos did not exceed .1 per cent C, which corresponds to 1.2 Hz/°C for the 1200 Hz sync oscillator, 1.5 Hz/°C for the 1500 Hz black oscillator and 2.4 Hz/°C for the 2400 Hz white oscillator.

Circuit Description. Referring to the schematic in Fig. 9-17, U1, U2 and U3 are 555 timers used as 2400 Hz, 1500 Hz and 1200 Hz reference oscillators. The component values specified in Table 9-4 have been selected so that the oscillator frequencies can be adjusted ±16-19 per cent from nominal to allow for normal component variations. The pattern generator pin assignments can be found in Table 9-5. To improve the oscillator's ability to be set, R2, R5 and R8 can be changed to 2k and R4, R7 and R10 trimmed (selected) for the correct frequency. There is nothing critical about the values specified and they can be changed as required, but keep the duty cycle in the 47-49 per cent range. Table 9-6 provides nominal values of Rb for preferred values of C when Ra equals 1k. Nominal values include ½ the value of the series trimpot selected. The outputs of U1, U2 and U3 are brought out to pins C, B and A.

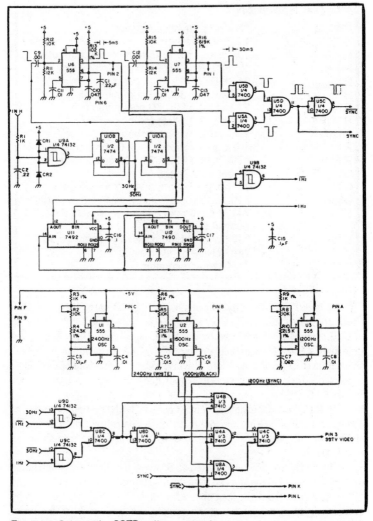

Fig. 9-17. Schematic, SSTB pattern generator.

Table 9-4. Parts List.

C1, 2	.22 uF ±10% 250 V
C3, 4, 6, 8, 11, 14	.01 uF ±10% 250 V
C5	.015 uF ±10% 250V
C7	.022 uF ±10% 250 V
C9, 12	.001 uF disc ceramic
C10, 13	.047 uF ±10% 250 V
C15, 16, 17	.1 uF ±10% 250 V

Note: With the exception of C9 and C12, all caps are metallized polyester Mepco/Electra Series C280AE/A or equivalent (.4″ lead spacing).

CR1, 2	1N914 or equivalent silicon
R1	1 k ±10% carbon
R2, 5, 8	10k ±10% 89PR 15 turn cermet trimpots;
R3, 6, 9	1k ±1% RN55/60C 50 ppm metal film
R4	24.3k RN55/60C 50 ppm metal film
R7	26.7k RN55/60C 50 ppm metal film
R10	21.5k RN55/60C 50 ppm metal film
R11, 14	12k ±10% ½Watt carbon
R12, 15	10k ±10% ½Watt carbon
R13	102k ±1% RN55/60D 100 ppm metal film
R16	619k ±1% RN55/60D 100 ppm metal film
U1, 2, 3, 6, 7	NE555V IC Timer
U4	7410 triple 3 input positive NAND gate
U5, 8	7400 quadruple 2 input positive NAND gate
U9	74132 quadruple 2 input positive NAND Schmitt trigger
U10	7474 dual-D type positive edge triggered FF
U11	7492 divide by twelve counter
U12	7490 decade counter

U4 is a TTL 7410 three input positive NAND gate. U4A and U4B alternately gate the 1500 Hz black and 2400 Hz white oscillators to the input of U4C, except when inhibited by the SYNC signal from U5C. During the time that U4A and U4B are inhibited by SYNC, U8A is enabled by the SYNC signal from U5D, gating the 1200 Hz sync oscillator to the input of U4C. U4C is as used a three input inverter, its output, the SSTV video output, brought out to pin 3.

U5 is a TTL 7400 quadruple 2 input positive NAND gate used as a sync adder, combining the 5 ms horizontal and 30 ms vertical sync pulses from U6 and U7. The outputs from U5D and U5C (SYNC, SYNC) are brought out on pins L and K.

U6 and U7 are 555 timers used as 30 ms vertical and 5 ms horizontal sync generators. U6 is triggered from the negative edge of U10A (15Hz) and U7 is triggered from the negative edge of U11 (⅛Hz). Pin 6 may be externally grounded to increase the horizontal sync pulse width from 5 ms to 30 ms. This provides a 30 ms wide sync pulse at a 15 Hz rate for easily

Table 9-5. Pattern Generator Pin Assignments.

1	Vertical Sync Pulse, 30 ms wide
2	Horizontal Sync Pulse, 5 ms wide
3	SSTV Video
6	Contact closure to ground increases the width of the horizontal sync pulse from 5 ms to 30 ms.
9	Ground
A	1200 Hz
B	1500 Hz
C	2400 Hz
F	Power in, +5V ±.25 @ 150 mA
H	6.3-12.6 Vrms AC @ 6-12 mA
K	Sync (combined horizontal/vertical)
L	Sync (combined horizontal/vertical)

adjusting the monitor's vertical sync separator. The outputs from U6 and U7 are brought out on pins 2 and 1.

U8 is a TTL 7400 quadruple 2 input positive NAND gate. U8C combines the output of U9C and U9D and in conjunction with inverter U8D, alternately enables/disables gates U4A and U4B. U8B is not used.

U9 is a 74132 quadruple 2 input positive NAND Schmitt trigger. U9A squares the 60 Hz AC input pin H providing fast output rising/fall times, the positive edge triggering the first dual-D FF U10B. U9B is used as an inverter for the 1 Hz output of U12.

Opposite phases of the 30 Hz output from U10B and 1 Hz output from U12 are connected to U9C and U9D, which alternately gate the 1500 Hz and 2400 Hz oscillators every 30 picture lines and 16.67 ms producing a 4 × 4 checkerboard pattern.

U10 is a TTL 7474 dual-D type positive edge triggered FF triggered by the positive edge of U9A. U10A and U10B divide the 60 Hz output from U9A by four, producing the 15 Hz SSTV horizontal scanning frequency. The negative edge of U10A triggers the 5 ms horizontal sync generator U6 and

Table 9-6. Nominal Values.

CAP.	1200 Hz	1500 Hz	2400 Hz
.01 μF	59.5k	47.5k	29.5k
.015 μF	39.5k	31.5k	19.5k
.022 μF	26.8k	21.32k	13.14k
.033 μF	17.68k	14.04k	8.59k
.047 μF	12.27k	9.71k	5.88k
.068 μF	8.32k	6.56k	3.9k

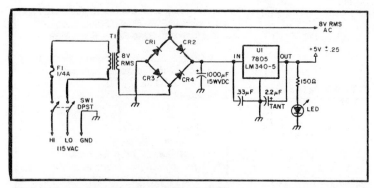

Fig. 9-18. This power supply satisfies the requirements for the pattern generator, but since the generator only requires 150 mA +5 V, it can easily be borrowed from an existing supply.

U11. Both phases of U10B (30 Hz, 30 Hz) are connected to the inputs of U9C and U9D.

U11 is a TTL 7492 divide by 12 counter and U12 is a TTL 7490 decade counter connected to divide the 15 Hz horizontal scanning frequency by 120, generating a 30 ms vertical sync pulse at the end of 120 lines or 8 seconds. Each counter contains a ÷2 element combined with a ÷5 element (7490) and a ÷6 element (7492). The schematic is not typical of two cascaded counters as follows: 7492– ÷6, 7490– ÷5, 7490– ÷2 and 7492– ÷2. This connection provides the vertical scanning logic which produces the symmetrical 4 × 4 checkerboard pattern when combined with the horizontal logic in U9C and U9D. In truth, the 4 × 4 checkerboard pattern is not symmetrical as displayed on a monitor, as 5 ms of the leading edge is blanked by the horizontal sync pulse; that is, the 1500 Hz and 2400 Hz oscillators are overridden by the 1200 Hz sync oscillators during the horizontal and vertical sync pulses.

Construction: Printed Circuit Board. The components are mounted on a double-sided 5½″ × 3″ glass-epoxy circuit board fabricated to fit a standard 10 pin card edge connector with .156″ spacing. You may prefer not designing double-sided circuitry because of the problems encountered making them at home and the increased costs involved, but the artwork density using twelve ICs made it impossible for this hobbyist to design single-sided circuitry which fits a 5½″ × 3″ circuit board. The artworks were prepared at home using commercially available artwork aids, taped 2:1 and photographically reduced at a local photo shop. Using direct positive photo-resist coated boards available from the Vector-Electronic Co. (CU70/45WE-2RN, 7″ × 4½″, 1/16″ double-sided glass-epoxy), the boards were exposed, developed and etched following the instructions that came with the boards. After cutting and drilling (#65 drill for IC pads, #60 drill for others), he tin plated the finished boards with Shipley LT-25 chemical plating solution. After assembly, the component side is top soldered as required and jumpers soldered top and bottom in the three feedthrough holes.

234

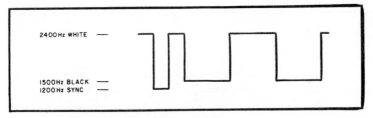

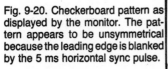

Fig. 9-19. Output of the WB9L VI low pass filter.

Adjustment. Obviously the easiest way to set the 1200 Hz, 1500Hz and 2400 Hz reference oscillators is with a frequency counter, and the values specified in the parts list should allow you to adjust the oscillators without reselecting R4, R7 or R10. If a frequency counter is not available, the oscillators can be set with a triggered scope with a *calibrated* timebase. Set the timebase to .1 ms/div, internal trigger and adjust the 1200 Hz oscillator for a width of .83 ms, the 1500 Hz oscillator for a width of .67 ms and the 2400 Hz oscillator for a width of .417 ms. Switch to line trigger and readjust the oscillators for a stable waveform (i.e., a waveform which is not slowly drifting from right to left or left to right across the screen). This adjusts the oscillators against the 20th harmonic (1200 Hz), 25th harmonic (1500 Hz) and 40th harmonic (2400 Hz) of the 60 Hz power line and is accurate within .1 per cent provided your timebase is calibrated, as it is very easy to adjust the oscillators against the wrong harmonic.

Operation. The pattern generator requires +5v ±.25 at 150 mA and 6.3-12.6 Vrms AC, and the supply illustrated in Fig. 9-18 satisfies the power supply requirements for the generator.

Syncing the scope on the positive edge of the horizontal sync pulse will be adequate for most troubleshooting designs to repair an ailing monitor with the generator. The 1200 Hz output can be used to tune the sync circuitry,

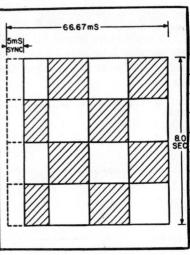

Fig. 9-20. Checkerboard pattern as displayed by the monitor. The pattern appears to be unsymmetrical because the leading edge is blanked by the 5 ms horizontal sync pulse.

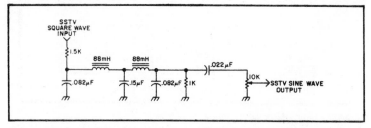

Fig. 9-21. This low pass audio filter used by K4EEU requires no power supplies.

and the 2400 Hz output can be used to adjust the discriminator. The horizontal and vertical sync pulses may be used as horizontal and vertical discharge pulses to drive the triggered deflection circuitry, and the checkerboard pattern can be used to optimize the gain(s) of the low pass filter for a square wave response with just enough overshoot (ringing) to enhance picture detail. The video output can be attenuated with a potentiometer to determine the minimum input for limiting, determined by limiter gain. Figure 9-19 is the output of the WB9LVI low pass butterworth filter with the generator connected, and Fig. 9-20 illustrates the checkerboard pattern as displayed by the monitor.

Although designed primarily for tune-up, calibration and repair, the generator can also be used for transmission by using low pass audio filter as shown in Fig. 9-21 to convert the square wave video to sine waves. This filter requires no power supplies, but an alternative is the filter used in Fig. 9-22.

Conclusion. Compared to the cost of an SSTV monitor, the pattern generator represents a modest investment. Naturally, a monitor can be constructed and aligned without this, but it should save countless hours of work should a snag develop.

Master Sync Generator

This master sync generator is designed to produce interlaced sync for a black and white camera at minimum complexity and cost. The heart of this circuit is an IC made by Texas Instruments for the '74 series of Sylvania color TVs.

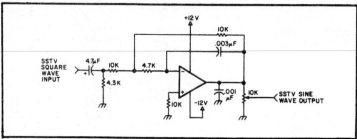

Fig. 9-22. WLMD low pass audio filter.

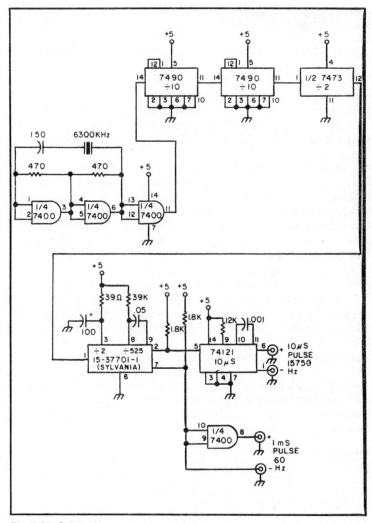

Fig. 9-23. Schematic.

This IC, Sylvania part number 15-377701-1, is a ÷2, a ÷525 and a single-shot all-in-one 14 pin dip IC. The list price is $5.40, and the net price is proportionately lower. This IC was designed so that the TV set generates its own sync signals if off the air snyc is temporarily lost.

By putting in a 31.5 kHz signal, and adding a singleshot to the horizontal output lead, interlaced sync results. One method of getting 31.5 kHz is shown in Fig. 9-23. With a surplus 6300 kHz xtal in the junkbox, this hobbyist followed it with a ÷200 set of flip-flops. Many other combinations of xtal and dividers would work fine, such as 3150 kHz and ÷100, 504 kHz and ÷16, etc., or you can pay the long dollar for a 31.5 kHz xtal.

Digital Signal Source

There seems to be a lack of a suitable signal source in this day and age. One hobbyist tried several things, including squaring circuits on the wide-range rf generator and several pulse/function generation units, with only marginal success. Recently, while working on a frequency synthesizer, an idea occured to this experimenter that *after breadboard construction* seems to be the answer. Development of this idea will result in a square wave signal source covering from 20 MHz down to sub-audio in fully tunable decade steps. It has proven to be an ideal unit for experimenting with amateur radio applications of TTL and CMOS logic.

Theory. The basic idea as presented in Fig. 9-24 is for a tunable oscillator in the 10 to 20 MHz range with switchable decade dividers for the range selection and switchable binary dividers for band selection. The resulting frequencies and time constants are listed in Fig. 9-25.

Construction. Figure 9-26 is the oscillator that this hobbyist used. Others would do as well. Try your favorite—just be sure the output is adequate to drive the digital buffer. Drive requirements to the first TTL stage can be cut down by biasing that stage into its linear range with a 2.2k resistor to ground as indicated in Fig. 9-26.

Figure 9-27 represents the power supply circuitry and is self-explanatory. Figure 9-28 includes the dividers and output circuit. Construction is straightforward with few precautions. It would be wise to keep all divider-to-switch wires separated from each other slightly (just don't bundle them all together in a cable harness). Run a separate power and ground lead for each IC. That way you need only one bypass capacitor on the common +5 volt and ground point.

This final version has a calibrated dial and tunes with a 20:1 VFO drive. However, you'll find it more convenient to cable the output to your digital counter for direct frequency readout.

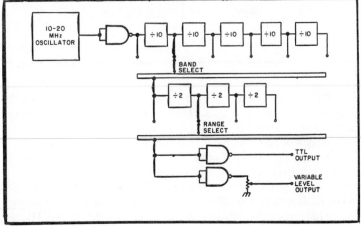

Fig. 9-24. Tunable oscillator.

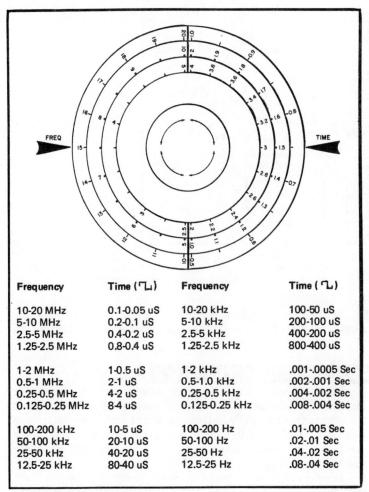

Frequency	Time (⌐⌐)	Frequency	Time (⌐⌐)
10-20 MHz	0.1-0.05 uS	10-20 kHz	100-50 uS
5-10 MHz	0.2-0.1 uS	5-10 kHz	200-100 uS
2.5-5 MHz	0.4-0.2 uS	2.5-5 kHz	400-200 uS
1.25-2.5 MHz	0.8-0.4 uS	1.25-2.5 kHz	800-400 uS
1-2 MHz	1-0.5 uS	1-2 kHz	.001-.0005 Sec
0.5-1 MHz	2-1 uS	0.5-1.0 kHz	.002-.001 Sec
0.25-0.5 MHz	4-2 uS	0.25-0.5 kHz	.004-.002 Sec
0.125-0.25 MHz	8-4 uS	0.125-0.25 kHz	.008-.004 Sec
100-200 kHz	10-5 uS	100-200 Hz	.01-.005 Sec
50-100 kHz	20-10 uS	50-100 Hz	.02-.01 Sec
25-50 kHz	40-20 uS	25-50 Hz	.04-.02 Sec
12.5-25 kHz	80-40 uS	12.5-25 Hz	.08-.04 Sec

Fig. 9-25. Suggested dial layout.

Wire the VFO portion using point-to-point technique on insulated standoffs. The divider is on perfboard and wired with wire-wrap pencil.

Summary. While this unit is just the basic generator, additions might enhance its operation. Some of the possible additions and their applications include the following:

- FM the VFO with an audio oscillator and voltage variable capacitor for working with FM receivers, phase detectors or PLL circuits.
- Switch the divider chain from the internal VFO to an external input jack. This would allow signals from external sources to be divided.
- Switch the divider chain input between the internal VFO and a crystal oscillator to generate harmonically-related standard frequencies.

239

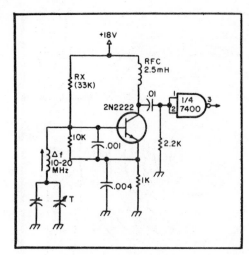

Fig. 9-26. Oscillator.

- Add a second buffer and output circuit for opposing polarity outputs.
- Run the output through a one-shot multivibrator for thin-line pulse generation.

In case you use this as an rf generator for general purpose work, you might be interested to know that the square wave output generates strong harmonics beyond 2 meters!

Bargain Audio Frequency Source

About a month ago one hobbyist had occasion to check the frequency response of an audio filter he had built. Unfortunately, his modest test bench lacked an audio oscillator. He had been unable to complete several projects because he didn't have such an oscillator. Buying one was out of the question. Being a college student, his financial status was only slightly above that of a medieval serf.

One day while visiting a local Radio Shack, he walked by the record rack and looked over the selections. All of a sudden there it was—a solution to the audio oscillator problem. In the corner of the rack sat Stereo Test Record

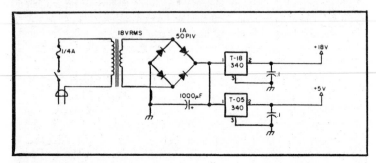

Fig. 9-27. Power supply circuitry.

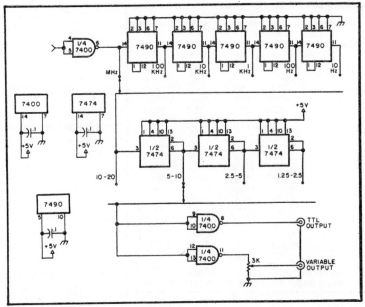

Fig. 9-28. Dividers and output circuit.

number 50-1971, which, in addition to many stereo system tests, contained twenty different test tones. It was only $1.49.

Inside of an hour, he had checked out the record on a friend's oscilloscope, which showed that the tones were pretty much sinusoidal. A little work with pen and paper showed that, if played at different speeds, the record could produce 71 unique tones. Granted, the record is hardly a replacement for a good oscillator, but it's certainly great in a pinch. It can even be used for checking the approximate bandwidth of phase-locked loops.

Table 9-7 lists the frequencies, to the nearest cycle, which can be generated when the record is played on a four-speed turntable.

The Audio Synthesizer for RTTY, SSTV and Whatever

Have you ever had a need for an audio generator for precisely tuning your SSTV or RTTY equipment? This audio frequency synthesizer generates highly accurate tones useful for tuning both SSTV and RTTY station equipment. The output frequencies are within .4 Hz of the exact desired frequency of eleven tones critical, or very helpful in the tuning of an SSTV or RTTY station. A single crystal is subdivided to derive the desired tones, and the complete unit can be built for less than $20.

The frequencies considered critical to SSTV tune-up were 1200, 1500 and 2300 Hz, representing Sync, Black and White respectively. In addition, 1700, 1900 and 2100 were considered useful for grey scale adjustments of SSTV, so these were added, resulting in six synthesized frequency requirements for SSTV.

SPEED (rpm)			
33⅓	16⅔	45	78*
30	15	41	70
70	35	95	164
100	50	135	235
200	100	270	470
300	150	405	704
400	200	540	939
440	220	594	1033
500	250	675	1174
800	400	1080	18	78
1000	500	1350	2348
2000	1000	2700	4696
3000	1500	4050	7043
4000	2000	5400	9391
5000	2500	6750	11739
6000	3000	8100	14087
7000	3500	9450	16435**
8000	4000	10800	18782**
9000	4500	12150	21130**
10000	5000	13500	23478**
15000	7500	20250**	35217**

FREQUENCY (Hz) is the row label for the frequency columns.

*Actual turntable speed is 78.26 rpm.
**Beyond the capabilities of most phonographs.

Table 9-7. Generated Frequencies.

RTTY has five frequencies which can be considered critical to tune-up. These are 1275, 1445, 2125, 2295 and 2975 Hz. These are the five tones which may be utilized on either wide or narrow frequency shift keying, using the low or high tone set.

Theory of Operation. The af synthesizer starts off with a simple crystal oscillator feeding three synchronous binary counter ICs. Since each IC has four stages, a possible frequency division of 2^{12} or 4096 exists before the counters carry to ϕ. However, if the proper outputs are gated together, the counters can be reset prior to 4096, thus establishing control of the exact frequency division. As shown on the schematic in Fig. 9-29, by sequentially labeling the outputs of the counters in powers of 2, the desired frequency division can be easily assembled. The problem then becomes one of getting the simplest gating combination for the desired accuracy.

A rather complex computer program was written to analyze the problem stochastically, and the relevant portion of the printout is shown in Table 9-8.

The computer did in a few minutes what would take years of manual calculation and comparison. Eleven NAND gate decoders are wired to detect a 1 bit at every input to produce a ϕ bit at the output of one of the decoders, the decoder of the selected frequency to be synthesized. This ϕ bit at the output of any decoder then produces a 1 bit at the output of the OR section, resetting the divider section and also toggling the output flip-flop.

The various NAND division decoders are selected by a switch which puts a + enabling voltage on one input of the desired frequency's decoder. If remote manual, or electronic selection is desired, such as a grey scale pattern for SSTV, or AFSK for RTTY transmission, the builder merely has

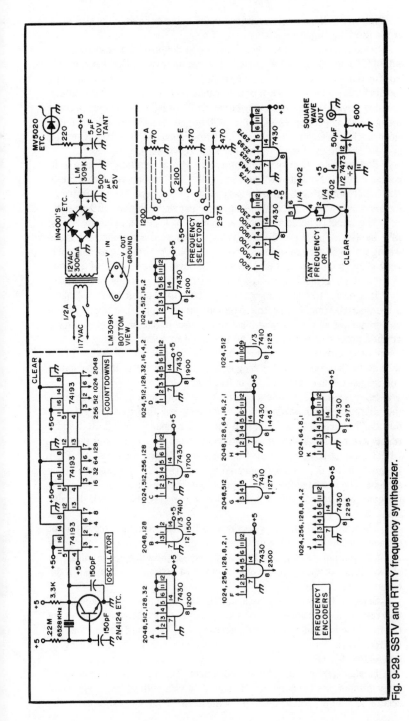

Fig. 9-29. SSTV and RTTY frequency synthesizer.

243

to supply a + voltage to the desired decoder selector input in the sequence or for the time required.

Since the calculated frequency has been doubled, then halved in the output flip-flop, the output waveshape is a symmetrical square wave, excellent for tune-up and calibration, but poor for transmission. If you wish to transmit this signal, build a low pass or bandpass filter to convert the output to a sine wave. Those not needing either SSTV or RTTY synthesis can omit the unneeded decoder NANDs, the 7430 or gate associated with the undesired block of decoders and the 7402. The remaining 7430 in the OR section is then directly connected to the 7473 and the Clear line.

Construction. One experimenter built the af synthesizer as a self-contained unit having its own power supply. The cabinet is a Radio Shack #270-252 measuring 4″ wide by 2-⅜″ high by 6″ deep. The power transformer is any 12 volt AC 300 mA or more unit that you may have around.

The actual IC board is a perfboard unit with .1″ center holes measuring 2½″ by 4½″. The ICs and other parts are inserted and then the tedious wiring with #30 wire and a small, fine tipped soldering iron begins. Since it is very easy to make a mistake, be sure to develop some kind of a wiring system, such as labeling the ICs and wiring similar sections sequentially.

It is easier to build if an ordered process is followed which will allow progressive testing. Complete and test the power supply first. Next wire up power pins of all the ICs and build the crystal oscillator. After you have verified proper oscillation, wire up the three frequency dividing ICs, and temporarily ground pin 14 of these three ICs. Look at pin 7 of the last 74193 with an oscilloscope. If everything is working correctly so far, you should see a square wave, 1593.75 Hz in frequency.

Start wiring the decoders by wiring the 7430 labeled "1200". Reading the abbreviated schematic in Fig. 9-29, you will find that the 1200 Hz decoder requires a 2048, 512, 128 and 32 bit input to pins 2, 3, 4 and 5 respectively. The 2048 and 512 bits come from pins 7 and 2 of the third 74193, and bits 128 and 32 come from pins 7 and 2 of the second 74193. Pin 1 of the 1200 Hz decoder is connected to "A" (the 1200 Hz position of the switch). The output of the 1200 Hz decoder (pin 8) is connected to the 7430 OR gate NAND associated with the SSTV frequencies, the 1200 Hz decoder input, pin 1.

Then wire the other connections to the OR section, except do not connect the OR inputs to the unwired decoders yet. Remove the temporary ground from pin 14 of the 74193s and connect the OR section output as shown on the schematic to the pin 14s of the 74193s and the input pin of the output ½ 7473.

Select 1200 Hz on the switch, and the output of the synthesizer should now read 1200 Hz, ± one digit, on a frequency counter. Now wire the rest of the frequency decoders and test each one. If any selected output does not read out correctly on the counter, you either have an error in your wiring, or a defective IC. Or maybe a bum counter!

Extending/Modifying The Af Synthesizer. The unit can also synthesize other frequencies between 800 Hz and 3264 kHz with varying degrees of accuracy. The lower the desired synthesized frequency, the greater the

Table 9-8. Computer Printout of Frequency Synthesis Combinations.

3264000							
	SSTV 2720.00	2176.00	1920.00	1717.89	1554.29	1419.13	
	RTTY 2560.00	2258.82	1536.00	1422.22	1097.14		
2720	1200.0000	2048	512	128	32	0	0	0
2176	1500.0000	2048	128	0	0	0	0	0
1920	1700.0000	1024	512	256	128	0	0	0
1718	1899.8836	1024	512	128	32	16	4	2
1554	2100.3861	1024	512	16	2	0	0	0
1419	2300.2114	1024	256	128	8	2	1	0
2560	1275.0000	2048	512	0	0	0	0	0
2259	1444.8871	2048	128	64	16	2	1	0
1536	2125.0000	1024	512	0	0	0	0	0
1422	2295.3586	1024	256	128	8	4	2	0
1097	2975.3874	1024	64	8	1	0	0	0

probability that the resultant output will be very close. A number of extra, currently unused, input gates in the OR section are shown on the schematic. These can be connected to additional decoders, and additional switch positions will allow selection of up to 16 synthesized frequencies.

Suppose you wish to add 1000 Hz output. First use the following formula to derive the frequency division needed:

$$\text{Division required} = \frac{3264000}{\text{Synthesized Frequency Desired}}$$

Entering our desired frequency of 1000 Hz we get:

$$3264 = \frac{3264000}{1000}$$

Next, we determine the binary divisions required. Sequentially subtract the highest binary number listed on the 74193 outputs, in descending order:

$$3264 - 2048 = 1216 - 1024 =$$
$$192 - 128 = 64 - 64 = 0.$$

This indicates that a decoder connected to the 2048, 1024, 128 and 64 outputs of the 74193s will give a 1000 Hz synthesized output from the af synthesizer when selected.

A problem comes in when the division required is not a whole number. In this case, the division required is rounded off to the nearest whole number, but some resultant inaccuracy will have to be tolerated, or a new computer analysis can be run, using this new desired frequency as another simulation constraint.

Table 9-8 shows several examples of how the computer listed the resultant error for this hobbyist. The first number on the top line was the stochastically selected master frequency. The SSTV and RTTY frequencies required the 11 division ratios shown to the right of the master frequency. 2100 Hz actually required a division of 1554.29. The computer rounded off to 1554 beneath, and then calculated the resultant frequency and the required binary divisions. Notice that the 1554 division results in a frequency .3861 Hz too high, but this is close enough for this application so the design was accepted.

Another consideration is that a 7430 has only eight input legs. Since one input goes to the switch, each decoder must use a maximum of seven input legs to the counters. This was another variable entered into the computer as a design constraint. Note on the printout of Table 9-8 that only one decoder (1900) required all eight inputs. Synthesized frequencies requiring only two counter inputs can use a ⅓ 7410 as shown, and those requiring only three can use a ½ 7420, in the interest of lowering the total IC count.

Conclusion. This project has presented an af synthesizer for SSTV and RTTY frequencies which is accurate to within .4 Hz, in the worst case. A computer was utilized to obtain the needed data for building a unit which gives the required accuracy, as well as minimizing the complexity of the unit. The output is a symmetrical square wave.

Build an Amazing Function Generator

The increasing popularity of a relatively new piece of test equipment (the function generator) has spurred at least two IC manufacturers to design special monolithic chips for this purpose. The Intersil 8038 and the Exar XR2206 are examples of such specially committed ICs. These new ICs offer great simplicity in function generator construction, but offer the user very little feel for what is actually going on in the process of waveform generation. Since the building of a piece of test gear is also a learning process, circuit flexibility and stage-by-stage analysis are important. For this reason, an older design for a function generator from a Motorola application note (AN510A) was the starting point for this project. The original AN510A output amplifier section is a real *klooge* by today's standards, and so it was replaced by a simpler all-IC substitute. The original power supply used two dual-winding power transformers, four integrated bridge rectifiers and four power IC regulators. This rather elaborate supply was replaced with one inexpensive transformer, one integrated bridge rectifer and two of the newer Raytheon ± regulator ICs. A feature in the new function generator is DC offset, a simple addition that is really worthwhile.

The circuit of the function generator is shown in Fig. 9-30. Note that four ICs as well as a number of discrete devices are used in the waveforming circuitry. In addition, two more power ICs are used in the power supply, shown in Fig. 9-31. A block diagram of the waveforming circuitry is shown in Fig. 9-32.

The integrator is composed of U1 and Q1, an op amp and an emitter-follower to lower the op amp output impedance.

The comparator is composed of U2, Q2, Q3, Q4, Q5, Q6, Q7, D8 and D9. U2 is an Emitter-Coupled Logic IC capable of extremely fast switching. Associated with, but not actually part of, the compartor are D10, D11 and D12 which serve as voltage regulators to provide U2 with +1.4 volts and −3.9 volts. It is worth noting that Q4, Q5 and Q6 were originally designated as Motorola MPS-L08 types in AN510A, but since this transistor type is now obsolete, appropriate substitutions have been made.

The *reference switch* is made up of D1, D2, D3 and D4. Note that D3 and D4 are dual diodes. Also note that RB in Fig. 9-32 is either R8 plus R10 or R9 plus R11, depending on the state of the reference switch.

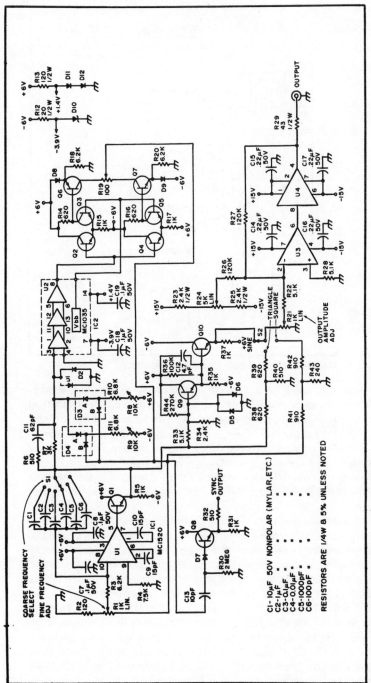

Fig. 9-30. Function generator, waveform circuits.

247

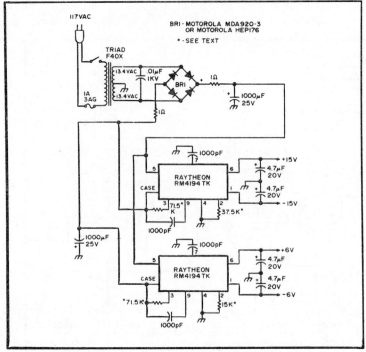

Fig. 9-31. Regulated power supply for function generator.

The *sync amplifier* is a simple differentiator, rectifier and emitter-follower. The square wave from the comparator is differentiated by C13 and R30. D7 allows only the positive-going spike to be passed to the base of Q8. This positive spike is then available at the emitter of Q8 for a sync pulse.

The *sine wave shaper* consists of D5, D6, Q9 and Q10. D5 and D6 act as soft clippers on the triangle wave, and produce a near approximation to a sine wave. Q9 and Q10 simply act as emitter-followers after shaping—one NPN and one PNP, so that their emitter-base voltage drops cancel each other.

Since it was necessary to attenuate the triangle wave (with the voltage divider R33-R34) to make it compatible with the shaping diodes D5 and D6, the resulting sine wave is smaller in amplitude than the triangular wave or square wave. To equalize the output levels of the three waveforms, simple "T" attenuators are placed in the triangle and square wave lines. These consist of R38, R39, R40 and R41, R42, R43 respectively.

Finally, the *output amplifier* consists of U3 and U4, a high slew-rate op amp and buffer amplifier. Note that the buffer is inside the closed loop of the amplifier. By adding in variable DC at the inverting input of U3 via a 120k resistor, a "DC offset" adjustment is easily obtained. This DC offset enables one to offset the three types of waveforms for testing of circuits which will accept only unipolar signals, such as logic circuits.

The power supply utilizes a Triad F40X (26.8 Vct 1A) transformer which combines low price, small relative size and high current capability. An integrated bridge and two Raytheon RC4194TK integrated circuits in the circuit of Fig. 9-31 provide ±15 volts and ±6 volts for the waveforming circuitry. The RC4194TKs are heat-sink mounted to the chassis with T066 mica wafers for heat transfer and electrical insulation. The F40X transformer is mounted under the chassis to keep it electrostatically shielded from the top-mounted waveforming circuits. The two 1000 uF filter capacitors are also mounted under the chassis because of the relatively large AC line ripple on them.

The function generator is built into an old 7″ × 8″ × 10″ steel equipment cabinet, to which an aluminum panel has been fitted. The aluminum panel was originally an old black-crackle finished relay rack panel which was stripped of paint and cut down to size. The left over portion of this same panel was made into the frequency dial, by rough sawing and turning down the outer diameter on a lathe. The large aluminum spinner knob for the center of the frequency dial was also turned from a scrap of bar stock on the lathe. The basic planetary drive for this control was a National Velvet Vernier salvaged form a surplus BC375 or BC191 tuning unit. Those with fancier dial systems available or without access to a lathe and the above surplus drive units can use other methods—even a plain large knob. One will note that the timing capacitors in the circuit of Fig. 9-30 is a 10 uF non-polar type. An old 10 uF 600 V transmitting capacitor was used. It was mounted on a homemade insulating mount, as otherwise stray capacity to the capacitor case was objectionable in this circuit.

The relatively small amount of circuitry of the power supply is mounted under the chassis. The power supply should be the first section checked out, preferably before connecting it to the waveforming circuits. Note that each RC4194TK has a 71.5k resistor from "case" to #3. This is the nominal value suggested by Raytheon. Sorting through one's 68k and 75k resistors will

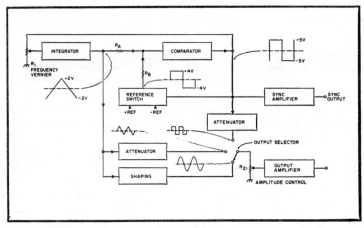

Fig. 9-32. Block diagram of function generator.

Table 9-9. Parts List.

D1,2	—Silicon signal diodes: Motorola MSD6102, 1N4454, or 1N914
D3	—Dual Si. signal diode:Motorola MSD6150, or two 1N4454, 1N914
D4	—Dual Si. signal diode: Motorola MSD6100, or two 1N4454, 1N914
D5-9	—Silicon signal diodes: Motorola MSD6102, 1N4454, or 1N914
Q1, Q8, Q9	—NPN Xstr: Motorola 2N4124 or HEP53, or 2N3643, or HEP-S0014
Q2, Q3, Q7	—NPN Sw. Xstr:Motorola 2N709 or HEP50, or 2N3646, or HEP-S0011
Q4, Q5, Q6	—PNP Sw. Xstr:Motorola 2N4260 or HEP720, or 2N3640, or HEP-S0019
Q10	—PNP Xstr:Motorola 2N4126 or HEP57, or 2N3644, or HEP-S0019
U1	—Motorola MC1420G or MC1520G
U2	—Motorola MC1035P or MC1235L
U3	—National Semiconductor LM318H, LM218H, or LM118H
U4	—National Semiconductor LH0002CH or LH0002H
D10	—Mototola 1N4730A or HEP-Z0403
D11, 12	—Motorola 1N4001 or HEP-R0050

give a few values close to 71.5k. Use these. The 37.5k and 15k resistors from pin #2 of the RC4194TKs will then be target values to give ±15 volts and ±6 volts respectively. Some juggling of these target resistance values may be required to give exactly the voltages desired.

Assuming that the power supply has been checked out, as above, and puts out ±15 and ±6 volts, the waveforming circuitry can be connected to it. The pots R1, R8, R9 and R19 should all be set at mid range, and the pot R21 set at minimum. The DC offset pot should also be set at mid range.

With the scope on the wiper arm of R19, adjust this pot until the waveform looks like a square wave (i.e., the positive portion is as long as the negative portion). Then put the scope on the emitter of Q1, and adjust R8 and R9 until the triangular wave observed there has a ±2 volt value. R19 should be retweaked and then R8 and R9 again. The scope can now be shifted to the output, and DC offset, level, frequency and sine-shaping checked.

This function generator is a useful piece of test equipment built up from odds and ends. While the semiconductor costs can be as high as $35.00, appropriate substitutions can trim this figure somewhat. To this end most of the diode (D) and transistor (Q) designations have several acceptable numbers given in the parts list of Table 9-9.

Build This Simple Noise Generator

For several years, there have been quite a few noise bridges available. The noise generators are all based on the use of noisy diodes (usually a 1N753, 6.8-volt zener, back biased) which go through two or three stages of transistor amplification and then into the various configurations of bridges. They all do a creditable job, but they use a great number of components and draw too much current. A noise bridge is a must for mobile operation, where the mobile antennas have a high Q and should be operated at very close to resonance. Using a bridge, you can QSY quickly and readjust the top section of the antenna for exact resonance. Plus or minus 10 kHz is about the tolerable excursion.

A transistor version of a white noise generator was built and tested. The results were superior to any of the systems tried with a noise diode.

Some diodes were so noisy that little amplification wes needed; others made so little noise that even three stages of amplification were not enough, to say nothing of the undesirable added current consumption.

How does this circuit generate such a wide band of frequencies? This oscillator may be described as a self-quenching device which is capable of generating square wave pulses with nearly ideal square wave. We know that a pure sine wave does not contain any harmonics. A sine wave with some distortion generates some harmonics. A *perfect square wave generator* will produce its fundamental frequency plus all the higher frequencies to infinity. With the materials used to produce this perfect square wave, we find that the usable bandwidth is actually far short of infinity. Tests indicate that the usable noise frequency is from somewhere around 500 kHz to far beyond 30 MHz, the amplitude remaining quite constant over the entire bandwidth.

Oscillation starts at 1½ to 2 volts and continues at a nearly constant level on all the HFs up to 20 volts. At 3 volts, the current drawn is about 200 microamps. At 20 volts, the current drawn is about 6 milliamps. This is ideal, whether you want to use the bridge with two AA cells, a 9-volt transistor radio battery or 13 volts from your car battery.

Several types of NPN trnsistors, high gain and in metal can or plastic, were tried. Most of them worked. However, it was found that 2N2222 transistors are the most easily obtained. They are cheap, also.

There are three components that must be given special notice. They are the RFC, the number of turns on the toroid and the manner in which they are wound and the disc ceramic capacitors. The .01 disc capacitors must be the 1,000-volt variety. Several dozen units have been built and sold with no problem, with these precautions taken.

The RFC consists of 24 to 26 turns of enamel #26 to #30 wire wound on a form about 4 to 6 mm (3/16″ to ¼″) in diameter and about 17 mm long (11/16″). The form to hold the wire may be any reasonable good insulator, such as a plastic knitting needle or a piece of #8 house wire with the wire removed from the plastic insulation and a wooden match stick inserted in place of the wire to make it rigid. A ½-watt ohmite-type of resistor, more than 100,000 ohms, also worked fine.

The toroid coil is wound on a T-50-2 (red mix) core. The core is rated for use between 500 kHz and 30 MHz.

For the primary winding on the core, which will connect to transistor collector and to the positive voltage, wind 24 turns as follows: Hold the

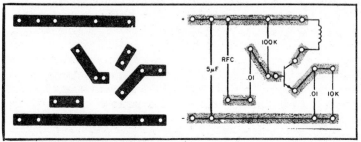

Fig. 9-33. PC board-actual size.

toroid in your fingers, in a horizontal plane, and make the first turn with the starting end of the wire sticking up out of the core about 5 cm (2″). Now make the first turn so that the long end of the wire will come out of the toroid on the right side of the starting wire. The next turn will come out of the toroid on the left side of the starting wire. The completed primary winding will have a bundle of 24 turns, or 12 turns on each side of the starting wire. By doing this, we have a coil with minimum stray capacitance coupling between primary and secondary.

The secondary, or output winding, will be 12 turns of bifilar, wound and connected to give a total of 24 turns. Cut two pieces of the same enamel wire about 40 cm (16″) long. Put two ends together in a vise, and twist them about one twist per com (2½ twists per inch). Wind 12 turns on the toroid, leaving about 5 cm (2″) of wire out of the toroid on both ends. Use a pocketknife to carefully remove all the enamel far enough back toward the toroid so that you can cut off the excess length not needed for making the connections to the coax connectors, bearing in mind that the two outside ends of the winding should be exactly the same length.

Select a wire from each end of the winding and test with an ohmmeter to find two wires that do not show continuity. Connect those two wires together. Now you should test to make sure that you do have continuity on the remaining two wires. This gives you the two bifilar windings in series, with a center-tap, or 24 total turns. Adjust these windings on the toroid core so that the primary and secondary windings have equal spacing between the two coils on both sides.

The toroid core with its windings should be mounted on a small plastic or wood pillar about 2mm (⅜″) long and cemented on the PC board. The white silicone cement available at nearly any store is ideal for this purpose.

Refer to the PC board layout in Fig. 9-33. Follow the pattern quite closely and all the parts will fit nicely into place. The values of all the components are indicated in Fig. 9-34. The diagram showing the Rx pot, the 360 pF variable capacitor, the 180 pF fixed capacitor and two coax connectors should be studied and used as a guide for making it a balanced system. That is, the lead lengths of the fixed 180 pF capacitor should be the same length as the lead of the 360 pF variable where they connect to a solder lug at the detector (DET) output coax fitting. The lead to the 100-ohm ohmite pot should be as short as possible. The PC board is shown actual size, and it may be used as a template to make small punch marks on the copper foil.

Chapter 10
Frequency Counters

Selecting a Frequency Counter

Thanks to microcircuits, there has been a major revolution in test equipment within the last few years. The prices of devices such as digital voltmeters, triggered sweep, oscilloscopes and others have dropped dramatically, a neat trick considering inflation! But more important, this excellent, low cost gear is getting into the hands of servicers and electronics enthusiasts where good equipment is both useful and necessary to cope with today's advanced products. The frequency counter has also benefitted from microcircuits and price cuts, but only recently has it begun to "take off" with electronic hobbyists and service people. This may be due to the past high prices, and a misunderstanding of what a frequency counter can do. But these things are going to change as more people are discovering what a counter can do for them!

We are going to show you some of the things to look for when you shop for a counter, and some of the not-too-obvious pitfalls to avoid. There's a lot more than specs to consider, too—all counters have special features that may be obvious or not in the advertising literature. In short, we are going to try to make your selection a better one by showing the more important general features and explaining them, so you'll know what to look for.

Some of you skeptics are probably thinking, "Why do I need a counter?" A good question, indeed. Have you ever designed/built an oscillator and couldn't find the frequency? Then spent hours pruning the circuit to the proper frequency? A counter would tell you where you are at a glance. Or have you ever aligned a filter or trap or i-f stage and found that the center frequency was a city block off? A counter would help you keep tabs on the calibration of your signal generator and get a better alignment in the bargain! Have you ever had trouble accessing the local 2 meter repeater? A counter can tune up that tone generator in a jiffy. Ditto the transmitter with a VHF counter. Or, perhaps you are a CBer and people complain that you "bleed" on several channels. This could be caused by a sick transmitter with a bad crystal(s). A qualified technician can easily check this out with a good counter. Anybody still skeptical? Remember that you can do far more with a counter if you put your imagination to work!

A Bit of History. You might be interested in how frequency counters evolved. The first method of frequency measurement evolved around the turn of the century when it was necessary to measure the frequency of radio transmitters. The gadget was called a *wavemeter* and it consisted of a paralleled coil, variable capacitor and spark gap (later replaced with a meter or light bulb). The capacitor had a calibrated dial, and it was adjusted until a spark appeared at the gap. The frequency was either read off the dial or extrapolated from a coil/capacitor chart if the dial read pFs. By the '30s, another form of frequency measurement came into vogue: the *frequency meter*. This unit had a built-in frequency standard and the unknown was mixed with it; the meter was adjusted until the signals "zero beated" in the headphones that were part of the unit. Old-timers will no doubt recall the BC-221 and LM frequency meters of WW II vintage with this mention—they were very famous! Digital electronics was the next development, and a digital frequency counter appeared in the early '50s. The first ones had 40-plus tubes, had neon lamps arranged in 0-9 columns for a readout, and were called EPUTs (Events Per Unit Time meters). The top frequency of those early counters was only a few MHz at most, although VHF range extenders appeared quickly. The early manufacturers were Berkley Scientific and Hewlett-Packard. The next step was in the 60's when ICs became available. Counters became smaller and much cheaper, too. Many companies are still using these warmed-over cicruit designs today, despite advances such as CMOS. Recently, one semiconductor company has introduced a *2 chip* counter, where all of the major parts of a six digit counter are on two LSI chips! Needless to say, the future has much to hold!

A Look at Basic Specs. Selecting the right counter takes some thought and an understanding of counter fundamentals. Without these things, you could end up with a three digit "toy" and have to align a generator that is 10 times more accurate, or a unit that is so versatile it does everything from making the morning coffee (extra cost option) to checking a phono oscillator. So, needless to say, the way to start is to sit down and analyze *your own needs*. Ask yourself such questions as, "What am I going to use a counter for?" and "What do I expect to be doing with my counter within a few years?" These questions will help you decide the *primary* features that you must have in the counter you select.

For example, suppose you are an experimenter and you like to work with audio circuits and TTL logic. This suggests that you should start looking for a counter with a 20 MHz maximum range, because that is probably the highest frequency you are working with (frequency limit of standard TTL). Accuracy probably isn't critical to you and you might be able to settle for a unit that is 0.03 per cent accurate.

Or perhaps you are a professional servicer and you are getting into CB repair. Your requirements are more strict. Since CB is 27.255 MHz maximum, you need one of the popular 30 HMz counters. Also, you need a counter that is 0.001 per cent or "10 parts per million" (abbreviated 10 ppm) accurate to satisfy FCC requirements.

Consider the future, too. If the experimenter upgrades to, say, CB, he'll need a faster counter. But there are devices such as *prescalers* to extend the range of counters at low cost. So, when the day arrives, he might only have to add a low cost black box to his counter.

On the other hand, if the servicer upgrades to commercial radio repair, he may have to replace the counter. Why? The frequency tolerance of commercial gear is often 0.0005 per cent and that requires a counter of 0.00001 per cent or 1 ppm. That spec is 10 times better than the old unit's! These are a few thoughts to keep in mind when you start your search for the right counter. Try to anticipate the future!

Let's list the primary specs of a frequency counter and then discuss them. Note: They are not necessarily listed in order of importance.

- Accuracy
- Input sensitivity
- Minimum and maximum frequency range
- Display
- Power supply
- Special features or options

Accuracy. The accuracy of a frequency counter is determined by a quartz crystal or sometimes the AC power line (60 Hz). Fig. 10-1 shows a block diagram of a simple frequency counter. As you can see, the crystal/60 Hz is divided in frequency to operate the rest of the counter; this entire section is called the *timebase*. The timebase (or TB, for short) is the heart of the counter and its accuracy determines the accuracy of the counter. Price is also a function of the quality of this section. You'll find counters that use the 60 Hz line as a frequency element that sell for up to $150, and you'll see units that have crystals in *ovens* (temperature stable enclosures) that sell for $6000 or more. The difference in accuracy is incredible: The 60 Hz TB counter will average 0.034 per cent accuracy, the average accuracy of the 60 Hz coming out of the wall, to 0.0001 per cent and better for the crystal oven unit! This is clearly one area where money talks. Typical counters run 0.005 per cent (50 ppm) to 0.001 per cent (10 ppm.) They use 4 MHz or 10 MHz crystals without ovens for low cost. These crystals are usually custom ground and the TB is carefully adjusted to get this degree of accuracy.

When you shop for a counter, try to get a counter with as much accuracy as you can to suit your needs. Remember you need at least a 10 ppm timebase for CB repair work, if that is your interest. The 60 Hz units

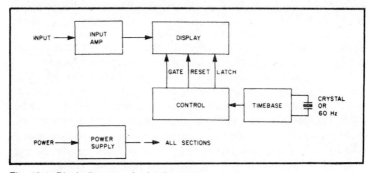

Fig. 10-1. Block diagram of a basic counter.

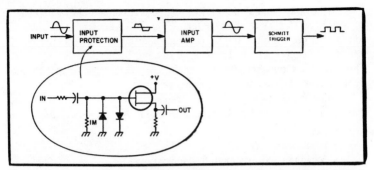

Fig. 10-2. Expanded drawing of input amplifier and a rough schematic of a typical input circuit.

are not for serious electronics work; 0.034 per cent accuracy is terrible in the counter world. Anyhow, the few units available with these may be off the market by the time you read this!

Input Sensitivity. Another sign of a counter's quality is the sensitivity it has to measure the frequency under question. Figure 10-1 shows an *input amplifier,* the section that has to do with the input sensitivity, and Fig. 10-2 shows a block diagram. The job of the input amplifier is simple: It amplifies the input signal and converts the signal to a corresponding set of pulses that is necessary to drive the counter's digital circuitry. The heart of the unit is the Schmitt *trigger;* it converts the incoming signal, which may be any kind of a waveform, into the necessary digital square wave. Generally, when you read input sensitivity specs, you must assume that the sensitivity refers to the minimum signal required at the maximum rated frequencies of the counter to get a steady reading. In other words, the specs are worst case. Generally, the sensitivity of most counters in not flat over frequency, hence the min-signal-at-max-frequency bit. Figure 10-3 shows the sensitivity plot of the Gary Model 301 Counter, a 32 MHz unit. As you can see, the input sensitivity is not flat! But in this case, a non-flat sensitivity curve can work to your advantage. The high gain at audio frequencies allows use of devices such as microphones and magnetic pickups for organ tune-ups and tachometers. The lower gain at 32 MHz reduces the chance of overload by radio transmitters, too.

Typical input sensitivity is on the order of 10 mV to 120 mV for most counters. It is desirable to get a counter with slightly more than enough sensitivity to do the job. In most labs, 100 mV worst-case does fine.

You will also find a maximum input voltage spec on some units. This refers to the maximum voltage you can apply to the counter without damage. This is always the *peak* AC signal plus any DC that may be present in the signal. For example, suppose you are measuring the signal of a homemade oscillator directly at the collector of the output transistor. You have, say, 9 volts DC at this point, plus, say, a 6v peak-to-peak signal: You are applying 9v DC + 3v peak or 12 volt worst case to your counter! Figure 10-4 illustrates this. This spec isn't really important to you unless you work around high voltage tube circuits and moderate to high power transmitters. In this case, there are counters built with range switches (usually calibrated ×1, ×10, ×100, etc.) for use around high voltages. Typical maximum voltages are 20 to 100 volts for counters without range switches and 500 volts for the ones with them. These voltages are rated at a counter's maximum rated frequency, where the maximum input voltage must be reduced to prevent damage to the input amplifier. At low frequencies, such as 60 Hz, most counters will handle 120 V ac without problem.

Min/Max Frequency Range. This is probably the most advertised feature of frequency counters. Everywhere you look you are bombarded with ads screaming, "30 MHz Counter," "80 MHz Counter," "Or How About A 250 MHz Counter..." and so on. Yet top frequency isn't all that important. A 30 MHz counter measures a 27 MHz signal as readily as an 80 MHz unit or even a 250 MHz unit or above. Yet, the price difference between these models can be at least several hundred dollars. All you are getting for the extra money of a faster counter is a unit with more capability. And that's fine if you sometimes need a counter to measure VHF signals; otherwise you can save money if you stick to a lower cost, lower frequency counter that will read the maximum frequency you expect to be working with. Remember, too, that low cost prescalers are available that divide the input signal by 10, so you can have your cake and eat it too!

An often ignored spec is the minimum frequency a counter will display. This depends upon the timebase and the sensitivity of the input amplifier. In order to get any accuracy from a counter when measuring slow signals, at least 10 Hz must be displayed on the standard MHz range switch set to kHz (actually 1 second gate time). Why? All counters have a built in ±1 count inaccuracy due to the circuitry. That's 10 per cent of 10 Hz—in other words, your 10 Hz is 10 per cent accurate! Also, TB errors show up in these digits, and can worsen or even improve accuracy at this point. The input amplifier sensitivity may drop at this point, too, making it hard to pick up low frequency signals. It's safe to say that the low frequency limit of most counters is 50 Hz for good accuracy (5x the minimum frequency is good test equipment practice) even though the counter will read lower frequencies. If

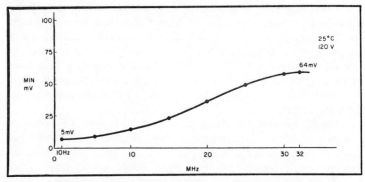

Fig. 10-3. Input sensitivity of a commercial counter. This plot shows the minimum voltage required to get a stable reading on the counter.

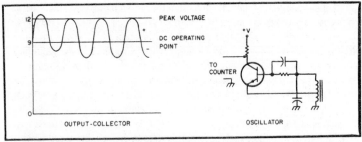

Fig. 10-4. Calculation of peak or worst case voltage that could appear across a counter input. Note sample circuit.

you do a great deal of measuring at low frequencies, you should consider a counter with a 10 second gate time, or a unit with a frequency multiplier, notably the Heath IB-1103. They will give you a display such as 60.1 Hz (10 sec TB) or 60.030 Hz (IB-1103).

Display. The display on a counter can have a powerful bearing on whether you buy a certain model or pass it up. Important things here are the appearance of the digits and the number of them. LED type displays currently reign supreme in counters, so this is mostly what you'll see. Nixie tubes are seen in some older AC-only counters, but their use is dying out. Liquid crystals show promise for the future, because they don't require the high current of LEDs and the high voltage of Nixies. When you shop, look for the largest and brightest display you can get, but do not let the appearance of the display bias your thinking. Accuracy and sensitivity are far more important in determining your choice.

The number of digits should be considered, also. As a rule of thumb, five to six digits are about the optimum number. Fewer (such as the three digits found on an inexpensive kit) require complicated range switching or more knob turning for you. And more digits can be confusing to read. Ever see 146.96 MHz read out on a large counter as 146960001 Hz? It's very confusing at times. Also, large digit counters consume larger amounts of power; consider that if you are buying an AC/battery model!

Power Supply. Some counters offer you different ways to power themselves, often at no extra cost. The most common power sorce is 120 volts 60 Hz. You'll find many units that can be wired for 240 volts AC, but you may have to check the operator's manual to find this out. There are also counters that have 120v plus 12v DC connections. You can often plug them into a cigarette lighter for working on mobile electronics. A few battery operated counters are becoming available, but at present none have the full features of the other models.

Special Features/Options. A whole plethora of special features are available on counters. Switching ranges from a simple kHz/MHz switch to a full blown range switch of 0.01 ms to 10 sec. Inputs range from simple jacks to complicated attenuators and trigger level controls. Input impedances may be different, too. You'll find the standard 1 meg plus 10 to 50 pF input and a straight 50 ohm input (for VHF) on some units. Some of the more expensive units have timers/stopwatches and time interval measuring features built in. And the list could go on. Base your selection on what features you really need and not what looks nice. It makes no sense to buy a $300 counter/timer, then use the timer once and spend the rest of the time measuring frequency!

Most options can be quite useful. Some popular options include frequency range extenders, high stability TBs, power cords, etc. Also, don't overlook good probes and cables. A ×10 scope probe is often used with the standard 1 meg input counter and will increase the maximum input voltage 10 times, preventing overload in most cases, and making connection to the equipment under test much easier. Needless to say, this is a very handy option and one that you should consider! If you feel that you must go for options, try to anticipate future needs. Sending a counter back to the manufacturer for modifications at a future date is usually much more expensive than ordering everything at the same time.

Some Pitfalls. Let's wrap this discussion of counters up by briefly mentioning some pitfalls that can trap you. Here are some of the areas to watch. Accuracy specs are sometimes hard to find on a counter data sheet, and when you do find them, they may not be complete! Although the data sheet may say "Accuracy: 10 ppm," nothing is said about the temperature or supply voltage, all of which affect the accuracy slightly, If accuracy is important to you and you are in this situation, it may pay to contact the manufacturer. Input amplifiers also have questionable areas. The input circuit of counters without attenuators should have an overvoltage protection network, usually diodes. Check. If the counter doesn't have one, avoid it or you'll be doing a lot of repair work on the counter! Finish up your evaluation by looking over the counter for any other features that you think will cause trouble.

If possible, try to get a "hands on" demonstration of the unit you select. Make sure that everything works to your satisfaction, and that the counter lives up to your expectations.

A Fun Counter Project

The goal in designing this counter was to produce an inexpensive, no frills, reasonably accurate, minimum parts design.

High impedance FET input, .2 to 30 MHz frequency range, sensitivity 60 mV RMS across most of its range, and compact packaging are some of its features. Also, two modes of timebase operation are possible: xtal controlled with an accuracy of ±20 PPM ±1 count and line frequency (±.05 per cent). The heart of the counter (less readout and power supply) is mounted on 2 PC boards (=3″ × 5″). The cost of the boards filled with parts (for line frequency operation) plus the six Minitron readouts will run you about $40. An additional $8 pays for the xtal controlled timebase, and of course you'll need a power supply. Any junk box parts substitute will naturally reduce your costs.

Operation. The block diagram of Fig. 10-5 shows the signal flow and control. Referring to Fig. 10-6, Q1 to Q4 shape the incoming signal to

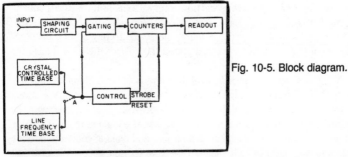

Fig. 10-5. Block diagram.

produce TTL compatible levels. A precise 100 ms gate (at IC19 pin 2) is then *ored* together with the signal (IC19 pin 3). This gated signal is counted down and displayed on the readouts. The trailing edge of the 100 ms gate triggers the one shot IC24. A pulse is generated that strobes the quad latches IC7 to IC12. The stored data is transferred to the decoders and readouts. The trailing edge of the one shot triggers another one shot IC25 to produce the reset pulse to counters IC13 to 18. All counters are reset to zero until the next gated input is counted. Thus the sequence is: gate, count, transfer, reset.

Construction. As mentioned earlier, the heart of the counter is mounted on 3″ × 5″ PC boards (actually 3″ × 5-3/16″). You can package these boards any way you like. Just allow sufficient ventilation in any enclosure you use. One of the drawbacks of laying out small boards is that you can't get all the circuit connections on them that you'd like. So we're going to have to add some jumpers. Refer to Figs. 10-6, through 10-8. After all the components

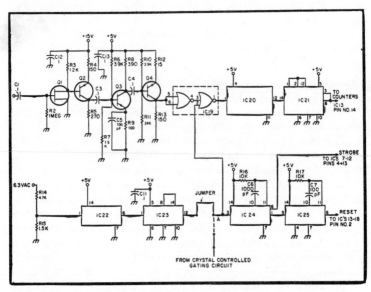

Fig. 10-6. Input and Control board.

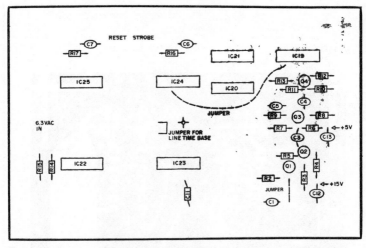

Fig. 10-7. Component layout, Input and control board.

are mounted, jumper IC19 pin 2 to IC24 pin 3. Also jumper point "A" to the pad right above it (for line base operation). That's all the jumping we need for the Input and Control board.

The Readout board will require the following jumpers (Fig.s 10-9 through 10-11):

- IC7 to 12-jumper all pins 4 and 13 (strobe).
- IC13 to 18-jumper all pins 2 (reset).
- IC13 to 18-jumper pins 1 and 12 on each IC only.
- IC13 pin 11 to IC14 pin 14.
- IC14 pin 11 to IC15 pin 14.
- IC15 pin 11 to IC16 pin 14.
- IC16 pin 11 to IC pin 14.
- IC17 pin 11 to IC18 pin 14.

Also, power to all ICs and all power grounds will have to be added. Refer to Fig. 10-9. Jumper grounds to the ground plane and between boards. Use Molex terminals to mount all ICs. The added cost is worth it.

The Input and Control board artwork (Fig. 10-8) is laid out for 60 Hz operation. If you're satisfied with an accuracy of approximately 500 PPM on your counter, use it as is. You won't need the LM340-5T in the power supply, so that's an additional saving.

The crystal controlled timebase should be added to get the most accuracy from your counter. Figure 10-12 is a suggested circuit. There is no artwork included for this additional circuitry. Layout is not critical and could be hand-wired in no time. If used, you won't need IC22 and IC23. To connect, remove the jumper at point "A" and connect the output on the last counter to point "A".

That's about it. If you got all the jumpers in right and your power is connected right (double check), you should be finished.

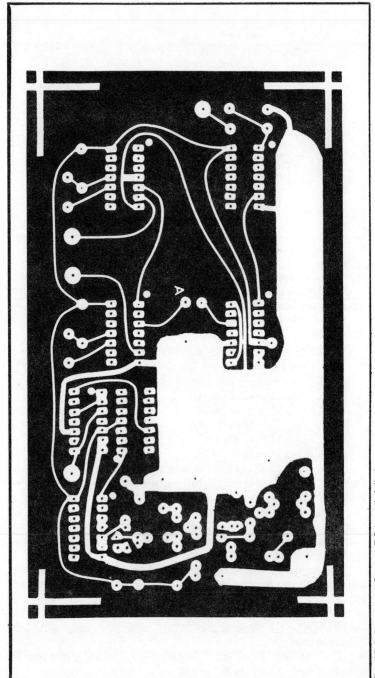

Fig. 10-8. Input and Control PC board, full size.

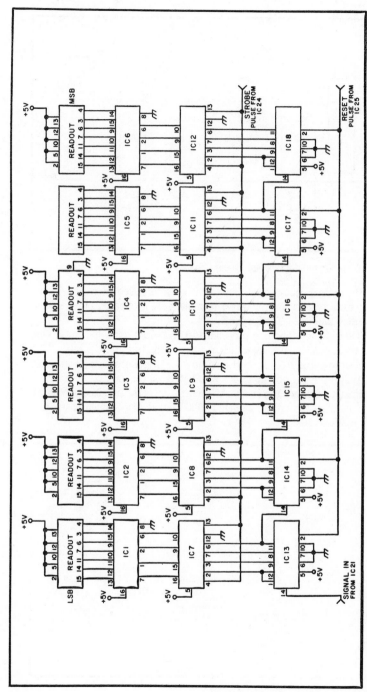

Fig. 10-9. Display board.

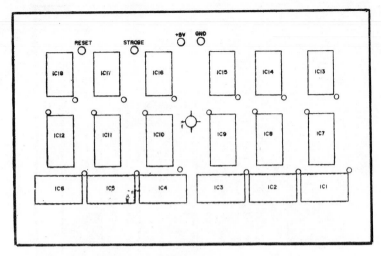

Fig. 10-10. Component layout, Display board.

For a final check, make sure all ICs are installed correctly. That's another reason for using the Molex terminals. It's a lot easier to reverse an incorrectly mounted IC with them.

There is no calibration procedure. It it's wired right, it'll work. One work of caution. If the counter fails to count all the way up to 30 MHz, IC21 could be the cause. It has to divide by 5 up to 15MHz. Substitute it with other 7490s until you get one that goes to 30 MHz on the input.

Finally, your counter is compatible with any ot the prescalers available on the market, extending its range to 300 MHz. For a complete parts list, refer to Table 10-1.

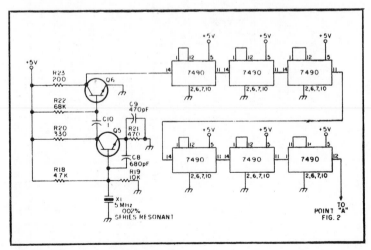

Fig. 10-12. Crystal controlled timebase.

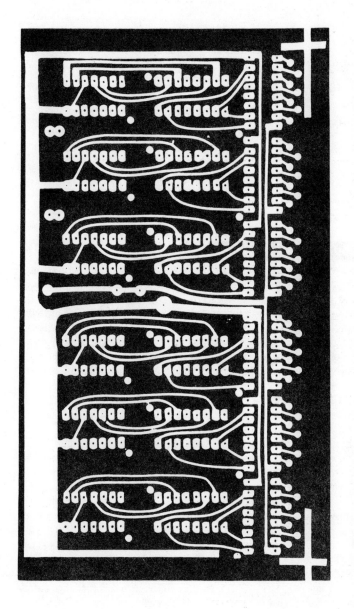

Fig. 10-11. Display PC board, full size.

Table 10-1. Parts List.

IC1 to IC6	7447	R11	24k
IC7 to IC12	7475	R12	15
IC13 to IC18, IC21	7490	R14	4.7k
IC24, IC25	74121	R15	1.5k
IC22	7413	C1, C3, C4, C11	.1mF 20 V
IC23	7492	C5	100 pF
IC19	7402	C6	.001 mF 20 V
IC20	74S73	C7	100 pF
R16, R17	10k ¼W	Q1	MPF 102
R2	1 meg	Q2, Q4	2N3638A
R3	1.2k	Q3	2N2222
R4, R13	150	Q5, Q6	2N2222
R5	270	IC terminals Molex Soldercon	
R6	3.9k	Series 90 Minitron Display (6)	
R7	1.5k	Note: Transistors, ICs, terminals	
R8	390	and display are available from	
R9	100	Digi-Key Corp. P.O. Box 126,	
R10	3.9k	Thief River Falls, MN 56701.	

A $50 Self-Powered Counter

To you this may look like another construction project. "A pretty neat little counter," you'll say. "Think I'll build one." Little will you know how long it took to get the bugs out and keep the cost down. This project is the pinnacle of savings. It is a 6-digit (that reads out to a full 8-digit capacity) counter, portable, hand-held, sensitive, rugged, LED readouts, requires no test equipment to build, and more, all for just a little over $50.

Costs are sliced down to an absolute minimum by cutting all the frills. For instance, many counters advertise no-flicker, or non-blinking readout. This one blinks, but you can only see it on the lowest reading scale, when you are reading signals to 1 Hz. The rest of the time the digits blink for 1 ms, and when was the last time you could see a 1 ms blink? Most hobbyists will use the MHz and kHz scales most of the time anyway so the "blink" will be for 10 us and 1 ms respectively. And, when you are showing the XYL or neighbor your latest creation, show them on the Hz range so they will be impressed by the blinking, counting readouts!

While cutting out the non-blinking feature does save some money, the greatest saver is cutting out 3 of the 6 digits! I know, you are wondering how it is that we get a 6-digit counter with 3 of the digits missing. At current prices, dropping 3 unnecessary digits saves nearly $25. Even at TWS a 3 DCU costs $18. So if you can liberate an extra $18 for this project, you can have all 6 digits! But it is all the more fun to drop 3 digits, build the unit in half the width and shift the readouts around to get a look at 8 digits of the incoming signal. Even in a 6-digit counter you have to shift the range switch around to get readout of say a 14.85 MHz signal to the nearest 1 Hz! With the handi-counter you have just one more shift to get the same capability with only 3 digits!

The basic "brains" for the counter come order a digital dial kit from TWS Labs, PO Box 357, Provo, UT 84601 for $45.95. When you order the digital dial, you get the same stuff you would get if you order the individual kit (a 3 DCU and a timebase), but you get a free AC power supply in addition. In fact, you can power your unit from this supply if you don't want to add nicads like in this version of the handi-counter. Or you can keep the power supply and have a good source of fairly well regulated 5v DC for your workbench. The fact that the basic counter comes in kit form makes this project all the easier. All you need to do is make some very simple modifications to the kit, add the digit switching circuit, a couple of extra circuits on perfboard and you have an instant 3 × 6 digit counter!

Operation. Let's look at how you use the unit before you build it. Actually, the principle of operation is very simple. There are several front panel controls. First is the on/off switch—a simple enough function. Second is the MHz/kHz selector (both these switches come with the kit); this one selects timebases and does some decimal point selection. We could have labeled it Hz/kHz like its 6-digit brothers; it would only have required putting the decimal points in some other places that were not convenient (and cheap) with this kit.

The third switch, the one in the middle of the panel, we can call the "display selector" switch. This one comes from your corner Radio Shack and is p/n275-405, a 4 PDT switch that currently is priced at 69¢. Other than the box and batteries, this is the only expenditure that is not available in a kit. You could have a complete counter that would only read to 100 Hz so far.

Back to the 4 PDT switch. It is used to select which 3 digits of the 6 available digits are viewed at any one time. When you throw the switch to the left, you see the left most significant digits and to the right, you see the 3 right most significant. The decimal point jumps around every time to keep you thinking straight. Figure 10-13 shows how it works. For an example, we used a 40 meter signal that is actually 7,224,362 Hz. This is how the counter sees it: Note the placement of the decimal point; also note that in kHz left and MHz right, the readout is the same, but the decimal point is shifted. That is because, if you look at the original signal you can see that there are: 7.2

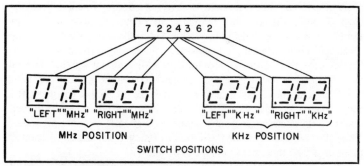

Fig. 10-13. Here is how the digital display in the Handi-Counter shows the eight complete digits in this forty meter signal. You simply move the digital display around and "look" at different portions of the signal at selected times.

MHz, .224 MHz or 224 kHz (see, the readout means the same thing) and finally .362 kHz. That last 0.362 kHz is the same as 362 Hz, but would demand another timebase switch position, so we merely switch a decimal point and call it a fraction of a kHz.

Suppose you hook the sensitive input circuit of the counter to an oscillator you built. You want a look at the output frequency of the oscillator. Maybe you see 21.8 on the MHz left setup; that means the oscillator is running at 21.8 MHz. If you were only interested in roughly knowing the MHz range of the oscillator, you would stop right there. If you wanted the kHz range you would leave the display left, and throw the timebase selector to kHz. That would give you a frequency check to the nearest kHz. Suppose you wanted further accuracy, or to know how much it drifts; then flip to the right digit display (still in kHz) and read out in Hz. You can see the flexibility of the whole thing! You have 3 digits in a lot smaller package, for a lot less cost and you are reading a 21 MHz signal out to eight places!

Figure 10-14 is the block diagram of the counter and this will help you understand what is going on in there. The easiest process takes place in the kHz position. Here, the timebase selector and display selector switches choose to provide a one second (right) or a 1 ms (left) signal. This is high-accuracy stuff with the timebase being divided by a million to get the one second clock; any errors or drift in the timebase oscillator (already crystal-controlled) are divided by the million too. At the very beginning of the count time (1 sec or 1 ms), the timebase gives a reset pulse that starts the counter stages at 000. Then the counter merely counts the input pulses that occur during the selected time period. On the kHz left position, for example, the counter is cleared and enabled for one second. When counting a 60 Hz line input, the counter would start to flash (the part that will impress the average observer) until the end of the counting cycle when it will sit there displaying .060 kHz, or 60 Hz.

When the display selector switch is pushed left, we apply a 1 ms gating signal to the counters and count the events that occur in 1 ms, or in other words, we count kHz. A 30 kHz signal will show up as 030. Note that there is no decimal point. There is no need, as the timebase selector is in the kHz position so it reads out as a whole number, not a decimal fraction. From the block diagram, when one switches to the MHz range, the clock now comes from 1 ms and 10 us. The 10 us clock is fed to the counter circuits in the "MHz left" position and merely allows the counter stages to count for 10 us. On this scale, the chance for error from the 1 MHz crystal is greater and the fact that the timebase is crystal-controlled becomes the greatest asset. At the 10 us point the clock signal is only divided by 10. With 10 us gating the counter can count up to 99.9 MHz; in real life the limit is imposed by the fact that the counter circuits will only operate to 30-40 MHz without a prescaler. Since most prescalers divide the display by ten, you will display MHz on all three digits, so the decimal point can be left out.

It is probably a good idea to explain why there is a different reset circuit than the one with the kit. The reset circuit with the kit is designed to give high accuracy with fast input signals and operate at a 10 ms and 1 ms gating output. While the kit's reset is more accurate than the "extra one" built for the handi-counter, it causes some problems while running at slow speeds. The reset actually resets a couple of clock pulses before the gating signal enables the counter. That works great at 1 ms and 10 ms and means that you are not depending on the characteristics of the input impedance of a gate and some resistors and capacitors for your reset pulse; the pulse is clock controlled. The trouble comes when the clock is one second. The strobe circuit causes the counter to sample the input signal every second and a half or so and you can extend it to a few seconds by increasing the size of the timing capacitor in the unijunction oscillator circuit. Well, at one second clock speed, the counter resets to 000, waits a couple of clock pulses (seconds), then enables the counter stage and the stages count the input signal. After the one second counting period, it is time for another sample and the display resets again. The result is you don't have the display lit long enough with the number to even read it. You may choose to merely extend the sample time by putting more capacitance in parallel with the timing capacitor (33 μF) in the unijunction circuit that comes with the kit, or you may use the reset circuit shown here.

The reset circuit built for the handi-counter works like this: When the enable signal goes active to allow the counter to count the input signal, a very fast RC differentiating network creates a narrow spike that resets the

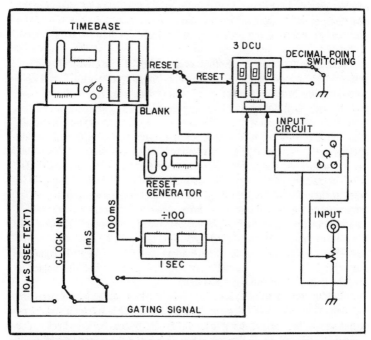

Fig. 10-14. This is the block diagram of the unit showing some of the switching that is accomplished with only two switches in the unit, a DPDT that comes with the timebase kit and an "added" 4 PDT.

stages. You can see that since this spike occurs because of the enable signal, it actually resets the circuits when they are supposed to be counting. This is no real problem at slow speeds, but at higher speeds, if the reset pulse is very wide, it can knock off a pulse or two and give an inaccurate reading. For the 10 us signal knocking a few pulses off can throw the MHz readout way off. In most counters that use this method, they merely adjust the clock signal so that the counters are enabled a little longer, but few of those counters use a gating signal of only 10 us, so that's why we have the two reset signals.

The counter input amplifier shown is pretty sensitive. It works well just picking rf up with a little wire antenna. Its major problem is that you can blow the FET pretty easily. So be careful about input signals.

To calibrate the counter you can use a known accurate signal, or beat the oscillator against WWV.

Construction. Building the counter is really not hard. Obviously, the first step is to collect all the parts, then do the layout of the front panel of the case, and finally, mount the pieces in the box. You can use any box for your counter. The key to switch mounting is to keep the switches out of the way of the 3 DCU and timebase circuit boards. The input jack and the sensitivity control are mounted on top of the case. You will probably want a jack for charging your nicads without removing the cover from the box. If you decide to make your unit nonbattery-powered, you will still have a lot of room in the box because all the power supply components mount on the top of the timebase PC board, and the transformer is small enough to sit on the bottom of the case.

Mounting Batteries. The bottom of this hobbyist's box wasn't flat; it had some design raised in there, presumably to add strength. This was a problem in mounting the batteries and he ended up being able to get only one bolt through each holder and that only thumb tight. That only allows the batteries to stay in place, and offers very little strength against drooping, etc. The batteries are mounted to leave a "cavity" so the crystal, which sticks into the case pretty far, has room to keep from getting crushed. This cavity also makes a convenient place to put the three external circuit boards.

The battery mounting bolt leads stick out of the bottom of the case to mess up the effect of the little "feet" on the box. It is easy enough to glue flat grommets to the bottom, with the hole over the bolt heads for better feet than were on the box in the first place. Note: Later experiments showed that it is easier to buy a plastic 4-C cell holder and use it. The holder mounts in a smaller space than the individual cell holders and there is plenty of room for the crystal and external circuits. An added benefit is that the holder is cheaper than four of the individual units.

Building the Circuits. It is easy to build the kits; all the instructions are included. Trim leads very close to the PC boards. While the kit instructions do not stress this, we are going to mount the PC boards back-to-back under some pressure and we don't want any extra long or sharp leads sticking out of the boards. Even trim the leads on the crystal. They don't mention it in the kit, but it helps in sandwiching the boards closer together. Also take care in

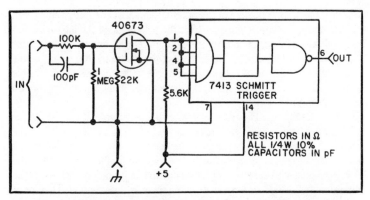

Fig. 10-15. Schematic diagram of the input circuit used with the Handi-Counter.

running the external wires on the board. They should be laid out so that they do not cross any pins or leads on the back of the board. Under pressure, these can puncture the insulation on the wires and cause a short.If you are using a different physical approach and have plenty of depth in your box, use the mounting method described with the kits and you eliminate many of the aforementioned problems.

Building the Input Circuit. The input circuit is built on a phenolic board, although you can make a tiny PC board for it if you want. Or, if space is at a premium, you can get the timebase extender board that comes in kit form from TWS; it is roughly 2″ × 3″ and has all three of the external circuits right on the one board. It was mounted near the input selector switch, while in the final version, longer wires were added and it fit in the cavity formed by the batteries, and near the input level control and input jack. Make sure your wires are long enough hookup wire with the kit for most installations. After the circuit is built, put some tape on the bottom of the board so the circuit will

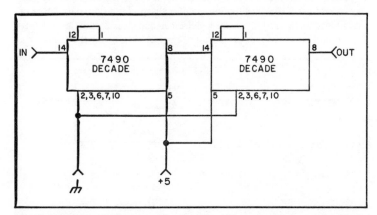

Fig. 10-16. Schematic diagram of the divide by 100, one second extender circuit. This circuit allows the counter to look at signals in the hundreds of cycles range and provides readout to 1 Hz.

not touch the battery holders or the other circuits that are stuffed along with it in the battery cavity. Figure 10-15 is the schematic of the input circuit.

Divide by 100, Extender Circuit. The 10 ms clock from the kit drives this circuit shown in Fig. 10-16. The output of the divide by 100 circuit is one second clock used in the timebase generator to make the one second enable signal for the counter. The circuit is very simple, and consists only of two decade dividers in series.

Reset Generator. The reset generator is very simple; it consists of the RC network and a 7400 gate used as an inverter. Figure 10-17 is the schematic of the circuit. The output of the gate is a normal logic 0. When the sample signal comes along (actually the inverse of the signal called the *blanking* signal with the kit), it causes a positive pulse output and resets the counter stages. Because of the nature of this reset, as mentioned before, it is best not to use it (use "kHz left") when looking at very fast signals. On slower signals, there will be no difference in reading between this and the "MHz right" readings.

Interconnecting the Circuits. Figure 10-18 shows the major interconnections and the hookup of the input circuit. First mount the 3 DCU counting and display unit on the front panel. Then run the power and interconnections as given in Fig. 10-18. Run 1″ wires from the terminals on the 3 DCU card. These will connect later to the switch and the timebase board. Run the decimal point wiring to the switches on the front panel as shown in Fig. 10-19.

Figure 10-20 shows how to connect the reset circuit to the timebase and 3 DCU boards. Figure 10-21 shows the hookup for the interconnecting wiring that does the timebase switching, including the divide by 100 circuit.

When the wiring is completed, put an insulating card (a piece of 3″ × 5″ file card works great) between the 3 DCU and the timebase card. First mount the whole thing up and tighten down the mounting bolts. Then take the thing apart again and inspect the piece of card for punctures where a wire from the circuit board was too long or too sharp. Trim the faulty wires or pins, reinforce the card with a black electrical tape strip around the puncture (if any), add another layer of insulating card, and remount the boards.

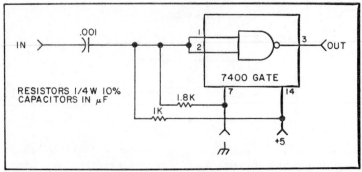

Fig. 10-17. Reset generator schematic. This added circuit provides a reset for the counter in the low speed (1 Hz) readout position.

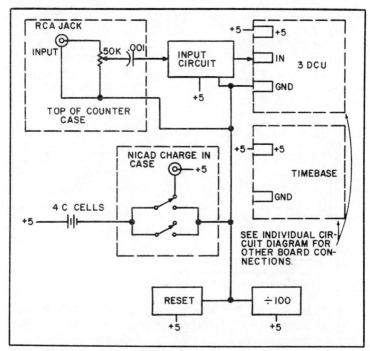

Fig. 10-18. Major power hookups and input circuit hookup. Note the nicad charging circuit. With a power supply (charger) plugged in, the nicads will charge when the unit is off. You may wish to allow them to charge at all times. If you use the power supply supplied with TW's kit, you turn the unit on and off by cutting the ac from the ac cord to the transformer input.

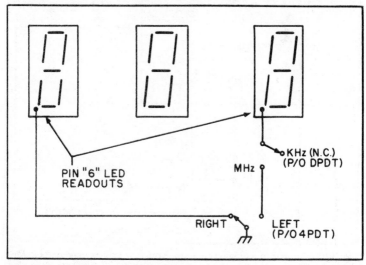

Fig. 10-19. Decimal point wiring diagram.

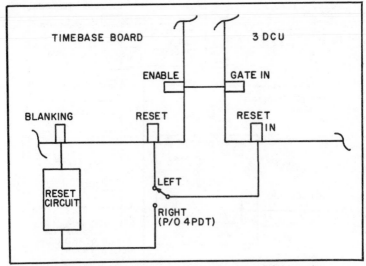

Fig. 10-20 Reset circuit hookup to the timebase and 3 DCU boards.

Put in the batteries, drop the input circuit, divide by 100 and the reset circuit boards in the cavity. Don't forget to install the input jack and pot. Wire the power connections to the batteries. The displays should light.

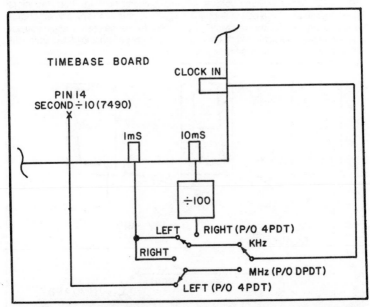

Fig. 10-21. Timebase switching diagram. Note that 1 ms is used on both MHz right and kHz left. The difference is the decimal point, and the source of reset pulse. (See other figures for details.)

Checkout. When power is applied, the readouts may show any number reading at all. This is normal. After a second or two, when the oscillator is running and all the circuits *settle*, the random numbers should clear. The display should flash "000." Now you are ready to use your counter and look for all manner of signal. Make sure the thing resets in all switch positions and you can tell if your new reset circuit is working properly.

Points to Remember. The unit will only run for a couple of hours from fully charged nicads, so don't leave it sitting around "power on" unless you are using it. The crystal oscillator is extremely stable and will work with little or no warm-up so don't worry about leaving the power on. Build a nicad charger and keep the nicads fresh. Some will want a trickle charger that will always keep the unit "ready to go." You may want to build up the power supply external to the unit, let it trickle the nicads (there is 8v unregulated in there). You can also use cigarette lighter power by using a suitable dropping resistor. For that purpose, it is possible to use the regulator circuit and either positive or negative ground 12 volts. Instructions for mobiling come with the kit.

The Calculating Counter

Traditional counters provide very limited resolution of audio frequencies associated, for example, with organ and piano tuning. The usual one-second gate time yields a reading varying between "59" and "61" for 60 Hz. To determine this frequency with any accuracy, you must measure the period of one cycle (16,666 microseconds), reach for your handy pocket calculator, and calculate the reciprocal (or wait 100 seconds perhaps!).

We decided to explore the possibility of teaching a counter to calculate frequencies from period measurements. It soon became apparent that

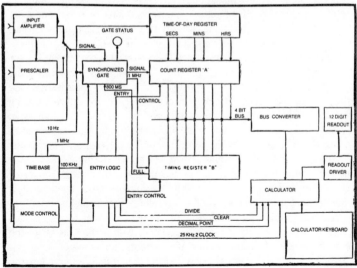

Fig. 10-22. Block diagram.

designing a circuit to calculate reciprocals was no simple task. A ready-made solution to this dilemma presented itself in one of the "calculator chips" widely advertised at attractive prices. The "One-Chip Calculator" advertised as a type 5001 turned out to be both inexpensive and practical. The chip comes in one of two voltage ratings (for desk or pocket calculator use). While we used the low voltage model, which requires −8.5 and −13.5 volt power supplies, the higher voltage unit should be directly interchangeable, with minor power supply modifications.

We wished to construct an all-purpose counter, and our final design includes a basic 30 MHz counter, a high frequency prescaler, a reciprocal calculator, a desk calculator, a digital clock, an interval timer and an incident counter. All logic circuits are TTL, except for the MOS calculator chip and ECL prescaler. Discrete devices are used for TTL/MOS interfacing and for driving readouts.

Design, construction and debugging of a project of this magnitude is virtually impossible without a test-breadboard facility. It is necessary to breadboard each function, and test it in turn. A suitable power supply is also essential, and we recommend construction of the power supply first, but note that use of different readouts or other components may require modification of the power supply.

Theory of Operation. Overall data and logic flow may be followed by reference to Fig. 10-22. The frequency to be counted is connected either to the input amplifier (high impedance) or directly into the prescaler (50 ohms). The prescaler is automatically activated in the MHz mode. Note that the high impedance input may be used in all count modes, and it provides good sensitivity through the two meter band.

The synchronized input gate directs input pulses to register A, an 8 digit decade counter, and timing pulses to register B, a 6½ digit counter. Gate time in kHz and MHz modes is one second (plus or minus one input pulse period), and in the Hz mode varies between 800 milliseconds and two seconds (800 milliseconds plus one input pulse period).

When the input gate closes, entry logic is enabled, and in the kHz or MHz modes, the contents of register "A" are entered serially into the calculator, which displays the entry just as if it had been made from the keyboard. In the Hz mode, the contents of register "A" are entered, trailing zeros are padded to improve resolution and the "divide" function again activated. This sequence causes the number of input pulses stored in register "A" to be divided by the number of timing pulses stored in register "B". The result (with the decimal point appropriately positioned) yields the frequency of the input (with up to four decimal places!). The bus converter interfaces the TTL registers with the MOS calculator, and changes the BCD coding of the TTL to conform to the 5001 chip's requirements.

The time-of-day register constantly contains the time, updated each second. Register "A" utilizes decade counters with preset inputs. In the time-of-day mode, once each second, the contents of the time-of-day register are strobed in parallel into register "A" and then serially entered into the calculator as in the kHz mode.

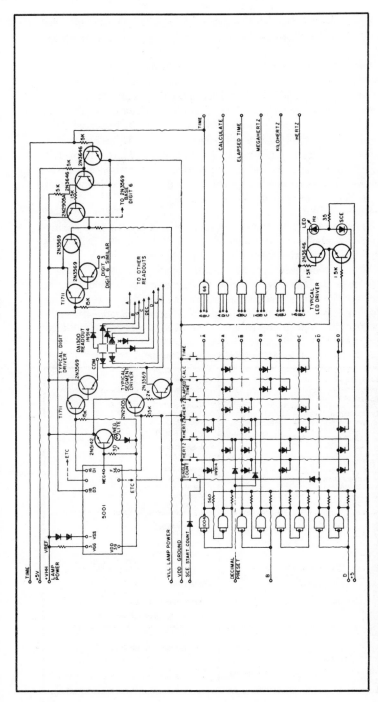

Fig. 10-23. Readouts and Mode Switching. The neg. light is the G-segment of the left-most readout.

277

The accumulate and period modes cause the input gate to be controlled either manually or externally, and either timing or input pulses are routed to register "A" as appropriate.

Readouts. Selection of readouts is guided by desired digit size, price, availablility, power supply and driving requirements. Readouts for use with the 5001 chip must have decimal points to the right of the digit.

Various readouts are suitable for use with calculator chips. The wide array of LED devices available suggests these would be preferred. Most calculator chips utilize multiplexed readouts, where like-segments of all digits are connected to a common bus line. Each digit is selected (turned on) for a brief period in turn. This driving method reduces wiring and driving hardware, because only eight segment drivers (including the decimal point) and one digit driver per digit are required.

LED readouts may be driven by the newly available 75491 and 75492 Darlington driver arrays; however, as they are NPN devices, PNP predrivers may be required. Drivers shown in Fig. 10-23 are also suitable for driving LEDs, with the addition of limiting resistors.

Incandescent readouts require drivers capable of handling substantial inrush currents. Digit drivers should be selected to handle 500 mA. The author's choice of incandescent readouts was influenced by cost at the time, and size of display. Present LED prices make them much more attractive. Selection of the 2N3569 for drivers was influenced by our junkbox stock, but

Table 10-2. Calculate/Count Program.

Count Mode—kHz	Elapsed Time Mode
Decimal Point Preset at "3"	Route Clock (MS or uS) Pulses to Register "A" (gated)
1. Open Input Gate	Route Input to Control Input Gate
2. Close Input Gate (1 second later)	Update Once Each Second
3. Clear Display (CE)	Clear Register "A"
4. Enter 5 High Order Digits From Register "A"	1. Clear Display (CE)
5. Enter Decimal Point	2. Enter 5 High Order Digits From Register "A"
6. Enter 3 Low Order Digits From Register "A"	3. Enter Decimal Point
7. Clear Registers	4. Enter 3 Low Order Digits From Register "A"
Count Mode—MHz	**Accumulate Count Mode**
Decimal Point Preset at "3"	Route Input Pulses to Register "A"
Prescaler Activated	Update Once Each Second
1. Open Input Gate	Clear Register "A"
2. Close Input Gate (1 second later)	1. Clear Display (CE)
3. Clear Display (CE)	2. Enter 8 Digits From Register "A"
4. Enter 6 High Order Digits From Register "A"	
5. Enter Decimal Point	**Time-of-day Mode**
6. Enter 2 Low Order Digits From Register "A"	Blank 2 Digits
7. Clear Registers	Update Once Each Second
	1. Clear Display
Count Mode—Hz	2. Load Register "A" Time-of-day Register
1. Open Input Gate	3. Enter 8 Digits From Register "A"
2. Open Clock (Register "B") Gate	
3. Close Input Gate (When Register "B" Full)	**Calculate Mode**
4. Close Clock Gate	Lock Out Data Entry Sequencer
5. Clear Display (CE)	(Keyboard Always Live)
6. Enter 8 Digits From Register "A"	
7. Enter Three Zeroes	**Continuous Count Mode (Hz, kHz, MHz)**
8. Enter "Divide" Command	Enable Automatic Count Restart (input gate)
9. Enter 4 High Order Digits From Register "B"	on Register Clear Signal
10. Enter Decimal Point	
11. Enter 3 Low Order Digits From Register "B"	**Single Count Mode (Hz, kHz, MHz)**
12. Enter "Equals" ("Divide") Command	Manual Count Restart (input gate)
13. Clear Registers	
(Delay New Count Cycle to Provide 2 Second Update)	

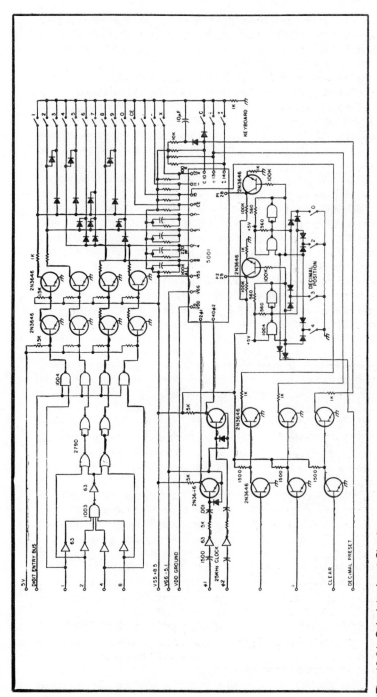

Fig. 10-24. Calculator Input Circuits.

279

the 2N2905 (PNP) is probably more suitable. The calculator data sheet shows PNP drivers, and use of the 2N3569 (NPN) requires the circuits of Fig. 10-23 to make the NPNs think they are PNPs. Diodes are required for isolation between each segment and the segment bus.

Driving circuits for Nixie and Numitron type readouts are included with the calculator data sheet. Table 10-2 is the calculate/count program.

The 5001 chip provides 12 digits, which is handy for calculating, but not really needed for counting. Although fewer than 12 can be used, you must then either provide overflow detection or take a chance on errors. Note also that there are restrictions upon the size of the arguments that the 5001 can handle without overflow. These restrictions cause an overflow if you attempt to calculate the frequency to too many decimal places. (You have the choice of zero, 1, 2 or 4 decimals.)

The Calculator. The 5001 Cal-tex chip widely advertised is probably the most suitable for this application. Data input is modified BCD, and is readily driven from TTL Circuits. The 5005 and several others use multiplexed inputs, which require more complex input interfacing. Although no investigating was done, the 4016 CMOS switches should be suitable for use with multiplexed inputs. The 5001 is available in 2 voltage versions. If you are not sure which version you have, it would be sensible to use the lower voltage supply to test it. This one turned out to be the 13 volt version. If yours works at the lower voltage, it is a low voltage unit and should not be tested at the higher voltage.

TTL and ECL are devilish creates unless adequate power supply by-passing is done. The surplus boards noted also yielded some very handy power bus strips. These are composed of two ⅜" wide strips of copper foil separated by a thin dielectric. Feed terminals are provided at convient locations which permit short leads to ICs. Even with use of these bus strips, 1.0 mF (miniature electrolytic) and .1 (miniature ceramic disc) capacitors were liberally distributed from +5 volts to ground. A .1 discap across the supply terminals of every TTL and ECL IC is recommended, if you don't use bus strips. Failure to provide these bypass capacitors will result in miscounts at best, and total non-operability at worst.

Figure 10-24 shows the various input circuits required for the 5001 chip. Pins 5,6,7 & 8 are 1,2,4 & 8 BCD lines, with the exception that a zero is read in as a "10" (lines 2 & 8 low). A low reading on any line results in a digit entry. The "digit entry" converter at the top of the diagram converts a zero entry into a "10." The "digit entry" bus is driven high whenever a digit is to be entered, opening the 4 NAND gates to the 2N3646 level converters.

Level converters are also provided for the decimal point entry, divide function and the clear function. Although nonsynchronous calculator clocking could have been used as shown on the 5001 data sheets, we chose synchronous clocking to minimize cyclic errors, and tapped a 25 kHz signal from the time-base coundown chain. Electronic, rather than mechanical decimal point switching, was chosen to permit automatic decimal point placement when selecting functions.

Figure 10-23 shows outputs from the calculator chip. A Quasi-Darlington connection is used for digit drivers. The digit drive transistors must be maintained in full saturation, otherwise an "8" display will result in a dimmer digit than a "1", not to mention possible destruction of the transistor. Digits 3 & 6 (from right) are blanked in the time-of-day mode.

Mode switching uses 8 NAND gates connected as 4 flips-flops. These gates serve as an instruction register and their outputs control the various logic circuits. LEDs are provided near each mode switch to indicate which mode has been selected.

Input Gating. A simple gate which opens for a fixed period of time is not adequate for use when calculating frequency from period measurements. The synchronized gate used here ensures only full pulses are counted, as opposed to a random gate which may open or close in mid-pulse. This substantially improves accuracy in the frequency calculate (Hz) mode, and reduces hunting of the last digit which is common to all counters.

The front end of the input board (Fig. 10-25) includes a wideband amplifier and a high speed prescaler both described later. Gates in the signal path (1D, 2A, 3A, 3B, 3C), flip-flop 21 and the input divider of register "A" should be selected for high speed performance.

Assume for the moment that decade divider 8 is enabled (Fig. 10-26), and we are in the kHz (Standard Count) mode. The first 1 Hz pulse arriving at the count input of FF 21 A causes the Q output to go high (gates 2C & 2D are enabled via gate 10B and the J & K inputs of FF 21B are conditioned to permit it to toggle "on"). The input signal is routed to the count input of FF 21B via gates 1B, 1D, 3B and 2A and the first subsequent input pulse causes FF 21B to turn on. Gate 3A is now enabled, and following input pulses are

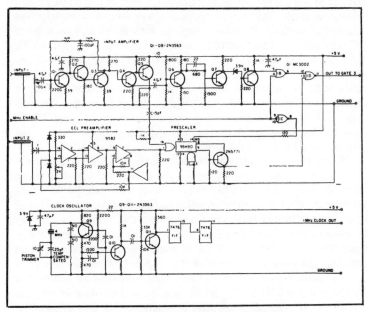

Fig. 10-25. Details: Input Amplifier, Prescaler, Clock Oscillator.

routed to register A via gates 3A and 3C. FF20A was also set to toggle "on" by FF 21A and the first 1 MHz pulse at the count input of FF 20A causes it to turn on, and gates 5A and 5C cause the display-start line to go high. This readies the display entry logic.(1 MHz pulses are routed to register "B," but are not used in this mode.)

After exactly one second, FF 21A switches off, and conditions FF 21B to switch off. The next input pulse causes FF 21B to switch off, and gate 3A closes. Register "A" now contains the count of input pulses during a one second interval. FF 20A promptly switches off (about one microsecond later), the display-start line is brought low, and the digit entry logic is initiated. When all digits have been entered (about one-fourth second), the 10/0 line goes low, and the one-shot generates a 1 ms reset pusse, which clears registers "A" and "B" and simultaneously sets decade divider 8 to a count of 9, through gate 4D. About one-tenth of a second later (in the continuous count mode), divider 8 reaches a full count, transmits a pulse to FF 21A, and a new count cycle begins.

Single Count. When line D is high, after the count/display sequence, gates 4C and 4D are locked up by the reset pulse, and divider 8 is held at a count of 9. The start (SCE) line is brought down each time the single count push-button is operated, restarting divider 8, a new count, and resetting gates 4C and 4D.

Calculate Frequency (Hz) Mode. A logical 1 on the Hz line enables gate 10D, prevents FF21A from resetting from its count input and prevents (gate 10C) the 10/0 line from initiating a reset pulse. Gate 3A is enabled via FF 21A and FF 21B, similar to the kHz mode. It should be noted that FF 20A (which follows gate 3A by one microsecond) routes one microsecond pulses via gate 7A to register "B" in all count modes. In the Hz mode, when register "B" has counted 800 milliseconds, it sends a pulse via gate 10D to reset FF 21A. The next input pulse closes gate 3A via FF 21B, and gate 7A closes about one microsecond later. Register "A" now contains the number of input pulses which were counted during the time (in microseconds) stored in register "B." The display-start line causes entry of the number stored in register "A" a "divide" function, the period measurement in register "B" and a second divide (equals) function (which calculates the frequency), and initiates a reset pulse via the 16/0 line.

Elapsed Time and Event Counting Modes. When the elapsed time line is high, divider 8 is disabled via gate 5D, and either input pulses or timing pulses are routed to register "A," depending upon the status of "D" (gates 9B, 9C and 10B). Microsecond or millisecond timing pulses may be selected by a back panel switch. In the elapsed time mode ("D" is low), the gate may be operated via the regular counter input, or via the manual or external gate inputs. Pressing "Single Count" and "Elapsed Time" together selects event counting, and input pulses are routed to register "A" when the gate is opened manually or externally. In these modes, display-start occurs once each second (gates 5D & 5B), and the reset generator is disabled via gates 1A, 4A and 4B. A single reset pulse is generated each time this mode is selected, which permits each count to start from zero.

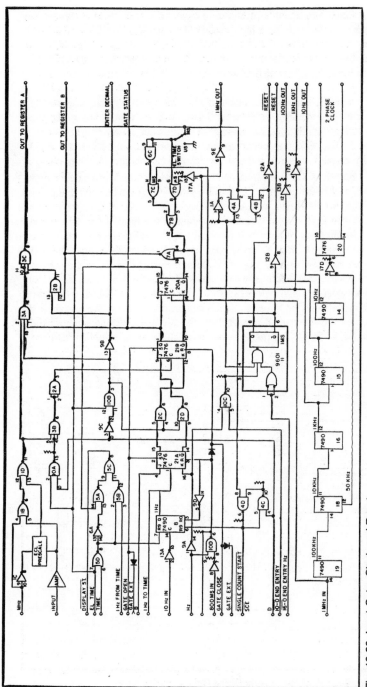

Fig. 10-26. Input Gate, Clock and Reset.

283

Clock and Countdown Chain. In order to assure count accuracy, the clock oscillator must have high stability. The circuit chosen has proven to be very stable. There is little sense in providing 7 or 8 digits or readout unless the clock oscillator is stable to 10^7 or 10^8 or better. When the digital clock is adjusted to keep time to one-half second in 10 days, accuracy is better than 10^6.

All devices in the clock chain, and the time-of-day register are provided with battery standby. Interfacing between these devices and other logic devices was made via open collector buffers. Provision of pullup resistors to the regular 5 volt line, and powering these buffers from the battery, avoids the possibility of glitches when the counter circuits are switched off.

Registers. Data registers and associated logic are diagrammed in Figs. 10-27 and 10-28. Referring first to Register "A," it can be seen that pulses presented to the input are counted during the interval the input gate is open. When the gate closes, counting stops and Data Entry Logic is enabled. Divider 30 (Fig. 5) is caused to count up, at a 20 millisecond rate, from "zero" though 8, and then resets to "zero." Decoder 29 changes the BCD output of divider 30 to a "one of 10" output. Each group of four gates to the data bus is enabled in turn, which causes the BCD output of the associated decade counter (high order first) to be presented to the BCD data bus (gate 24 presents divider 16, gate 23 presents divider 15, etc.). When divider 30 reaches the count of 8, a signal on the "stop entry" line stops further counting of divider 30, resets it to "zero," and in turn prevents further data entry.

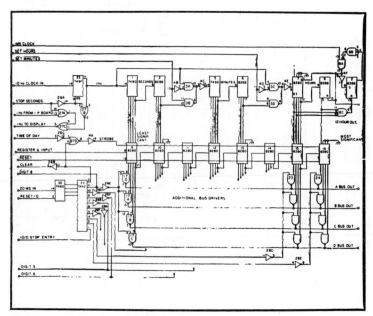

Fig. 10-27. Register "A" and Digital Clock.

At this stage, the calculator is displaying the contents of register "A." In the kHz and MHz modes, no further data entry occurs, and a new count cycle begins.

The time-of-day register utilizes decade and divide-by-12 dividers to determine the current time. Gates 26A, 27A, 27B, 27C and divider 25 are provided to permit setting the clock. Dividers 1 and 2 divide the 1 Hz pulses by 60, providing one minute output pulses, and storing the "seconds" of the time. Dividers 5 and 6 similarly divide by 60 providing "minutes", and an hourly output pulse. Divider 7 and the divide by 2 (flip-flop) portion of divider 6 provides a 12 hour count register. When a count of "13" is sensed by gate 8C, the hour register is preset to 1 through the action of FF2 and gates 8B and 8C.

The 8280 dividers used in register "A" have a very valuable feature. When the strobe input is raised to logic "1," the divider is preset to the value found on its BCD inputs.

Time -of-day can then be entered into display, just as if it were a "count." Note that the digits between the "seconds" and minutes, and between the "minutes" and "hours," are blanked in the display, so a dummy entry must be made in these two positions of register "A."

Data Entry. Register "B" (Fig. 10-28) does not require parallel entry, so 7490s were used here (more readily available and cheaper).

In the "Hz" modes, divider 18 is caused to count from "zero" up through "15" to zero. Two 7442 decoders provide a one-of-16 output from the BCD output of divider 18. Table 10-2 shows the data entry "program" for the various modes. Flip-flops 15, 16 and 17, along with associated gates, provide for entry control of both registers, and flip-flops 15A and 15B enter a "decimal" point in the appropriate place during register "A" entry.

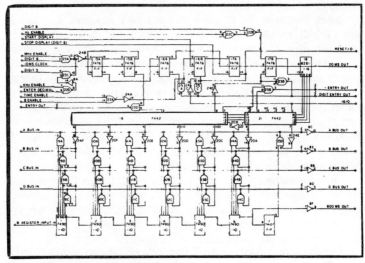

Fig. 10-28. Register "B" and Entry Logic.

Flip-flop 16A is normally enabled, and provides 20 millisecond pulses to the two data entry sequencers. Gate 25A signals the bus converter that there is valid data on the bus, and the converter in turn presents this data to the calculator for 10 milliseconds. The action of flip-flop 16A is inhibited (and no digits are therefore entered) during decimal point entry. Although the 5001 data sheet states 4 ms is required for data entry, anything faster than the 20 ms data entry cycle rate chosen was found to be unreliable with the bargain price 5001. Flip-flop 16B switches "on" in response to a pulse on the start-display line (input gate closure). Flip-flop 17A is then permitted to toggle "on," which unlocks the register "A" data entry sequencer and the "digit entry" line (gate 25C). Register "A" entry sequencer then counts up (entering digits), and when it reaches a count of "8" the stop-display line permits flip-flop 17A to toggle off, which resets the entry sequencer to "zero" and stops further entry of register "A."

In the "Hz" mode, a count of "8" on the register "A" entry sequencer permits flip-flop 17B to toggle "on" which enables the 8281 register "B" entry sequencer (18) and unlocks the digit entry line at gate 25C. Register "B" entry sequencer then counts up, entering register "B," and at count 15 gate 24D enables the flip-flop 17B "K" input. The next count pulse causes sequencer 18 to return to a "zero" count, and simultaneously toggles flip-flop 17B "off," which disables divider 18.

Input Amplifier and Prescaler. Details of the input amplifier and pre-scaler are shown in Fig. 10-25. The input amplifier uses low-cost high speed 2N3563 transistors throughout. The input amplifier has a passband from ½ Hz through at least 150 MHz. Some resistor changes were found necessary to optimize response. The 9582 ECL preamplifier was found to have less sensitivity than the discrete preamp, and stability proved a problem. In retrospect, we would eliminate the ECL preamp and use the discrete preamp for all input signals.

Clock Oscillator. Time base stability is very important if accurate counts are to be expected. The oscillator circuit was chosen after much experimentation. The crystal is a 4 MHz HA type (high accuracy) ordered for room temperature use. This frequency seems to be about optimum for thermal stability, and use of a series/parallel combination of NPO and negative temperature capacitors for the crystal series capacitor permits stability to be adjusted. The 10 pF capacitor is a glass piston trimmer can be anything smaller than 10 pF (4 pF probably is about right if you can find one). Q9, Q10 and Q11 are 2N3563 or 2N4274. Use of this oscillator permits stability to one part in 10 period. Only a very slight difference in time-keeping ability was noted when the unit was run "warm" for a similar period.

Mode Switching. Use of 8 open collector TTL 2 input NAND gates (or inverters) connected as 4 S/R flip-flops permits a completely electronic mode switching system. The configuration of these flip-flops is set by pressing one of the mode push-buttons. The flip-flops then "remember" what mode switch was last pushed.

Outputs at the first three flip-flops are decoded by six 3-input NAND gates. The fourth flip-flop is used to control the single count function, and also, together with the "elapsed time" mode, to provide an "accumulate" mode. LEDs indicate the mode presently in use. Operation of either the kHz or MHz button presets the calculator to 3 decimal places. This permits display of the decimal in the appropriate location. An extra decimal point LED was mounted between the fifth and sixth digits (from the right) and connected via a driver to the MHz function. This provides a reading of xx.xxx.xx in this mode. (Although it might appear that the MHz function should preset 2 decimal places, the decimal point is entered in the proper place by the data entry logic.) Germanium diodes should be used in the "decimal preset" line to ensure that TTL inputs are brought below their switching thresholds.

Power Supply. In order to avoid loss of power to the digital clock, a battery standby has been provided for the clock circuits. Referring to Fig. 10-29, a constant current regulator, comprised of a 2N3055 and a 2N3569, maintains the Nicad battery in a fully charged state. The voltage regulator using another 2N3055 and a 2N3698 is somewhat unusual. A voltage doubler driven by G1 and G2 provides sufficient current to the 2N3055 to maintain it in saturation when the input voltage drops to 5.2 volts or less. This ensures adequate output voltage to drive the TTL clock circuits until the battery voltage drops to about 4.7 volts. At this point, zener diode Z1 (with diodes D5 and D6) reduces the base current to the 2N3055, which results in a drop in output voltage. Lowered voltage to the TTL countdown chain causes a loss of clock drive to the voltage doubler, and results in a complete shut-down of the voltage regulator. Deep discharge of a Nicad battery would result in permanent cell damage.

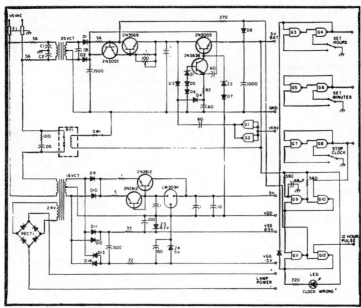

Fig. 10-29. Power Supply. *This LED signals a power off condition until the clock is reset.

Zener diodes Z1 and Z2 with their associated forward diodes must be chosen to provide about 5.7 volts at the base of the 2N3055, and about 5 volts at the base of the 2N3638.

The main 5 volt supply uses a PNP 2N2612 transistor to provide about 4 amperes of regulated DC, controlled by the 1 Ampere LM309K. The second 2N2612 is diode connected to provide a voltage drop similar to the base-emitter drop of the pass transistor, and as it is mounted on a common heat sink provides thermal stabilization. MOS voltage levels are provided by simple zener regulators. Note that Vdd is grounded, wheras the 5001 data sheet references voltage levels to Vss. Lamp power is unfiltered, which reduces readout brilliance changes with the widely varying readout load.

Three bounceless switches are provided for setting the time. Gates G9, G10, G11 and G12 increase the battery charge rate upon restoration of power after discharge. The high rate continues until first the "stop clock" (set seconds) switch is operated, and then a 12 hour pulse is received at midnight or noon. (That is, the high rate stops at noon or midnight after the time has been reset.)

Using the Calculating Counter. The completed counter has proven to be a very useful tool. By merely switching the unit on, the precise time is immediately displayed. Instead of those laborious parallel resistor computations with pad and paper, a precise calculation can be quickly made on the calculator. A short piece of wire attached to the input and placed near an antenna allows checking of transmitter frequencies. (A hand held business portable yielded a reading of 155.610.15 MHz-authorized frequency 155.610.)

As indicated earlier, while we wished to be able to measure frequencies through 2 meters, low audio frequency resolution was a prime target. All frequencies above 10 Hz may be measured to 5 or more significant figures.

An Inexpensive Modularized 50 MHz Counter

Need a relatively simple but effective counter? If it has to be easy to build and it has to be able to count to 50 MHz without using a prescaler you'll love this project. After giving the design approach considerable thought, we decided on the plug-in board method. Then we proceeded to break up the total circuit into the individual circuits that make up a frequency counter and design PC boards for each of them. Counter circuits are pretty basic and there are only so many ways to design one, so this one may look a lot like others. There are, however, a few new ideas.

One of the worst features of any counter is the awful current drain that the LEDs in the display manage to consume. Even with current-limiting resistors, 8 digits at 20 mils per segment can pull 1.12 amps. We got around this by scanning the display so that only one LED is on at a time. This reduces current consumption by one eighth. We also decided to get rid of that ridiculous number of resistors, since scanning the LEDs gave us a 10 per cent duty cycle. This meant that the average current through the LEDs would be 10 per cent of the maximum current without resistors. Checking the specs for the LEDs we were using (MAN8) showed that at 5 volts the

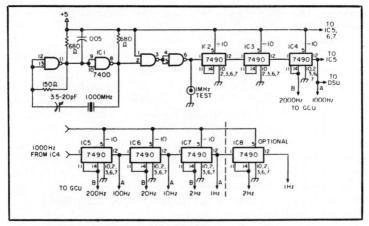

Fig. 10-30. Timebase Oscillator and Divider.

maximum current per segment would be 192 mils. But 10 per cent of that is 19.2 mils and is within the maximum of 40 for the LED. The maximum current drain is now reduced to 134.4 mils when the display is showing an 8.

Another change we made is the use of a 74LS90 as the first counter in the chain. The 74LS90 is pin replaceable for the 7490 and will count in excess of 50 MHz. Some designs use the 74196, which uses inverted logic as compared to the 7490. The 74LS90 uses the same logic so board design is simplified.

We also added the option of selectable gates, .01 sec, .1 sec, 1 sec and 10 sec.

Breaking the counter down into individual boards, there is the Time Base Oscillator and Divider (TBOD), Display Scan Unit (DSU), Decimal Counter Unit (DCU), Gate Control Unit GCU) and Preamp. All these units

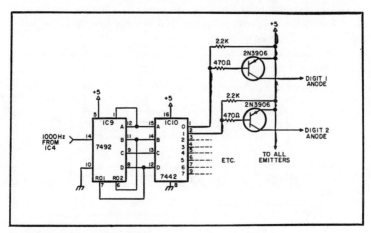

Fig. 10-31. Display Scan Unit.

plug into a master board which has all the inerconnecting circuit paths etched into it.

Approximate cost, less cabinet, is about $85 using all new parts, less if you have a well-stocked junk box and can make the boards yourself. Figures 10-30 through 10-48 illustrate this entire project.

Timebase Oscillator and Divider. The TBOD is constructed on a PC board that is 3.5" by 2". Due to the compactness of the circuit it was necessary to use jumpers for the frequency outputs. A double-sided board could be made and eliminate that need, but in the effort for simplicity we decided against it. The TBOD consists of a 7400 NAND gate for the oscillator and a series of 7490s wired to divide by 10 in the bi-quinary mode. This method gives a symmetrical square wave at the output, needed for proper gate timing. Also, the divide by 5 signal is brought to the edge of the board, as these frequencies are also needed. 1000 Hz is also used by the Display Scan Unit, so it has two outputs.

Curious as to the stability of this circuit since the crystal only has a tolerance of .005 per cent, we checked it against a 1 MHz signal with a known accuracy of 1×10.

Provision has also been made for bringing the 1 MHz signal out to the back panel of the counter for checking it against another signal.

This is also the most expensive of the units, costing about $15 with all new parts.

Display Scan Unit. The DSU is also built on a 3.5" by 2" PC board. The DSU has a 7492 divide by 12 counter wired to reset to 0 at the count of 10. The BCD ouputs of the 7492 are connected to the BCD inputs of a 7442 decimal decoder, which is used to scan the display LEDs by switching the Vcc on and off through a PNP switching transistor. The emitters of the transistors are connected to positive 5 volts and the collectors are routed to the anodes of the LED display. Pull-up resistors are used to keep the transistors biased off, along with current limiting resistors on the bases. If 5 volts does not provide enough brilliance from the LEDs, a slight modification on the board will enable you to use more than 10 volts, however. More than that and the LEDs may burn out.

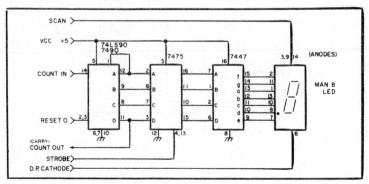

Fig. 10-32. Decimal Counter Unit.

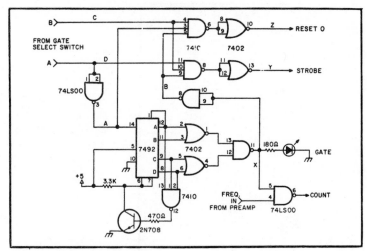

Fig. 10-33. Gate Control Unit.

The outputs of the 7442 are active low. That is, the output selected is at ground and all others are high. Grounding the base of a PNP transistor turns it on and switches Vcc to the proper LED. Because the 7492 is wired to divide by 10, the scan rate figures out to 100 Hz. This is fast enough to eliminate any flickering, but allows enough brilliance for normal room lighting.

With all new parts, the DSU costs about $6.25.

Decimal Counter Unit. The DCU is constructed on a 3.5″ by 1.7″ PC board. Except for the first DCU, all use standard 7490s as counters. The first DCU uses the 74LS90 by Fairchild, which was described earlier, for a 50 MHz count rate. Also, each DCU has a 7475 quadruple bistable latch, a 7446 or 7447 BCD to seven-segment decoder, and a socket for the LED. Any Monsanto LED may be used here as the pin-outs for most of them are identical. We used the MAN8, which is yellow, simply because we had them. However, the large .6″ MAN6 or the .27″ MAN7 will also plug in. Both of these are red.

For the LED socket, use the already preformed side mount socket or be cheap and bend the leads of a wire-wrap socket.

The 7475 is used to transfer the accumulated count of the 7490s to the display when strobed by the Gate Control Unit. A logic one is needed on the clock inputs to transfer the input information. When the clock is low, the latch will store the information until the next strobe pulse. If the input has not changed, the output won't either. If new information is present at the input, the outputs will change to agree with the inputs.

The outputs of the 7475 are connected to the inputs of the 7447 which decodes the BCD to the proper coding to display the corresponding decimal number on a seven-segment readout.

Four of the DCUs have provision for using the decimal point so that the display can be wired to show the frequency in either kHz or MHz. One need only wire the proper decimal point to ground through the gate select switch.

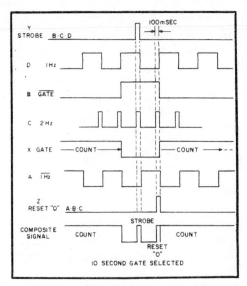

Fig. 10-34.
Gate Control Timing.

The cost of the DCU will be about $6.50 for the 50 MHz version, and a little less than $6 for the standard version.

Gate Control Unit. The original circuit for this was unsatisfactory, as the time needed for the strobe and reset 0 pulses was equal to the gate time. On the faster gates this was no problem, but on a 10 second gate it could be annoying having to wait 20 seconds for updating the display. So we redesigned it with the basic idea that we wanted a 10 second gate and the resets to occur within one second. A look at the timing diagram may help in understanding the operation of this circuit. Refer to the schematic for lettered lines. Note also that the board has a gate LED incorporated on it, eliminating the need to front panel mount one. It will show through the display window to the right of the digits.

The GCU is built on a 3.5″ by 1.7″ PC board. It consists of four ICs: a 7492, divide by 12; a 7410, triple-three input NAND gate; a 7402, quadruple two-input NOR gate; and a 74LS00, chosen also for the high toggle speed.

Let us assume we have selected a 10 second gate. Through the gate select switch, 1 Hz and 2 Hz signals are routed to the inputs of the GCU. One gate of the 74LS00 is used to invert the 1 Hz and apply it to the clock input of

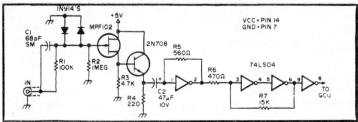

Fig. 10-35. The Preamp.

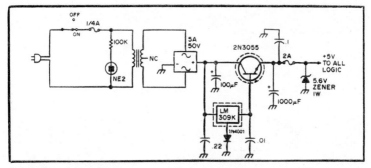

Fig. 10-36. Power Supply.

the 7492. Normally this IC would count to 12 and reset to 0, but with the 7410 gate connected to the A, C and D outputs, it will be forced to reset at the count of 11. Zero detecting the outputs with a NOR gate and NAND gate will produce a pulse that is high for 10 seconds and low for one second, which is the time between reset and the next input cycle. Another gate of the 74LS00 is used to invert this pulse.

By combining the inverted gate pulse with the 1 Hz and 2 Hz signals, and then inverting the outputs of the other two gates of the 7410, the Strobe and Reset 0 pulses are generated and transferred to the rest of the counter. Through trial and error it was found that there had to be a minimum amount of time between the two pulses and this circuit provides it. Unfortunately, due to the minimum pulse width needed to reset the 7490s in the counter (50 nsec), the fasest gate time allowable is .001 second. This is probably faster than needed anyway.

The current limiting resistor for the LED should be chosen for the particular LED being used. Generally, about 180 ohms should be right. Any color may be used.

This is also the cheapest unit, costing about $4.00 with all new parts.

The Preamp. We used a 74LS04 to obtain a 50 MHz working speed, a 2N708 for Q2 and a slightly different input scheme, in addition to the k20AW counter. Finally, sensitivity became livable. A scope and rf signal generator (HP 608D) showed a sensitivity of 10 mV from 10 MHz on up. At audio frequencies, about 50 mV was needed for reliable operation.

The reason for all that extra space on the board layout is that in the future, an onboard prescaler using the 11C90 by Fairchild will be incorporated. Provided, of course, that the circuitry is as simple as that using a 95H90. With the 11C90, the counter should operate in excess of 500 MHz.

Power Supply. Due to heavy current demand, about 1.3 amps, we decided on the circuit shown to regulate the 5 volt line, rather than use two LM309s. All components can be mounted on the rear panel with appropriate mounting hardware and solder lugs. Heatsink the pass transistor and LM309K. You can use a bridge rectifier module or individual diodes. They should be rated for at least 5 amps at 50 volts. The transformer is a 12.6v AC at 3 amps, or parallel two smaller rated ones. Use a 2N3055 for the pass transistor or a suitable substitute with similar ratings.

293

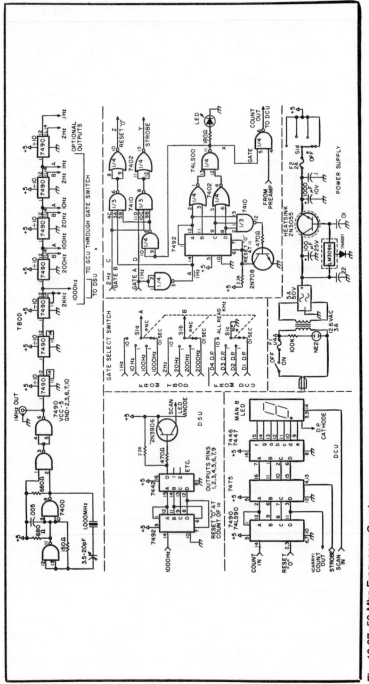

Fig. 10-37. 50 Mhz Frequency Counter.

294

Fig. 10-38. TBOD board (full size).

The fuses should be of the fast blow type. Fuse holders mounted on the rear panel provided the best method, but they can be soldered in inside the chassis. The power switch can be eliminated so that the counter is on whenever the line cord is plugged in.

General Construction and Testing. After etching and drilling all the boards, install the jumpers first because two on the GCU are under ICs. Next put in the resistors and capacitors, and then the transistors and ICs. The LED sockets on the DCUs come next; then install the LEDs. Install the sockets on the master board.

Wire the power supply in the case, along with the front and rear connectors and switches. Secure the master board to the case with small angle brackets, at least 4″ behind the front panel. This will allow the boards to be installed and removed easily. Or the master board may be hinged at the bottom to tilt back. Next do the main wiring from the power supply to the

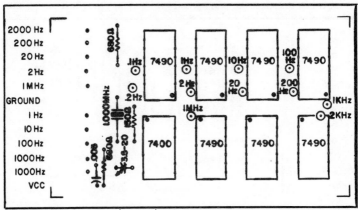

Fig. 10-39. TBOD component layout. Use jumpers to bring frequency outputs to edge. 1 board per counter. Dot indicates pin 1.

295

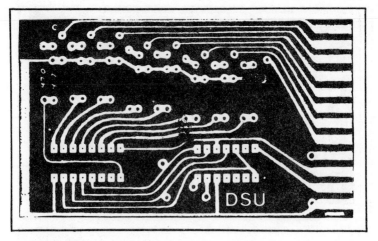

Fig. 10-40. DSU board (full size).

master board and the front panel switches. Use RG-174 for the counter input and the 1 MHz test output. Location of the controls and cutouts is entirely up to the builder and depends on the case used.

Before installing any boards, check the power supply for proper operation. If it's working, turn the power off and install the TBOD in the 12-contact socket on the left (from the front of the counter). An accurate frequency counter will be needed to set the 1 MHz oscillator, at least 1×10^{-8}. Turn the power on and with the counter connected to the 1 MHz test output, adjust the trimmer to read 1 MHz, plus or minus a few Hertz. If it won't adjust, try another 7400. This is an initial adjustment. After the rest of the boards are installed, and at least a one hour warm-up period, recheck the frequency. Before turning off the power, check the divider chain for proper frequency outputs. If everything checks out, turn the power off and install the GCU board. Select the one second gate and turn the power on. If the

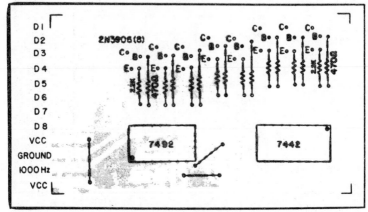

Fig. 10-41. DSU component layout. 1 board per counter. Dot indicates pin 1.

296

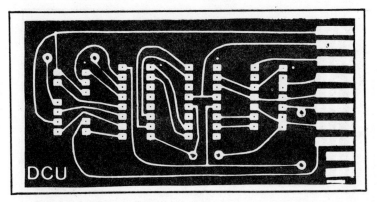

Fig.10-42. DCU board (full size).

GCU is working, the gate LED will blink on for one second and off for 10 msec. Check the Reset 0 and Strobe outputs with a scope for a 10 msec pulse. If the GCU checks out, turn the power off and install the DSU board.

Turn the power on and check each of the DSU outputs with a scope for proper switching. A frequency of 125 Hz should be measured. If the DSU is working, turn off the power and install the DCUs. Turn on the power and check that the display reads all 0s. If not, make sure all the boards are in the sockets tight or check for unsoldered connections, or bad ICs. If all 0s are displayed, turn off the power.

Install the Preamp board and turn the power on. The display should still read all 0s. If not, the DSU may need bypass capacitors on the Vcc line on the master board. Any signal on the Vcc line greater than about 20mV will trigger the Preamp and cause false counting with no input. Any input signal will have to exceed this by at least 10 mV to be counted.

If you get all 0s on the display, proceed to check out the whole counter by using a signal generator to check the frequency response and sensitivity. You may want to keep a graph or record of the results for future reference.

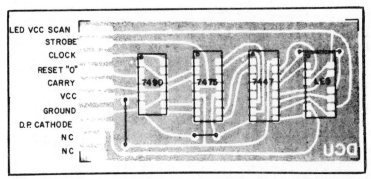

Fig. 10-43. DCU component layout. Use side mount socket for LED (see text). One DCU uses 74LS90 for 50 HMz count speed. 6 or 8 boards per counter. Dot indicates pin 1.

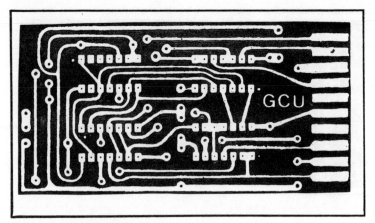

Fig. 10-44. GCU board (full size).

All that's left to do is recheck the TBOD frequency and button up.

Troubleshooting. If you run into difficulty getting the oscillator to zero on exactly 1.000 MHz, try another 7400. Some will oscillate better than others. It has something to do with the characteristics of different batches. You can use a scope or another counter to check the divider chain for proper division. On the GCU, a dual-trace scope is nice because you can check and compare the waveforms at more than one point and reference them to another. Most problems here are caused by loose ICs in the socket. Actually, most problems can be cured simply by trying a different IC. If you still run into difficulty, look for solder bridges, bad connections, wiring errors, or even the possibility of a leaky or bad transistor.

One can save in construction costs by not using sockets for the ICs, although it's a good idea to make certain the IC is good first. It's not fun unsoldering them. Do use sockets for the LEDs, though. The preformed

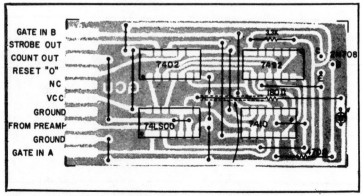

Fig. 10-45. GCU component layout. Use sleeving to insulate resistor lead. 1 board per counter. Dot indicates pin 1.

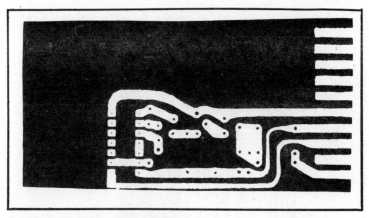

Fig. 10-46. Preamp board (full size).

side mount ones are best, but a wire-wrap socket will do just as well if you carefully bend the leads with needlenose pliers.

Try for a trade-off between price and visibility on the display. Sure, those large 6″ LEDs are easy to read, but expensive. The cheaper, .27″ ones will work just as well, and they can be had in different colors: red, green, yellow or even orange.

The Amphenol PC card sockets are available from Cramer International, Newton, Mass., or the local office in your area.

The cabinet will have to be at least 3½″ high, 10½″ long, and 7″ deep. Make sure you leave at least ½″ between the LEDs and the back of the front panel for removal of the plug-in cards.

If you don't need more than one gate, eliminate the switch and hardware the circuit. The switch used here is shorting miniature rotary, Centralab #PS-11. The extra position and pole made it possible to turn the power on when selecting the timebase. For a complete parts list, see Table 10-4.

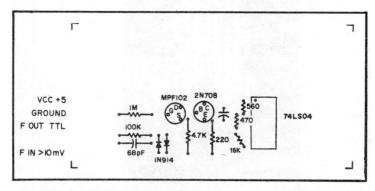

Fig. 10-47. Preamp component layout. Resistors are ¼ Watt. R5, R6, R7 are mounted vertically. 1 board per counter. Dot indicates pin 1.

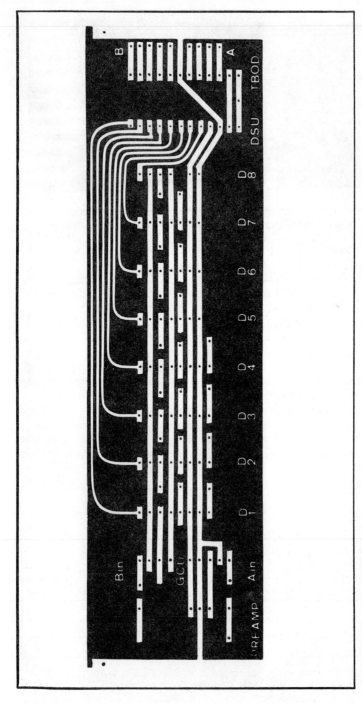

Fig. 10-48. Master board (full size)

Table 10-4. Parts List for Frequency Counter.

	TBOD	1	.1 μF Disc
		1	.22 μF Disc
1	7400	1	1N 4001
7	7490		
1	PC Edge Connector		**DSU**
	Amphenol #143-012-03		
8	14 Pin DIP Sockets	1	7492
2	680 Ohm ¼Watt	1	7442
1	150 Ohm ¼Watt	1	PC Edge Connector
1	3.5—20 pF Variable		Amphenol #143-012-03
	E.F.Johnson #274-0020-005	1	14 Pin DIP Socket
1	.005 μF Disc	1	16 Pin DIP Socket
1	1.000000 MHz Crystal	8	470 Ohm ¼Watt
1	PC Board	8	2.2k ¼Watt
		8	2N 3906 PNP Transistors
	DCU	1	PC Board
1	74LS90 (50 MHz Version)		**GCU**
7	7490		
8	7475	1	74LS00
8	7446 or 7447	1	7402
8	PC Edge Connectors	1	7410
	Amphenol #143-010-03	1	7492
8	Seven-Segment LEDs (See Text)	1	PC Edge Connector
8	14 Pin DIP Sockets		Amphenol #143-010-03
8	14 Pin Wire Wrap Sockets	4	14 Pin DIP Sockets
16	16 Pin DIP Sockets	1	470 Ohm ¼Watt
8	PC Boards	1	3.3k ¼Watt
		1	2N708 NPN Transistor
	Power Supply	1	LED
		1	@ 180 Ohm ¼Watt
1	12.6 V ac, 3 Amp Transformer	1	PC Board
1	50V, 5 AMP Bridge Rect.		**Miscellaneous**
1	2N3055		
1	LM309K		Gate select switch
1	100μF, 25 V		BNC
1	1000 μF, 10 V		connectors,plastic window,
1	.01 μF Disc		fuse holders,hardware, knobs,
			cabinet, etc.

300 MHz Frequency Scaler

This project is about building an IC frequency counter good up through the VHF range. That design has a built-in frequency scaler, and is good through almost 300 MHz. For those who have a low-frequency counter and just need a scaler to extend the range, here is a simple design which can be built for about $20.

The heart of the unit is a Fairchild μ6B95H9059X divide-by-10 scaler IC, which costs about $16 in unit quantities; we call it a 95H90 for short, and it's IC1 in Fig. 10-49, the schematic diagram. The output of this IC is fed into a 2N5771 PNP transistor which amplifies the output and provides a voltage change. If you have a counter which will go up to about 15 MHz, then this part alone will extend your range to 150 MHz. But if you have a slower

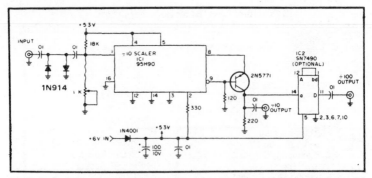

Fig. 10-49. VHF scaler schematic diagram. All .01 capacitors are disks. All resistors are ¼W, 10%.

counter, then you will need IC2 as well. This is an inexpensive ($1.30 or so) SN7490N TTL decade counter, which does a second division by 10. Both IC's together will take a 150 MHz signal and divide it down to 1.5 MHz.

The IC's need a voltage between +5 and about +5.3 volts. While you could build a power supply, that only increases the cost. The easiest way to provide power is with four D cells or a 6-volt lantern battery, which provides + 6 volts. The 1N4001 diode, or another silicon power diode, drops this to about +5.3 volts. The scaler needs about 150 mA, so a set of four D cells will last about 10 or 20 hours with intermittent use.

Figure 10-50 shows a printed circuit board layout which can be used to mount the scaler, and Fig. 10-51 shows the parts placement on the board. Leave the copper border around the board, to act as the ground connection for the board. Use phono jacks or rf connectors, and bypass the +6 volt lead. The purpose is to let rf into the box only via the input lead.

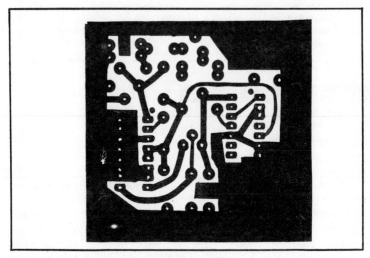

Fig. 10-50. Copper side of PC board, actual size.

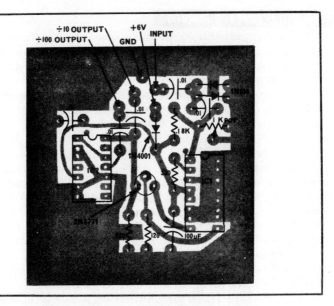

Fig. 10-51. Parts layout diagram (component side).

In use, avoid over-driving the scaler. Start with a low input level, and feed in just enough rf to get a counter reading. With a 19″ length of wire connected to the input lead you should be able to activate the scaler with a 1 or 2 watt hand-held unit from a few feet away.

Increased Speed for the k2OAW Frequency Counter

This project involves the construction of a frequency counter with a low frequency (0-20 MHz) input and a divide-by-10 prescaler capable of extending the upper frequency limit to 300 MHz. The counter is built using mostly surplus ICs and junk box parts and it works the first time you plug it in. By sorting the ICs according to speed, we were able to obtain a stable upper frequency limit of 32 MHz in the low frequency position. Much experimentation was accomplished on the front end of the counter, including the 40673, the 7413 and the 7400 input selector. Maximum attainable speed remained in the 32 to 34 MHz range.

The main problem appeared to stem from the limited speed of the 7473 flip-flop which comprises the divide-by-two function of the first decade. Simultaneously with the discovery of the apparent problem with the 7473, low cost Schotty TTL ICs became available from some of the surplus companies. (They are available from Solid State Systems, P.O. Box 773, Columbia, MO 65201.) A 74S73 flip-flop was obtained and plugged into the IC5 position on the circuit board and things started to happen. Using a four turn wire loop on the counter input and a grip dip oscillator as a signal source, the counter's upper limit was found to be 62MHz! Subsequent checks proved that the initial test was not a fluke and the counter was now stable from 20 to better than 60 MHz.

The 555 integrated circuit can easily be made into an inexpensive linear frequency meter covering the audio spectrum. The 555 is used in a monostable multivibrator circuit. The monostable puts out a fixed time-width pulse, which is triggered by the unknown input frequency.

Referring to Fig. 10-52, transistors Q1 and Q2 are used as an input Schmitt trigger. The unknown frequency input is clipped between 9 volts and ground by these transistors. Positive feedback is used to insure the waveforms have fast, clean edges. The output of Q2 is a square pulse with the same frequency as the input signal. The output of Q2 is differentiated by C2 and R2 to provide a short pulse for the 555. A small signal diode is connected across the differentiator to insure that the 555 input never exceeds 9 volts. The 555 is connected in the standard monostable circuit. Since a Schmitt is used to trigger the monostable, square, sine and ramp type waveforms may be used at the input to the frequency meter. A nominal voltage of 1 volt rms is required to trigger the Schmitt circuit.

The range scale timing resistors R3, which determine the monostable pulsewidth, are small potentiometers mounted directly on the circuit board. These pots are used to calibrate each frequency scale.

The output of the monostable is a fixed width pulse. Every time a zero crossing of the unknown frequency occurs, the monostable is triggered. Thus, as the frequency of the trigger pulse increases, the monostable output has a greater and greater duty cycle. The frequency limit on any one range is determined by the R3C3 time constant re-triggering, and a constant 9 volts will appear at the output.

The monostable output is a pulse with a duty cycle dependent upon the input frequency. Thus, by integrating or averaging the output waveform, a DC voltage is developed. This voltage is directly related to frequency. Resistor and capacitor R4C4 is used as a pulse averager, important on the

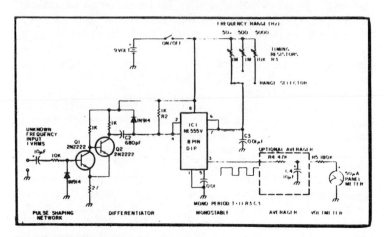

Fig. 10-52. 555 timer frequency meter.

lower range setting. As the input frequency increases, the panel meter itself can act as a waveform averager. Input frequencies greater than 50 Hz will be averaged by the meter fairly well; however, at lower frequencies the meter will respond to each cycle of the unknown frequency input. The meter is used as a high impedance voltmeter. A 1 mA meter could be substituted with a change of resistor R5. For a 1 mA meter, R5 should be about 9.1k and R4C4 should be changed appropriately. R4 should be a factor of 10 less than R5, and the same R4C4 time constant should be kept. This would make C4 a very large value, so R4C4 could be left out if the lower frequencies are of little interest (less than 50 Hz).

The range scales are set up in decades. To calibrate each scale, a standard input frequency is connected to the input. About 1 volt rms is needed to trigger the first stage. The monostable must be less than the maximum input frequency to be measured on each scale. With the maximum input frequency applied, each range potentiometer should be adjusted until the value of the input frequency corresponds to the full scale meter reading. The duty cycle of the monostable should be roughly 90 per cent at the maximum frequency input for each scale setting. If, for a 90 per cent duty cycle, the meter will not read full scale, meter resistor R5 should be lowered accordingly. Each scale setting should be calibrated by adjusting the respective R3 potentiometer.

During operation, when the scale reads off scale on any range, the scale should be changed to the next higher setting. Once calibrated, this frequency meter should read within 5 per cent of full scale. The useful frequency range of the meter is from tens of Hertz to well over 50 kHz. Although decade ranges are shown, the ranges between the decades can easily be added to give as many frequency ranges as deemed necessary.

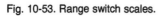

The Speedy Audio Counter

The need for fast and accurate measurement of the low frequency tones in an electronic organ provided the basic drive to produce this counter (Figs. 10-53 through 10-55). An earlier version was slightly less costly, but far too slow and lacked the needed resolution. This design works well, and should find acceptance with those wishing for precise measurements for most audio frequency applications.

Most electronic organs contain a master oscillator board that uses 12 slug-tuned coils for generating the highest tones, or the top octave. These are followed by a divider chain that provides successively lower octaves. It was decided to measure one of the lower octave ranges that spans from C6

Fig. 10-53. Range switch scales.

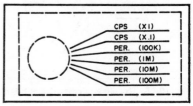

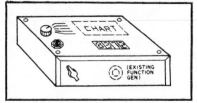

Fig. 10-54. Oblique view showing general outward appearance.

at 1046.50 Hz to C5 at 523.25 Hz. With a conventional counter, a 10 second gate would provide tenths of Hertz resolution at best, and a longer gate time is out of the question. The only reasonable approach is to count the number of higher frequency pulses that occur during one or more cycles of the frequency being measured. For example, if the number of one microsecond pulses were counted during one cycle of A5 at 880.00 Hz, you would get 1136. A 10 cycle count gives 11363, and 100 cycles will read 113636. The period of one cycle of A5 is .00113636 seconds, so 100 cycles requires just over .1 second. You get the required resolution and fast update to the counter. To calculate the required display reading, just divide the reference frequency by the input frequency in Hertz.

A look at Fig. 10-56 shows that CMOS integrated circuits are used throughout. A four stage display counter consisting of 4026s count and decode to seven segments in decade fashion. These were used in preference to 4033s, as they have a display enable input that permits us to blank the display while resetting or counting. No latches are used as it was found a "blinking" display is actually easy to use. A new display means an updated count. 4049s are used with 680 ohm resistors for segment drive to MAN-1 readouts. Three 4518s are used for the timebase divider to provide 100 kHz pulses, or .1 second and 1 second timebases. A 4518 is also used to divide the input to give ×1, ×10 and ×100 ranges. Two sections of a 4001 are used for the crystal oscillator and buffer and two sections of another 4001 comprise the input amplifier. The input amplifier is biased slightly above or below the threshold point to prevent random triggering. A 4013 and three

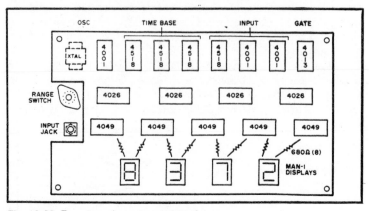

Fig. 10-55. Top view of component board.

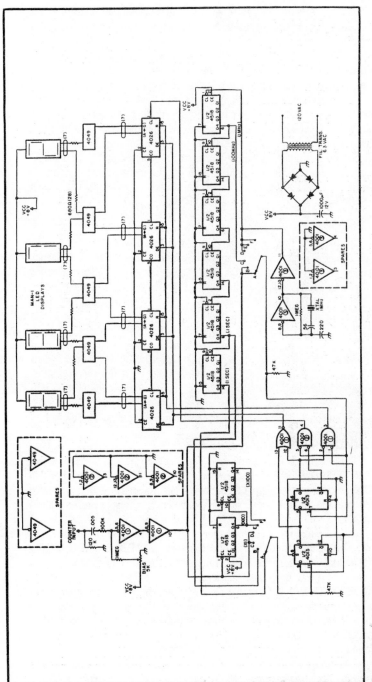

Fig. 10-56. Schematic.

sections of 4001 are used to provide synchronization and division of the input, reset, display gating and clock pulses to the counter chain. This gating system requires a minimum of parts, yet provides a 50/50 gate and display cycle. The reset is one-half cycle of the reference frequency that immediately procedes the count. As the 4026s require less than .5 usec for reset, it may be possible to use a higher reference frequency.

A two section, six position switch permits switching between the two count modes. Position A gives a direct reading in Hertz per second. Position B gates for .1 second. These two scales give a positive reference for gross tuning adjustments or quick identification of an unknown frequency.

Switch position C provides pulses at a 100k rate and D at a 1 microsecond rate. Positions E and F bring in the input divider to give an effective ×10 and ×100 in resolution. An example of expected readings of A5 at 880.00 Hz would be: A-0880, B-0088, C-0113, D-1136, E-1363 and F-3636. The more significant digits drop off on the higher resolution scales, but it was felt that an overflow indication was not needed.

Construction was on a piece of perfboard with holes spaced .1 inch, mounted under the top cover of an existing function generator. The cabinet measured 7″ wide, 5″ deep and 3″ high. The perfboard was slightly smaller with a section removed to permit the switch and input jack to be mounted on the top. The LEDs, ICs and resistors were mounted on the top of the board, and point-to point wiring used for connecting the components. The power supply is mounted on one end of the box. The crystal, originally from a BC-221 frequency meter, was removed from the metal cover and mounted on the board. If you have enough room in your cabinet, it would be better to leave it as is, and mount it with a clip or octal socket. A slot was cut out to reveal the LEDs and covered with plastic laminate for protection. A switch position scale and a chart of frequencies and their respective periods were typed up and pasted to the top, covered with more of the plastic laminate.

How Accurate Is Your Counter?

There is much misunderstanding among the ranks as to counter frequency accuracy, how a counter should be calibrated and the relationship with one's transmitter output frequency. This project's intention is to shed light upon the multiple ramifications involved in frequency counter calibration.

A most likely starting point, therefore, is the counter time base oscillator. This usually is comprised of a 100 kHz, 1.0 MHz, 5.0 MHz or 10.0 MHz crystal oscillator which may or may not be temperature stabilized with an oven. Typical time base accuracy may be as good as one part in 10,000,000 ($1/10^7$) or less per 24-hour period. If measurements with reasonable repeatable accuracy are expected, the counter will require a minimum warm-up time of one hour with 24 hours' time being recommended. If the counter is being calibrated to factory specifications, a 24-hour period becomes minimum and 3 months is not unheard of.

A laboratory-type crystal reference oscillator used to calibrate the basic time base oscillator in a counter generally does not reach its stabilized aging rate until a minimum "ON time" of 3 months. This type of standard has its complete oscillator circuit encased in a heavily constructed double oven (an oven within an oven). The typical size of such an oven may be as small as 8″ long and 3″ in diameter and ovens twice this size are quite common. Such oscillators, when calibrated against WWVB or WWVL, 60 kHz and 20 kHz, respectively, have accuracies as great as one part in 100 billion ($1/10^{11}$), with 1/10 transmission from Boulder, Colorado is $1/10^{10}$, one part in 10 billion. Because of signal path distortions, however, the accuracy is degraded by about two orders of magnitude or possibly more. This is known as "The Doppler Shift." The received accuracy is now $1/10^8$ or less. Under the most optimum of receiving conditions and using the best calibration techniques, including an overall time period of 6 weeks for the daily corrections of the crystal oscillator frequency, one could achieve a counter time base accuracy of one part in 10 frequencies of 2.5, 5.0, 10.0 or 15.0 MHz because of the limited accuracies available. The low frequency transmissions of WWVB and WWVL are used as they are not subject to the same types of transmission frequency distortions. In addition they can be and are transmitted with much higher accuracies. The transmitter oscillators of WWVB and WWVL do not use crystals. They are controlled by atomic references.

The Cesium Beam standard is an atomic resonant device which provides access to one of Earth's invariant frequencies in accord with the principles of quantum mechanics. It is a true primary reference and requires no other reference for calibration. A cesium atomic beam resonator controlled oscillator is used and is the nation's primary frequency reference. The accuracy of transmission is $1/10^{13}$, one part in 10 trillion.

The transmissions of WWVB on 60 kHz are intercepted on receiving equipment of very narrow bandwidth, usually a few Hz wide. The time interval between a local reference oscillator and the frequency of WWVB are compared minute by minute on a strip chart recorder which records the phase differences. The resolution of this comparison is typically 1 us. One part in 10^{10} takes 10^4 sec. to achieve a 1 usec. error (somewhat less than 3 hours).

The transmission of WWVB is coded in the binary coded decimal system. The characters are formed by variations of the carrier in ± 10 dB levels. This presents time-of-year information each minute: the minute, hour, day of the year and the millisecond difference between the broadcast signal and UT2 time.

The crystal oscillator, although not the most stable, is still the most economically available to the hobbyist. It is also compact in size and fits easily into the geometry of a compact frequency counter. The frequency counter does have a number of inaccuracies and the crystal reference frequency oscillator is the major one.

The crystal is the time reference base for a divider chain which basically sets up start and stop trigger pulses in the counter. The counter has a number of these, 1Hz, 10 Hz and 1 kHz being typical. Besides the time base inaccuracy there is also an inherent trigger error which causes a ± 1 count in the readout. The crystal when first turned on has a large drift error. This is overcome by use of an oven and a warm-up period. Thereafter the crystal has an aging rate. This is a constant periodic upward change in frequency with time. All crystals have it. After about 90 days of ON time this rate becomes quite predictable and can be compensated for if the oscillator is equipped with a dial capable of a 1,000 to 1 or better resolution. On the finer laboratory oscillators this vernier readout can be related to time in microseconds.

The average hobbyist is most apt to set up the counter's reference oscillator by zero beating the counter's time base reference oscillator against WWV's high frequency transmissions. This is generally performed aurally after a warm-up period of 1-5 hours. Most often, the inexperienced hobbyist may wind up by beating against the audio of the transmitted signal in which case the crystal will be set to the wrong frequency. WWV usually uses 440 Hz for a period during the hour, then shifts to 600 Hz. If the oscillator has been zero beat to one of these, the error will become apparent when the modulation frequency is changed as the apparent zero beat will have disappeared.

A comment quite often heard by lab technicians and radio experimenters contains the following substance: "I have the audible capability to zero beat a signal emanating from the speaker of my receiver." If this is so, then these individuals have indeed been endowed with a very special human ability, especially when one considers that the human ear cannot hear frequencies below 30 Hz regardless of their amplitude. This also brings into view the necessary unusual hi-fi characteristics required of the receiver's output transformer which probably does not respond below 100 Hz—to response. What one actually is hearing is a difference in the amplitude in the noise frequencies accompanying the signal.

A recommended procedure that affords quite good zero beat capability and is readily available to most radio experimenters is as follows:

- Tune in the WWV frequency to be used. The reception of the signal should be free of fading.
- The oscillator signal to be calibrated should be coupled so that its amplitude closely equals that of the incoming WWV carrier level.
- Zero beat aurally as closely as possible, then continue refining the adjustment while observing the receiver "S" meter indication. As zero beat is achieved the "S" meter will reach either a maximum or minimum stabilized indication. This reading will be maximum if both signals are of the same phase relationship or minimum if the phase is out by 180 degrees. It is of little concern which is obtained just so the meter needle has stabilized its position.

If the crystal is of high stability quality and is in an oven, one might want further refinement in the long-term stability capability. In such case it will be required that the crystal oscillator be left on continuously. Further refinements in the zero beat setting will be required each 24 hours, first to correct for the initial phase error setting and then to correct for the crystal aging rate. These further refinements could result in parts in 10^7 accuracy.

For long-term stability a 1.0 MHz or higher crystal is recommended. Zero beat becomes more difficult the higher the WWV referenced frequency. This follows because the resolution of adjustment is reduced and you are working to a finer tolerance.

Assume that the received signal is accurate to 1 part in a million ($\pm 10^{-6}$) and that the adjustment you have just made was set to zero beat with $\pm 10^{-6}$ of the received signal. This will produce a short-term accuracy of $\pm 2/10$ of ± 294 Hz depending on which side of the WWV carrier frequency you are zeroed in on. This accuracy will decay with time elapse between the act of calibration and use of the counter.

For those fortunate enough to have the means to perform more accurate measurements, the following is supplied.

- Counter minimum ON time: 90 days.
- Compute oscillator drift rate for single measurement spans. The average fractional error of frequency is equal to the fractional time error which is given by:

$$\frac{\Delta f}{f} = \frac{t2 - t1}{T}$$

Where

$\dfrac{\Delta f}{f} =$ Average frequency error

$t_1 =$ Initial time comparison reading

$t_2 =$ Final time comparison reading

Example (Time comparison reading at 9 am May 1 = 4.64 ms. A reading on 9 am May 4 = 1.70 ms.):

$$\frac{\Delta f}{f} = \frac{4.64 \text{ ms} - 1.70 \text{ ms}}{3 \text{ days}} \times \frac{1 \text{ day}}{8.64 \times 10^7}$$

$$= \frac{2.94}{3 \times 9/74 \times 10^7} = \frac{29.4}{25.92 \times 10^6}$$

$$= \frac{1.19}{10} \text{ or } 1.19/10$$

That is, the average oscillator error during this period (or assuming a constant frequency drift, the instantaneous error at 9:00 am on May 2) is 1.19 parts per million high. The average frequency of the oscillator during this measurement interval is given by:

$$f \text{ av} = f \text{ nom.} \left(1 + \frac{\Delta f}{f}\right)$$

Where

$f \text{ av} =$ Average frequency
$f \text{ nom.} =$ Normal Oscillator frequency
$\dfrac{\Delta f}{f} = \dfrac{\text{Average frequency}}{\text{error}}$

Thus, continuing with the above example and an oscillator with a nominal frequency of 1.0 MHz,

$$f \text{ av} = 10^6 \left(1 + \frac{1.19}{10^6}\right) = 1,000,000.119 \text{ Hz}$$

Using this method of continuous corrections and recording continuous data one is able to iron out the propagation anomalies of WWV and approach a precision better than 1 part in 10^8.

An Incredible Counter Calibrator

Wouldn't it be exciting if it were possible to have all of the following in one simple small package?

- The maximum readout accuracy a frequency counter or reference oscillator design will permit.
- A calibration device so simple that it requires only a level set control, a momentary switch and two BNC coax connectors on its panel.
- A compact package, all solid state and no warm-up time required.
- The cost of building under $60.00, if everything is purchased off the parts house shelf (under $25.00, if you know where to shop).
- Fast calibration, once the counter or reference standard to be calibrated has reached stability.
- Fun and excitement for both operator and spectators while in the process of calibration.
- Accuracy as great as one part in 10^{-11} (if the counter could only have this kind of stability!).

The only other equipment required is a family color TV set.

A wild dream? Not at all. Such a device is available and, in addition to the above features, it beats the long, drawn-out WWVB methods used by calibration (metrology) laboratories throughout the country. This longawaited method is called the *color bar calibration method*. Once you possess this device, calibrating your counter timebase oscillator or other reference standard becomes as exciting as using the instrument for its intended purpose.

The method to be described has been developed by Dick Davis of the National Bureau of Standards, and additional information relating to frequency measurement methods can be obtained by writing their Time and Frequency Services Section.

NBS consulted with the heads of the four broadcast networks (NBC, CBS, ABC and PBS), suggesting that they could provide a public service by precisely controlling the frequency of their TV color burst. This would require a modest one-time expenditure. The networks obliged by purchasing rubidium signal sources so modified that they included a synthesizer to generate a 3.579545454...MHz signal to generate their color-burst subcarrier. The rubidium frequency oscillators have accuracy beyond one part in 100 billion, but are offset from the exact frequency by about -3000 parts in 10.

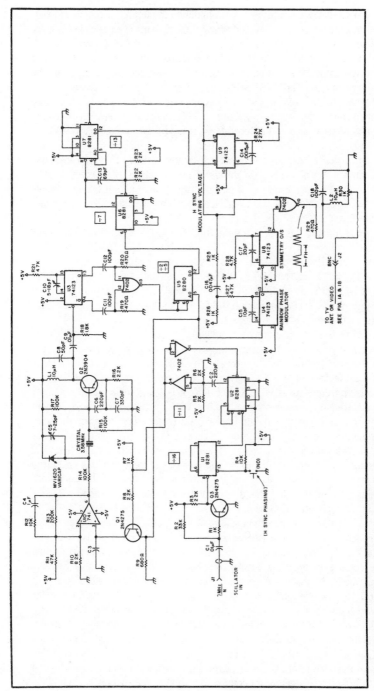

Fig. 10-57. Schematic diagram of the color bar calibrator.

313

With that out of the way, you should be moderately excited, at least. But first the question is, "How does this accuracy manifest itself in my color TV set?" All color television receivers lock onto the color subcarrier signal, and, if the color set is tuned to a network program, its internal 3.58 MHz oscillator generates a replica of the atomic oscillator signal back at the network studio. This 3.58 MHz signal from the color receiver is not a substitute for the oscillator in the frequency counter. It is a calibrating signal that can be used to set the oscillator. In only 15 minutes, one can match the results of days of data gathered from the NBS radio stations WWVL or WWVB.

In order to interface the counter oscillator with the TV set, a small box will have to be built (Figs. 1-57 through 1-60) containing 12 inexpensive ICs, a $1 color TV quartz crystal, 3 cheapie transistors and a few other simple components. This small box is called a *color bar generator*. All of its components fit on a 4″ × 7″ PC board.

Essentially, this box takes the 3.579545454 MHz network signal and phase-compares it with the 1.0 MHz, 5.0 MHz, 10.0 MMHz, etc., frequency standard or crystal timebase you have in your workshop.

The block diagram in Fig. 10-61 illustrates one of two ways the equipment is interfaced with the TV set. Basically, the color bar generator produces a wide vertical bar on the face of the picture tube simultaneous with receiving a network telecast.

With the crystal timebase oscillator of the counter coupled into the color bar generator, the bar generator's level control is adjusted for a comfortable bar presentation level. The bar will be moving horizontally across the face of the TV screen. If moving to the right, the counter oscillator frequency is high; to the left, it is low. Adjusting the counter crystal frequency in the proper direction will slow the horizontal movement of the bar until it stops moving. Pressing the momentary switch on the color bar generator positions the bar on the screen. Just keep pushing until the bar positions itself where you want it. Now, going back to the counter's crystal frequency adjustment, further corrections of the frequency will cause the spectrum of colors to roll through the bar presentation. If the colors are rolling from left to right through the bar, then the frequency of the oscillator is on the high side; if moving to the left, the converse is true.

When the colors are moving slowly through the bar, the frequencies or divisions thereof are on the same frequency, but not in phase with one another. (The frequency of the counter timebase is already much closer than one could ever adjust it by any aural beat frequency method.) Further adjustments in phase are necessary in an attempt to cause the rolling of colors to slow even further. When a single color can be retained for about seven seconds, the timebase accuracy has parts measurable in 10^{-8}. If a single color can be retained for about 15 minutes, then you have attained a crystal timebase stability measured in parts in 10^{11}. One thing to remember is that, if the crystal oscillator and associated circuitry does not have this inherent stability, then it will never be attained no matter how long a warm-up period is given the counter. However, one thing is certain: This

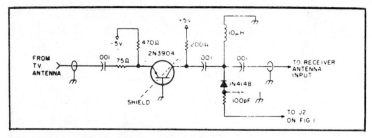

Fig. 10-58. Schematic diagram.

system will squeeze out every bit of accuracy of which the counter is capable. Now that you know what this device does, and if you have the appreciation of this advancement in the state of the art of frequency technology, your adrenalin should be running at top speed.

As stated earlier, the illustration in Fig. 10-61 shows how to interface the equipment by going into the antenna system of the TV set. This method has at least one drawback. Coupling the signal through the TV antenna causes the signal to pass through more of the TV circuitry than is needed, thus producing a visual beat note. This shows up on the TV screen as an annoyance.

A second method uses only the video circuits needed, eliminating this beat note. This method requires getting into the inside of the set and adding a small coupling capacitor, a length of RG-174 50Ω coax and a BNC connector. Essentially, you must tap into the TV chrominance bandpass amplifier stage. It in no way degrades the normal operation of the TV set. If you have the technical competence to expose the innards of your TV set, mother, father or spouse willing, then it is recommended you do so. Otherwise, the antenna input method will suffice.

What makes the color bar generator work is best illustrated by the schematic in Fig. 10-57. Let's see just how this thing takes an odd frequency like 3.57954545 and makes it harmonically related to a 10 MHz crystal and/or any submultiple, that is, 10 MHz divided by N where N = 1,2,3, up to 100. This allows one to compare frequency sources of many frequencies.

The crystal oscillator used in the frequency counter as a timebase reference or, for that matter, any oscillator used as a reference frequency

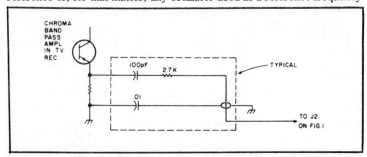

Fig. 10-59. Scehmatic diagram.

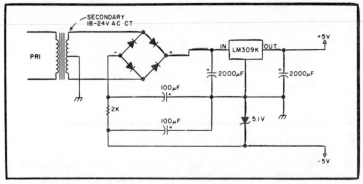

Fig. 10-60. Schematic diagram.

standard, is fed into the J1 input to Q3. The signal is divided by 16 and then by 11, for a total of 176, and drives the base of Q1, which is a phase locked loop comparator.

The 741 operational amplifier (connected as an RC integrator) drives a voltage-controlled crystal oscillator operating at the color subcarrier frequency of 3.58 MHz. The other input to the comparator comes to the Q1 emitter from the loop output circuit. The output from the VXCO drives a 74123 oneshot circuit for pulse-shaping of the oscillator output.

Two signals are taken from the one-shot, and the positive-going transitions are coupled through a 7402 NOR frequency doubler, the output of which is fed into the inputs of a 8280/74176. Part of this signal is divided by two and fed back to Q1 for phase lock.

The result of phase lock is that you have a crystal-controlled oscillator operating at the subcarrier frequency phase locked to a local standard. This permits you to inject a signal into a television receiver and compare it to a network color subcarrier. The second part of the injected signal to the 8280/74177 binary counters is further divided by 7 and 13, providing a total

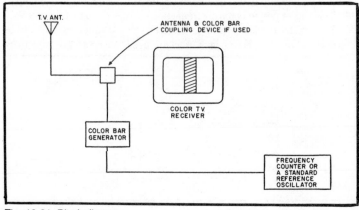

Fig. 10-61. Block diagram.

division of 455. The resultant frequency of 15,734,265 Hz is equal to the horizontal oscillator signal of the color TV receiver. This will produce a stationary vertical bar on the TV screen. This horizontal rate signal is used, with the output of the modulator, to drive the receiver. The block diagram of Fig. 10-61 shows the signal being interfaced through the antenna terminals of the TV receiver. Using this method requires the addition of a single transistor circuit and a pair of terminals to accept the TV antenna. This additional circuitry connects between the TV antenna and then antenna terminals on the TV receiver. See the schematic in Fig. 10-58.

The second method does not require removing the antenna from the receiver and placing it onto the bar generator, but does require obtaining a schematic of the receiver so that a point in the chrominance bandpass circuit can be located. A typical chroma interface is shown in Fig. 10-59.

A 15 MHz oscilloscope will be helpful in determining that all of the ICs are operating and dividing properly. There is only one circuit in the unit that requires tuning, and that is the 3.58 MH crystal stage; C10 requires adjusting to produce doubling of the frequency at pin 13 of the 7402 IC. Check pins 4 and 13 of U3 for equal amplitude signal. Now connect the color bar generator to the TV set, tune in a color program, preferably a network program but not absolutely necessarily at this point and adjust the level control R1 for an acceptable presentation of the vertical rainbow color bar on the screen. Adjust the 7-25 pF C1 crystal trimmer capacitor until the movement of the bar slows sufficiently. This will place the crystal frequency so that it is within range of the 741 op amp voltage swing and so that the varicap can lock the crystal on frequency when the local frequency standard approaches phase resonance.

A VTVM at pin 6 of the 741 op amp is another good test point. Variations of C1 will cause changes at pin 6. When everything is working properly, the device for calibrating the frequency standard is ready to be used.

Everything is now ready for that very exciting moment, except for an extended period of warm-up time for the local standard. This period depends upon how accurate a calibration one is looking for. Obviously, if you are seeking the ultimate, then it is recommended that the oscillator, once turned on, remain on indefinitely. For a starter, I would suggest a 24-hour warm-up be considered as minimum. Then start a program of measuring the frequency every 24 hours, preferably at the same time each day. The stability will be erratic for about a week. If a careful record is kept, you will note that this stability will improve over the next 90 days, when it will settle out with a predictable aging rate thereafter. So there is no misconception at this point, it should be understood that, if the standard oscillator is one of mediocre quality (has no oven, not a low drift crystal cut, etc.), then one cannot expect accuracy of better than perhaps parts in 10^{-7} or worse. In this case, the unit should be warmed up for a period of five to 24 hours, calibrated as best you can, and used as soon thereafter as possible. If the crystal has an oven, the expected accuracy will be better. The quality of the crystal and the oven are of major consideration.

Chapter 11
Oscilloscopes

Build an Engine Analyzer

If you are anything like most hobbyists, you hate to pay someone else to do something you can do yourself. That's the way it is with many experimenters and their automobiles. They are self-professed experts, yet have no way of taking on the complexities of the common Kettering automobile ignition system.

For this project, get your hands on an automotive timing light. It has a bright flash and operates from the car battery system. Hook it up to your car.

The instructions say to hook the red and black wires on the light to the positive and negative terminals of the car battery and then clamp the induction pickup around the number one spark plug wire. Elementary, so far. With the engine running, and being careful to watch that those dangling wires don't drop into the spinning fan blades, gently squeeze the trigger on the gun and watch the light spring to life.

One hobbyist tried this and since he had fiddled around some with automotive problems and knew that the timing marks are found on the side of the front pulley he figured that all that had to be done was to rub some chalk into those marks. Then you can see them easily, and, with the timing light aimed at the spinning pulley, press the trigger and watch the strobing action as the number one cylinder fires the timing light.

However, he did overlook a few small details. The vacuum advance line to the distributor must be pulled and plugged according to the manufacturer's directions. Also, "Timing must be adjusted with the engine running at manufacturer's specified rpm. If necessary, use a tachometer to set idle rpm."

Not having a tachometer, the first thought that went through this hobbyist's mind was to run out and buy one. Somehow that didn't settle too well with him. He needed to figure how many revolutions per minute his little Pinto engine was turning over however, and with a fair degree of accuracy.

We've all seen the modern, automotive electronics shops, with their big engine analyzer scopes all nicely calibrated, but who among us is going to rush out and buy one of those! What this hobbyist did have was a pretty fair B and K model 1461, 10 MHz, triggered oscilloscope, with 18 calibrated sweep ranges.

The problem was interesting. He began by thinking in terms of how the combustion engine works. It takes a fuel/air mixture into the cylinder on a downstroke, compresses it on the upstoke, where it begins burning the mixture by sparking the plug somewhere before top dead center. The resultant explosion gives the power downstroke. Finally, the cylinder on the last upstroke exhausts the by-products of burning. The problem is to fire the plug at just the correct time on the first upstroke before top dead center and do this timing with the engine running at a specified number of revolutions per minute. The timing light flashing on the timing marks will show the answer to the first problem, but that rpm problem must still be figured out. Remember, that cylinder fires only once for every *two* engine revolutions.

What we must do is get a good, *stationary* display of all cylinders firing on the scope, so we can measure the duration of all cylinder firings in time. With an externally triggered scope, this is a cinch. Take a clip lead and loosely couple it around the number one spark plug wire. Just use an ordinary clip lead with an alligator clip on one end. Clipping this around the plug wire gives plenty of induced pulses to easily trigger the scope. Switching to external trigger, the scope will now make one sweep, from left to right across the tube, for every firing of that number one cylinder. Then, by coupling the vertical input of the scope to the high tension lead coming out of the center of the distributor in the same manner, your display will show the firings of all cylinders in exactly the sequence they actually are firing. In the case of the Pinto, it will be, first, number one cylinder, followed by three, four, and finally, number two. It's a simple matter to immediately see if all plugs are firing, and also to see the relative amplitude of the spark voltage to each cylinder. The vertical gain control, along with the vertical positioning control, can be used to bring the voltage peaks of all firings onto the scope face. Just remember, we are only looking at *induced* voltage through the insulation of the spark plug wire. We have not connected our scope directly to any bare wire, as the plug wires can carry well over 10,000 volts of AC. In some cases, it may help to put a 2200 ohm resistor and .05 capacitor across the input of your scope to dampen out much of the high frequency informa-

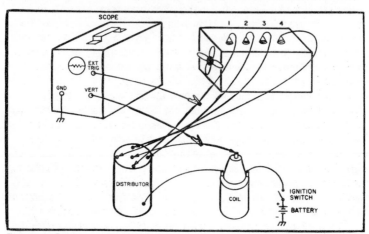

Fig. 11-1. Scope connections.

tion we are not interested in. Some experimentation is called for with the exact values. Nothing is very critical in this department. Time in seconds is the reciprocal of frequency in Hertz, and frequency in Hertz is the reciprocal of time in seconds.

Remember, we want to measure engine revolutions in time—specifically, revolutions per minute. Because, as stated above, frequency in Hertz is the reciprocal of time in seconds. All we must do is measure, with the scope, the time for all cylinders to fire, take the reciprocal of this time in seconds to get frequency in Hertz, and then multiply by 120, thereby getting revolutions per minute. (Remember, that cylinder fires once every other revolution; therefore we must multiply by 120 rather than 60.)

If we look at a calibrated sweep oscilloscope, we see that sweep time is usually measured in milliseconds or microseconds per division on the graticule over the face of the tube. All we must do is count the number of divisions, generally centimeters, multiply by the indicated number of milliseconds or microseconds per division of the sweep time scale of the scope, and take the reciprocal to find frequency. At this point, a small calculator is an immense help, unless you like to do long division with a pencil.

As an example, suppose we have connected our scope as shown in Fig. 11-1, and we are driving a four-banger. Our sweep time is set for 5 milliseconds per centimeter. The time between firings is 6.6 centimeters. Multiplying this by the sweep time of 5 milliseconds per centimeter, we find that time between firings is 33 milliseconds, or 132 milliseconds for four cylinders. Taking the reciprocal of 132 milliseconds and multiplying by 120 reveals engine revolutions to be 909 revolutions per minute.

Although the scope could have been set up for a display of all four cylinder firings, a little more accuracy is possible by using an expanded sweep and measuring the time for one cylinder firing, rather than by multiplying by the total number of cylinders. There probably isn't much difference, so it will boil down to what each individual feels most comfortable with.

To set the curb idle speed of your car, it is always best to refer to the manufacturer's specs, either in the owner's manual or in a local library, in a good automotive manual. Generally, it's a matter of adjusting the correct screw on the carburetor. Curb idle speeds will vary, and the specs may call out different rpm for such cases as cars equipped with or without air conditioning, etc. Once the idle speed has been properly set, the timing can be adjusted with the light. This involved loosening the lock nut under the distributor and gently turning the distributor, while watching the timing marks on the front pulley in the strobing flash of the timing light. Timing will also increase or decrease the engine rpm, so you may find yourself going back and tweaking the curb idle adjust again.

A word of caution is called for here. Adjustment of engine timing and curb idle speed will affect the emissions of your car. Go slowly the first time, consult your manuals and set your car up by the book. Don't forget to reconnect the vacuum line back onto the distributor when you are finished.

At today's prices for automotive analysis and tune-up, it won't take long before this simple equipment will pay for itself.

The oscilloscope is one of the most versatile instruments in the hobbyist's workshop and it would probably take an entire book just to list its many uses. Quite a few of the lower priced instruments, however, are handicapped by the fact that, though they can faithfully display the shape of the waveform under examination, they cannot be used to accurately measure the amplitude or the frequency of the signal. This is true not because there is anything wrong with the instruments themselves, but simply because most of them do not have build-in calibration circuits, or they have only very rudimentary ones.

Amplitude calibration is normally carried out by applying a signal or voltage of known amplitude to the vertical input of the oscilloscope and adjusting the vertical gain control until the displayed signal spans a given number of divisions on the reticle. You then have a known calibration value in volts per division (V/div.), and the amplitude of any unknown signal or voltage can be read directly from the screen.

The scope's vertical input attenuator can be used to accurately measure higher voltages. For example, if you calibrate the scope to read 1 V/div., then, by simply turning the attenuator switch to the X10 position, it will be calibrated at 10 V/div. You can go the other way, too. For example, start out with the attenuator in the X10 position and calibrate the scope to display 1 V/div. Then turn the attenuator back to X1, and the scope will be calibrated at 0.1 V/div. (i.e., 100 mV/div.).

Frequency measurements are handled in a similar fashion, but by using the horizontal sweep frequency and/or gain controls to calibrate the horizontal axis such that the signal of known frequency is displayed with a given number of cycles per division. The frequency of an unknown signal can then be measured by noting the number of cycles per division it occupies on the screen.

Some oscilloscopes have a built-in calibration signal, but often consist of nothing more than an extra winding on the power transformer to supply 1 volt peak-to-peak at 60 Hz. Consequently, any line voltage changes would cause the amplitude of the calibration signal to change by the same percen-

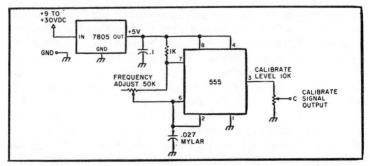

Fig. 11-2. Circuit diagram of the N5KR oscilloscope calibrator. It uses only seven components, all of which are readily available.

tage, which is entirely unacceptable for accurate measurements. The frequency is that of the line itself and, consequently, extremely accurate. However, 60 Hz is such a low frequency compared to the signals normally being measured in the audio range that the measurements which are possible in theory are not possible in practice because of the great percentage difference (that is, the ratio) between the reference signal and the unknown signal.

The solution to all these problems is the circuit shown in Fig. 11-2, which generates a stable wave calibration signal and costs less than $3 to build. The component values shown have been selected to provide a symmetrical 1-volt p-p signal at 1000 Hz when the two trimpots are calibrated. The 555 timer generates the signal at a stable frequency, while the 7805 regulator ensures a fixed amplitude. The entire circuit contains only seven components, which are mounted on a small PC board as shown in Fig. 11-3. The 1v p-p signal enables fast and simple calibration of the scope in the vertical (amplitude) axis, and the 1000 Hz frequency is a nice round reference frequency for analysis of most audio signals. The circuit board is small enough to mount inside of the oscilloscope itself, with the few milliamps of power required being supplied from the scope's existing power supply.

Construction. The circuit is simple enough to build on perfboard, and parts layout is not critical. It can be built on a small PC board, the layout for which is shown in Fig. 11-4. Parts placement, as viewed from the component side, is given in Fig. 11-5.

Component tolerances of 20 per cent are satisfactory, since the two trimpots are used to make the final amplitude and frequency adjustments. In fact; the .027 μF capacitor can be substituted with any value from .02 to .047, if you don't have a .027. But it would be better to go lower in value rather than higher, if you have a choice. The important thing is that this capacitor be of mylar or other temperature-stable type. The PC layout shown in Fig. 11-4 has two different solder pads (one on an extension) on the ground side of this capacitor to accommodate whatever capacitor you select, since its physical size may be different from the one used here. There are also a couple of extra solder pads for making an external ground connection, if desired.

The 7805 regulator may be substituted for an LM340T-5, which is the same device. On rare occasions, these regulators have been known to go into self-oscillation. Should you encounter this problem, replacing the .1 μF disc capacitor with a 1μF tantalum capcitor should clear it up.

Mounting and Power Connections. The circuit is so small that it can be mounted almost anywhere in the scope. In fact, you can trim the PC board down a lot smaller than the one shown in Fig. 11-3, which has a lot of blank space around the edges. You should, however, make it a point to mount it as far as possible from major heat sources so that the calibrator's frequency stability won't be impaired.

The input voltage can be anywhere from +9 to +30 volts DC, and you won't have any trouble finding a supply voltage somewhere in this range in a solid state oscilloscope. If your scope is tube type, you might still find a

Fig. 11-3. Photograph of the finished board ready for installations in the scope. The trimpot at the lower right sets the frequency, while the one toward the left sets the amplitude of the calibrate signal.

voltage of the proper value, but if not, you can obtain it from a higher voltage through an appropriate voltage divider. The supply voltage doesn't have to be regulated, but, on the other hand, you want to avoid one that fluctuates widely. The negative (ground) connection can be made through the mounting bracket, if desired.

If your scope has an existing jack on the front panel for a calibrate signal, just disconnect the wire from the existing calibrate signal and run a new wire from the jack to the calibrate signal output (C) on the circuit board. If you don't already have a jack or if you want to retain your existing calibrate signal, you will need to mount a small pin jack on the front panel for this connection.

Calibration. In order for the calibrator to be of any use, it must itself be calibrated in an accurate manner, and the circuit has been designed for simplicity in that respect. First, set the two trimpots to mid-scale. Now, with power applied to the calibrator circuit, use a VTVM or other accurate

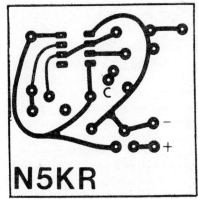

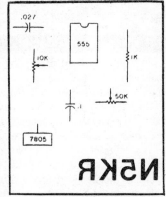

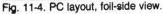

Fig. 11-4. PC layout, foil-side view.

Fig. 11-5. Parts placement, component-side view.

voltmeter to measure the calibrate signal at C. Set the voltmeter to read "DC volts," and adjust the 10k pot until the meter reads exactly one-half volt (0.5 V).

What you have done here is used the DC voltmeter to measure the average output voltage, and, since the signal is switching back and forth between 0 volts and 1 volt, the average value you read on the meter is 0.5 volts. This is true only for a symmetrical square wave, of course, and the component values in this circuit have been chosen such that any error in symmetry is less than 1 percent at 1000 Hz. You also have to be assured that the negative (low) half cycle of the signal is actually at zero volts, which it is, for all practical purposes, when the 555 is operated at the 5v input voltage supplied by the 7805 regulator.

The easiest way to set the frequency is to measure the calibrate signal with a digital counter. If you don't have access to one, an accurate audio signal generator can be used, with the calibrate signal frequency being adjusted to match the known signal while they are being watched on the scope. In any case, the frequency of the calibrate signal is adjusted with the 50k pot.

If you prefer a frequency other than 1000 Hz, you can set the pot for a lower frequency without hurting anything. If you set the pot to a higher frequency, though, the symmetry of the waveform will be upset. To go to a higher frequency, you should lower the value of the .027 capacitor instead, using the formula:

$$C = 28/F,$$

where C is the value in microfarads of the new capacitor, and F is the desired signal frequency.

Summary. This is one of those satisfying projects that you can put together in an hour or so, at a cost of only $2 or $3 and see the results immediately. It consists of only seven components, which are available from any retail or mail-order electronics dealer. It's a handy little device that will enhance the utility of lower cost oscilloscopes and you can sit back and enjoy using it for years to come.

Eight Trace Scope Adapter

Probably the most frustrating problem faced when designing digital circuitry is control of timing. After working out a design on paper, one usually breadboards the circuit to prove it out. In accordance with Murphy's well-known laws, there will be several logic errors which will then be apparent but very elusive. Depending upon the complexity of the design, the errors may be (but usually are not) easily located and corrected.

A number of tools are helpful in tracking down these problems—the logic probe and oscilloscope probably being the most helpful. A logic probe establishes the steady-state status of various points in the circuit, but tells nothing about pulse widths or repetition rates. The oscilloscope is used to visually illustrate these waveshapes, pulse widths and repetition rates. What most scopes do not show is the time relationship between pusles at different locations in the circuit. Sometimes this relationship is crucial in searching out a problem that may be caused by *glitches* (extremely short pulses caused by unexpected and unwanted time overlaps). Well-equipped laboratories use special multichannel logic scopes for this sort of work, but most of us are not equipped with the kilobuck pocketbook required to manage this. Even a dual channel high speed scope requires a considerable investment.

While such a scope would be most welcome in any experimenter's laboratory, most of us must settle for a relatively inexpensive general purpose scope. Fortunately, it is neither difficult nor expensive to build an adapter to display multichannel logic signals. The adapter permits viewing up to eight channels of logic signals simultaneously, and thereby examination of the relative timing between them. Although analog waveshapes cannot be

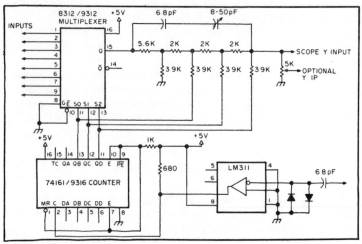

Fig. 11-6. Schematic.

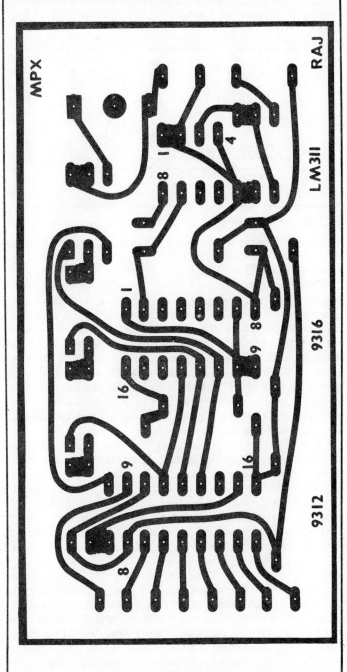

Fig. 11-7. PC board (full size).

327

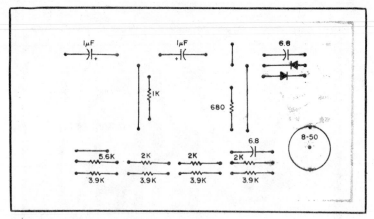

Fig. 11-8. Component layout.

displayed (you can use your scope without the adapter for this function), it will show the low or high states, in precise time positions, of any signals present in TTL or DTL circuits.

Almost any general purpose scope should work with this adapter, but it is recommended that it be equipped with a triggered sweep. The viewing of simple repetitive signals without a triggered sweep can be frustrating enough, but attempting to lock onto one of eight channels being displayed may be virtually impossible. If you are using a scope without this feature, consider adding a new triggered sweep, even if you do not build this adapter.

Scope bandwidth is not critical unless you are working with really high speed stuff, and a 4 MHz bandwidth will let you examine almost all you need to see. You must have a way to externally trigger the scope sweep, and you will have to find the sweep signal or blanking pulse to permit changing the input channels during the retrace interval.

The circuit itself is very simple. A small capacitor couples the scope sweep circuit to a voltage comparator (you may find it necessary to adjust the size of the capacitor for reliable trace switching). The sweep retrace causes a negative excursion at pin 3 of the LM311, forcing its output to go high. Each time this occurs, a 16 stage counter advances one count. Three output bits of the counter are connected to an eight-to-one 9312 multiplexer, which selects each input in turn and outputs to pin 15. If most of your work is at the lower frequencies, use the low order 3 bits of the counter, instead of the 3 high order bits (Fig. 11-6). When using the 3 high order bits, you may use the adapter with a fuel channel scope operating in the *alternate* mode.

A ladder network commonly used for digital to analog conversion is used to position each channel on the screen. The resistors should be well matched (i.e., 1 percent), but satisfactory results have been experienced with 5 percent units. If your display is not evenly spaced vertically, try swapping resistors in this network for best spacing. The variable capacitor is used to compensate for the scope input capacitance, and should be adjusted for best waveshapes. The output potentiometer will not be required in most instances, and should not be used unless essential.

Note in Fig. 11-6 that a 74161 or 9316 synchronous counter is recommended, rather than a 7493 or similar asynchronous type. It is unlikely that propagation delays in an asynchronous counter would result in viewable glitches on the scope in this application, but it is good design practice to always use a synchronous counter where the output states are decoded and fed back to the counter.

The adapter may be built on a small printed circuit board (Fig. 11-7) and installed inside your scope. However, it may be very conveniently enclosed in a small box which can be located near and powered from the digital project, and coupled to the scope via cables. You will need the usual vertical input cable and a sweep-out signal. Many scopes have an *Ext* jack for horizontal input, which is permanently connected to the input of the horizontal amplifier. When the sweep is running, this also happens to be the output of the sweep generator.

Should you experience difficulty in obtaining a stable trace, the sweep circuit may not be advancing the counter properly. Try a different spot in the sweep circuit first. You may find it necessary to invert the signal by using pin 2 of the LM311 (grounding pin 3) if the signal is reversed in polarity. The 74161 counts on a rising edge, and reverse polarity will cause the channel change to occur in mid-sweep, with obvious visible distortion. You may find experimenting with the size of the sweep input capacitor to be helpful, but be careful to avoid distorting the sweep. The scope will not be as bright as usual, as the trace is being timeshared among 8 signals. A slight adjustment of the brightness control compensates for this. The variable capacitor is adjusted for best waveform using a 10 kHz or higher digital pulse. A 74151

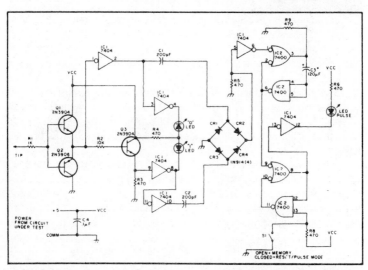

Fig. 11-9. The Superprobe. *Note: C3 is two small 60 uF caps wired in parallel.

329

```
IC1 —SN7404 IC
IC2 —SN7400 IC
Q1, Q3 —2N3904 transistor
Q2 —2N3906 transistor
CR1, 2, 3, 4 —1N914 diode
R1 —1k ¼ Watt resistor
R2 —10k ¼ Watt resistor
R3-9 —470 Ohm ¼ Watt resistors
C1, C2 —200 pF capacitor
C3 —120 µF capacitor (2 small
60   µF inparallel)
C4 —.1 µF disc capacitor
LED 1-3 —Any type/color LED
desired
```

Table 11-1. Parts List

multiplexer is functionally identical, but not pin compatible, with the 9312 unit. The LM311 comes in either a mini-dip or T0-5 package. As the pin-outs are identical, either may be used with the circuit board shown (Fig. 11-8).

Using your multitrace scope is a delightful experience: You see all of those signals at the same time, and can really tell what is going on. Remember that you must trigger the sweep from the slowest signal you are viewing. Otherwise, you will not be able to sync the slower signals. Also, be aware that the inputs are not protected in any way, and connection to potentials outside of the proper logic levels will destroy the multiplexer IC. Protective diodes may be added on the input lines to give marginal security, but care, plus a socket for the 9312, are probably adequate.

The small investment required to construct this unit will be quickly repaid the first time you use it to track down a problem.

Superprobe

If the modern technician is going to be successful when working with digital logic, he must have means of looking inside the circuit. The most common way in the industry is with the use of an oscilloscope or logic analyzer. Both of these instruments are great, but the cost puts them out of reach for most experimenters.

The *Superprobe* (Fig. 11-9) was designed to be an inexpensive piece of test gear which will provide the necessary insight into the digital circuit under test. This probe is not a toy, and will provide the user with almost as much information as a $3,000 oscilloscope, when dealing with TTL or DTL logic.

The probe has several desirable features, aside from the expected "1" and "0" indication. The first is a pulse stretcher, and pulse memory. Any time a high to low or low to high transition takes place, the pulse LED will flash. The flash will be visible even with very narrow pulses, since the probe will stretch the pulse width to a visible flash. If the probe is in the memory mode, the pulse LED will remain lit until reset by the operator, capturing any stray pulse. Another important feature is high impedance input. It will not load the circuit under test. The input is protected by the zener action of the input transistors, with current limited by R1, in the event the probe is touched to high voltage.

Yet another feature of this probe is that if the tip is touched to an open circuit, or to a chain of floating inputs, no light will light, thus identifying this condition immediately. The entire circuit uses only two inexpensive TTL ICs, three transistors, and can be built into a handy, hand-held probe. The parts list is found in Table 11-1.

The Audible Probe Is Safer to Use

This project is a piece of test gear that gives the output in audio. The unit does not have to fit totally in the hand since it's for bench work. Therefore, creation of the sound (loudspeaker) won't be a problem.

You can now build a logic probe that works with CMOS, produces a high tone for a logic high, a low tone for a logic low and no sound for an open or floating string. In addition to this, the unit will detect narrow pulses and stretch them so the user is able to hear them.

The audio probe will detect high to low, or low to high pulses and will do anything any good probe will do, except its output is in sound. It works great!

Circuit Description. Transistors Q1 and Q2 form the input stage (Fig. 11-10). Under no signal or floating input condition, both transistors are biased on by the base resistors. In the ON state, the collector of Q1 is near ground and the collector of Q2 is near plus volts. It is stated as "near" because of the 3 volt zeners in the emitter of each transistor. The zeners will subtract their voltage from what is developed in the collector circuit. The purpose of the zeners is to provide a threshold level at which the transistor will key. For Q1 to turn on, the base must be greater than 3.5 volts, and for Q2 to turn off, its base must be higher than supply voltage minus 3.5 volts. This choice of zener value is not critical. The choice for this project was made based on the fact that all of the cmos circuits are run from 12 volts. With this value of zener, the probe rejects any signal not below 3.5 or above 9 volts. A functioning circuit will easily exceed these values, and if it won't, the circuit is bad. From this point on, think logic. We will call the collector of Q1 "A," and the collector of Q2 "B." If the probe tip is floating, A is low and B is high. If the tip is low, A is high and B is also high. If the tip is high, A is low and B is also low.

Three unique states—one for each condition. Now all that is necessary is to decode the states. Only the 1 and 0 states are decoded, since if the state is not a 1 and 0, it must be the open state and nothing happens during this state. If a high condition occurs, the 1 decoder triggers the pulse stretcher. This pulse stretcher is a bit different in that it uses an OR gate. If pin 1 goes high, pin 3 also goes high, feeding back through C1 holding pin 2, and itself high for the time determined by the time constant of C1 and R5. When C1/R5 time out, the output of pin 3 will drop low again *if* there is not a high still on pin one. The gate acts like a straight piece of wire except that it will stretch a short pulse. Any time the output of pin 3 is high, it will enable the high pitch oscillator, causing a high tone to be generated.

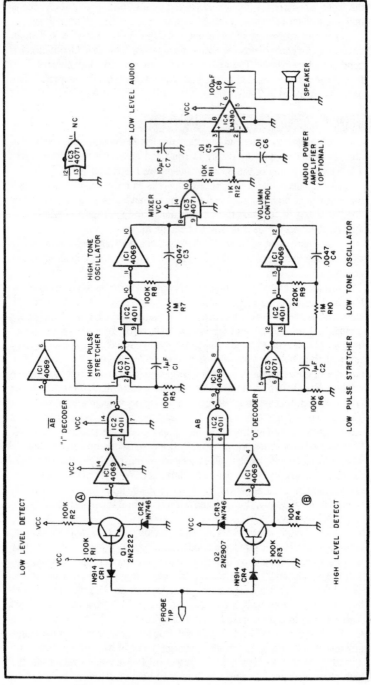

Fig. 11-10. Circuit description.

The 0 decode, pulse stretcher and oscillator work exactly the same as the circuit just described, except, of course, the tone generated is a low pitch. The two tones generated are mixed in the final section of IC3, with the inputs on pins 8 and 9, and the mixed audio on pin 10. The output at pin 10 is good for an earphone (used with a volume control) or a small speaker. You could even add a small audio amplifier if you work in a noisy environment. Of course, it could blast you out of the room. As mentioned, the unit will produce a high tone for a high and a low tone for a low. With a bit of practice, as with other types or probes, you can get a pretty good idea of what is going on by the sound. If it sounds like both tones are on, you have a square wave signal. If you have a high tone with a low tone *ticking,* you have a high with a low pulse, and so on. For a parts list for this project, see Table 11-2.

Mod for the Heath 10-102 Scope

The Heath IO-102 scope kit appears to be one of the "better buys for the money." The 30 mV vertical sensitivity, 80 nanosecond rise time, and 5 MHz frequency response seem to fill the bill. Actually, the vertical response is adequate for viewing 15 MHz or higher signals with sufficient trace height to be usable. Assembly is on four main circuit boards: vertical, horizontal, sweep and power supply. The finished unit is uncluttered, internal adjustments are no problem and most components are easily serviced.

Critical power supply voltages are zener regulated, a fact which contributes to the operational stability after warm-up. The power transformer, which is double shielded, and circuit board layout result in a trace like those drawn in textbooks.

The initial alignment and internal adjustments, although not difficult, *must* be performed as directed. An accurate VOM or VTVM is the only equipment needed. During this alignment, one wishes he had a screwdriver with a 10 to 1 drive. However, with a light touch, the desired results can be achieved.

Heath states that vertical drift during warm-up "for the first half hour or so" is to be expected. Experience with two of these units was as follows: From a cold start at 70°F, the trace was completely off the screen for the first 10 minutes. After 10 minutes, the trace appeared at the top of the CRT and drifted downward for 35 minutes, at which time it stabilized and no further drift was noted over several hours of operation. For someone who turns on the instrument and intends to use it all day, this may not appear as a problem. You may find juggling the vertical position control inconvenient, especially if your use is on an intermittent basis. The addition of four simple heat sinks produced two improvements:

- A usable trace on the screen in 3 to 4 minutes.
- A completely stable trace, without touching the position control, in 12 to 15 minutes.

Two heat sinks were made from ⅛" aluminum and attached to the vertical output transistor heat sink tabs. The tops of these sinks, which were bent 90°, may be secured to the CRT shelf with standoff insulators or RTV

Table 11-2. Parts List.

IC1	CD4069	Hex Inverter (RCA)
IC2	CD4011	Quad 2-input NAND (RCA)
IC3	CD4071	Quad 2-input OR (RCA)
IC4	LM380N	Audio Power Amplifier (National)
Q1	2N2222	NPN Transistor
Q2	2N2907	PNP Transistor
CR1,4	1N914	Signal Diode
CR2,3	1N746	3 Volt Zener Diode
R1-6,8	100k	¼ Watt Resistor
R7, 10	1 Meg	¼ Watt Resistor
R9	220k	¼ Watt Resistor
R11	10k	¼ Watt Resistor
R12	1k	Variable Resistor (Pot)
C1,2	.1μF	Capacitor
C3,4	.0047 mF	Capacitor
C5,6	.01 μF	Capacitor
C7	10 μF	Capacitor
C8	100 μF	Capacitor

cement. Two additional heat sinks, 1″ by 1″, were attached to the driver transistors with a small wrap-around wire.

Heath incorporates a 1 volt peak to peak calibrating signal at a front panel jack. A square wave calibrating signal and incorporated circuit are used. The calibrator provides a clipped sine wave signal of .1v, 1v and 10v. Three pin jacks were added next to the vertical input ground jack, to bring these signals to the front panel from the circuit which was mounted on the cover of the power supply transformer using existing screws.

Chapter 12
RF Test Equipment

The Benefits of Sidetone Monitoring

Sidetone is used in amateur circles to denote a tone from a CW keying monitor. But the term comes, of course, from the telephone industry, where a certain amount of the voice energy fed into the microhone unit of a telephone handset is deliberately coupled to the receiver unit in the headset so a person can barely hear himself talk. It was determined that transmission efficiency was improved by this scheme and a person using a telephone tended to maintain a steadier, more level talking level—especially in a slightly noisy environment. The method must have some value, since, besides in the telephone system, one will find audio sidetone employed in various commercial radio receivers, military transceivers and aircraft communications transceivers. This is *in spite of* the fact that these transceivers also employ various advanced forms of speech processing designed to keep the average modulation level as high as possible.

Amateurs might also consider the value of using audio sidetone in both mobile and fixed station installation. The use of such sidetone is mainly possible where a handset is used or a microphone/headset combination is employed. Otherwise, there is the possibility of audio feedback taking place unless a good directional microphone is used while the audio sidetone is monitored via a loudspeaker.

Simple Methods. A few simple methods for making the necessary audio sidetone interconnection in a station setup are as follows:

- If a separate transmitter and receiver are used, one can make an interconnection between the microphone amplifier stage in the transmitter and the audio preamplifier stage in the receiver. This should be done after the volume control in the receiver and utilizing a separate potentiometer to control the level from the microphone amplifier output in the transceiver. Figure 12-1 shows the basic idea. The audio stages in the receiver must remain activated during transmit periods for this scheme to work or the receiver transmit/ receive function suitably rewired to accomplish this. A relay is used to break the audio side-tone circuit when in the receive mode, since feedback can easily result if the high receiver loudspeaker output level in the receive mode gets back into the microphone circuit.
- If a transceiver is used, it may or may not be possible to utilize a scheme similar to the above between the receiver and transmitter

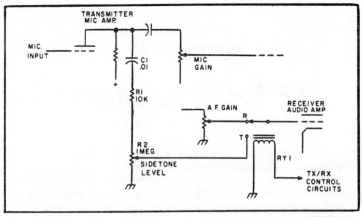

Fig. 12-1. Typical audio sidetone interconnection between separate transmitter and receiver or sections of a tube-type transceiver. The only added components are C1, R1, R2, and RY1.

portions of the transceiver. Some transceivers utilize complicated switching schemes for transmit/receive functions and the dual use of the low level audio stages both in the transmit function (as microphone amplifiers) and in the receive mode (as preamplifier after the product detector stage). In most transceivers, however, these functions are kept separate and one can switch the receiver section audio amplifier stages, via a relay, to pick up the output of one of the microphone preamplifier stages on transmit. A separate potentiometer should be used to control the level being fed into the receiver section audio amplifier stages and this potentiometer serves as the audio sidetone level adjustment.

The simplest scheme to achieve audio sidetone monitoring is via a separate small amplifier which is used between the microphone and the station loudspeaker or headset and is energized only during transmit periods. Any one of the number of simple amplifiers can be used, such as the solid state PA amplifier modules available from Lafayette Radio or Radio Shack for about $3. Although called *PA* modules, these modules will not exactly create enough undistorted volume to disturb the household. However, they are very suitable for audio sidetone purposes since they have a high impedance input which can be bridged across any microphone output without greatly affecting it, and their low impedance output can be used to drive either a loudspeaker or a low impedance headset connected across a receiver's output. At the low levels needed for audio sidetone, monitoring, their distortion is very low. They require ¾ volts for operation, and this can be obtained in tube/type transceivers by a simple half wave rectifier and filter across the filament line. Figure 12-2 shows a typical arrangement. For mobile transceivers they can be powered via a dropping resistor directly from the battery line.

The secret to obtaining maximum benefit from audio sidetone monitoring is in the adjustment of the sidetone level. You have to trick yourself, the same as the telephone company people do! Pick up your telephone and listen for the audio sidetone level as you speak to another party. It has been carefully chosen, as you will be inclined to *speak up* consistently rather than whisper into the handset. If the sidetone level were too loud, you would be inclined to speak more softly. If it is too comfortable to hear, you would be inclined to become intrigued with the reproduction of your voice, but not necessarily inclined to maintain a steady voice level. The sidetone level has been deliberately chosen so you *can just hear it,* and this is the adjustment to make for monitoring the audio sidetone level in a station. After a while, you will become unaware of the sidetone level, but you will try to speak at a consistent output level that allows you to detect the presence of the sidetone.

Audio sidetone monitoring is a relatively simple accessory to add to any station. The simplicity of the idea has probably been obscured by the usage of so many other speech processing devices such as audio compressors and clippers. However, if one can add only one or two dB more effective transmission efficiency to a station by audio sidetone monitoring and completely without any added distortion of the modulated signal, the idea seems worthy of a try.

Design Your Own QRP Dummy Load

A dummy load can prove to be a handy little gadget when it comes time to tweak-up the final in your handie talkie, QRP rig, or even CB set.

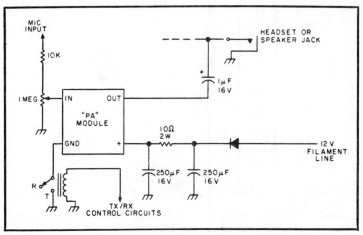

Fig. 12-2. Use of an inexpensive PA module amplifier provides audio sidetone with a minimum of circuit modifications.

Unfortunately, there doesn't seem to be anything on the market that lends itself to the task, unless you're willing to invest in an expensive kilowatt model. The only way to get around it is build your own, an easy accomplishment with a little forethought and a handful of components.

Choosing the Resistors. Ideally, the resistors in a dummy load should act purely resistive (no reactance) over the frequency range they are going to be used. The only inexpensive resistors that come close to meeting this requirement are the carbon (composition) type. Wirewound resistors, although cheaper and available in higher wattage values, exhibit too much inductance and are bothered by the skin effect. By contrast, carbon resistors are less subject to skin effect problems, so their resistance stays fairly constant as the operating frequency increases. In carbon resistors, reactance is mainly due to inductive effects in the leads and stray capacitance between the leads and nearby metal. These effects remain negligible up to around 100 MHz providing that the resistors are properly mounted and have values greater than 25 ohms.

Watts	R_{char}	R_{each}	N
4	50.0	100	2
4	75.0	150	2
6	50.0	150	3
6	73.3	220	3
8	50.0	200	4
8	75.0	300	4
10	48.0	240	5
12	50.0	300	6
12	73.3	430	6
14	51.4	360	7
14	72.8	510	7
16	48.7	390	8
16	77.5	620	8
18	52.2	470	9
18	75.5	680	9
20	51.0	510	10
20	75.0	750	10
22	74.5	820	11
24	51.6	620	12
24	75.8	910	12
26	52.3	680	13
26	73.3	1100	13
30	50.0	750	15
32	51.2	820	16
32	75.0	1200	16
34	76.4	1300	17
36	50.5	910	18
40	50.0	1000	20
40	75.0	1500	20

Table 12-1. Dummy Loads Using Standard 5%, 2W Resistors

The above design values will not result in a mismatch greater than 1.1:1 even when the tolerance is worst case.

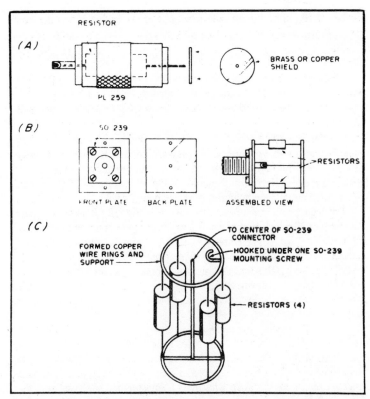

Fig. 12-3. At (C), the S-239 connector is mounted in the can lid. Make sure that the bottom ring does not make contact with the can.

Proper mounting dictates that leads should be kept as short as practical, and that there should be some separation between adjacent resistors.

Design. Because carbon resistors capable of handling more than two watts are expensive and difficult to locate, the only convenient way to achieve high power dissipation is to use several resistors in parallel. This increases reactive effects, but it offers greater power dissipation as an acceptable trade-off, unless you need an extremely accurate dummy load. Also, it should be remembered that the power rating of resistors is a continuous *free air* rating, so for short duty cycles, a dummy load can easily stand more. For example, with the 33 percent duty cycle of SSB the power rating may safely be increased by three, while the 50 percent CW duty cycle will allow an increase of two. A good rule to follow is to touch the resistors occasionally. If they are too hot to handle then they are dissipating too much.

The actual design of the dummy load is accomplished by starting with two parameters, the desired total power dissipation and the desired characteristic resistance. Assuming that all resistors have the same values and power ratings, the following equations may be used for designing a multiple-resistor dummy load:

Number of resistors needed = Desired power dissipation of dummy load
    Power rating of each resistor, and Value of each resistor
  = Total resistance of dummy load × Number of resistors needed

All too often the calculated value for each resistor will not be a standard resistance, a factor that will complicate matters. Table 12-1 helps to sneak around this problem by offering values based on the standard two watt resistors.

Construction. The actual construction depends upon the number of resistors that the above equation tells you are needed. If you can get away with just one, whether it be a ⅛ or 2 watt, then the design in Fig. 12-3A should do the job. Here, the resistor is soldered inside a PL-259 connect, with a small metal plate soldered to the end as a shield.

For higher power operation, the circuit in Fig. 12-3B allows compact mounting for 20 resistors or more. In this circuit, two copper or brass plates are used for mounting to eliminate capacitive coupling between adjacent leads. After the S0-239 connector has been mounted, a series of small holes are drilled in both plates for the resistors. A hole is also drilled in the center of the back plate and a wire soldered between the center conductor of the coax connector and the back plate itself. As always, leads should be kept as short as possible.

If shielding is what you need, and even higher power dissipation, then the circuit in Fig. 12-3C may be installed inside a pint or one half pint paint can. Transformer oil can be added to increase the short duration power handling capacity by a factor of three or four. A note of caution however: Don't use motor oil—because of its lower boiling point, it has a nasty habit of blowing off can lids, and sending hot oil all over the place.

In the interest of accuracy, it's worth the extra money to invest in 5 percent resistors for these circuits. This will give a tolerance of ±2.5 ohms for a 50 ohm dummy load, which results in a mismatch of only 1.05:1.

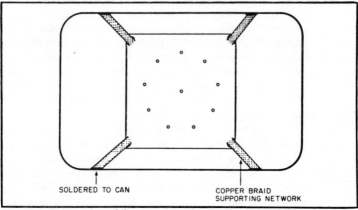

Fig. 12-4. Bottom view of the can.

In today's times of high prices, it is nice to know that home brew dummy loads are still within the price range of the poorest amateur. Half-gallon gallons can be constructed for less than $3 using readily available materials.

One-gallon paint thinner cans were prepared for the project by cutting off their bottoms with an ordinary can opener (Figs. 12-4, 12-5 and 12-6). This was followed by a thorough washing and drying procedure to remove any trace of the flammable contents. Next, twenty 470-ohm, two-watt 10 percent resistors (10 for each load) were sorted between two of us so that we had equal numbers of high and low values. This gave us parallel combinations that were very close to 50 ohms. Each group of 10 was then soldered between two 3" × 3" pieces of 1/16" brass plate predrilled to accept nine resistors in a 2"-diameter circle with the tenth one in the middle.

A hole was then punched on top of the can to accept a suitable connector. In this case, a flange-mount SO-239 was the choice. The flange of the connector was soldered to the can to make an oil-tight connection. The rear of the connector was also epoxied over to prevent oil seepage through the center conductor.

The resistor network was then mounted in the can, supported by a 2" length of heavy-gauge wire soldered from the center pin of the connector to

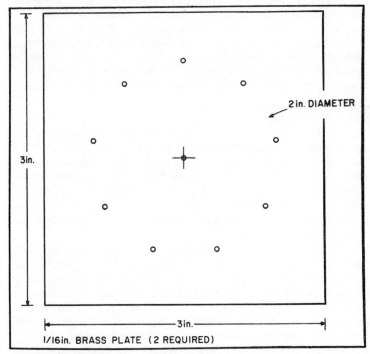

Fig. 12-5. Brass plate.

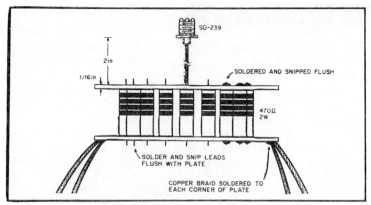

Fig. 12-6. Resistor network.

one side of the resistor assembly. Four pieces of copper braid were soldered to the bottom of the resistor network at each corner and then soldered to the can for support and to provide a good ground plane.

The can was sealed by cutting a piece of 1/16″ copperclad PC board approximately ¼″ larger than the base and soldering this to the bottom of the can.

Testing the load on a Hewlett-Packard network analyzer model 8407A showed that the loads were purely resistive at 50 ohms up to 32 MHz. The can was filled with a gallon of 30- or 40-weight motor oil (the cheapest brand available). The dummy load was designed for transmitters in the 200-watt class, but it is able to take at least twice that power. It is advisable to loosen or take off the cap of the dummy load when in use to vent any expanding oil or fumes that may come off when the load is used at high power.

Instant SWR Ridge

VSWR measurement is widely used by radio amateurs for adjusting and matching their antennas. One hobbyist saves a lot of walking and tunes his antennas. He started out with a model 190B Tektronix constant amplitude signal generator, which is easy to obtain now on the surplus market. He modified the signal generator slightly, built a 50-ohm resistance bridge and merged them to come up with an extremely stable and accurate vswr bridge. If you don't want to modify your generator, you could build only the resistance bridge portion.

The 190B signal generator is perfect for this vswr bridge because of its constant output level, which means that once the incident (forward) voltage is set, you can tune a very large frequency band without needing to readjust every few kilohertz or so. And there are other advantages. The frequency range covers 160 through 10 meters. With this hobbyist's generator-bridge arrangement, everything is complete in one package, so it's handy to use. Because of its low output level, there is no QRM radiated into space.

The following steps describe how to modify the signal generator and install the resistance bridge:

- Remove the external attenuator pad and its socket from the unit. (Note which pins the wires are on.) Fabricate a 1½″-square aluminum plate to mount an SO-239 rf socket, and attach the socket to the unit.
- Drill a 5/16″ diameter hole midway between the power switch and the SO-239 rf connector. Refer to Fig. 12-7. This hole will be for the vswr function switch.
- Remove V50, a 12AU7 tube used as the meter amplifier, and discard it.
- Build the 50-ohm resistance bridge according to the schematic shown in Fig. 12-8. Caution should be used in the wire dress because the frequencies will range up to 50 MHz.
- Install the resistance bridge in the 190B signal generator.
- Remove the wire from pin 3 of the attenuator socket (blue/white/yellow). It was a heater voltage supply on older units.

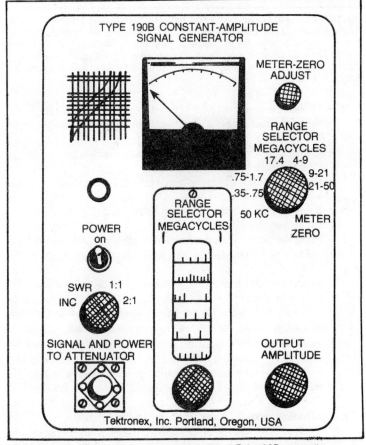

Fig. 12-7. Completed vswr bridge using modified Tek 190B generator.

343

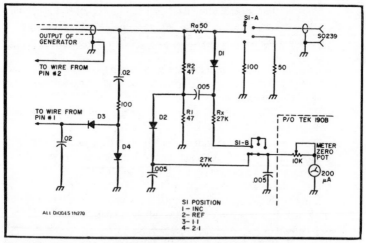

Fig. 12-8. This is the circuit of the 50-Ohm resistance bridge. R1 should match R2. RX should be trimmed so that incident (fwd) voltage equals reflected voltage. The 1:1 and 2:1 positions of switch S1 are simply 50-Ohm and 100-Ohm resistors to ground. When switched in, these positions give a quick self-check of the unit. D3 and D4 are the feedback diodes for the Tektroniz 190B generator. All diodes = 1 N270.

- Ground the wire from pin 2 of the attenuator socket (white/red) at the newly installed SO-239 socket. The coax shield also connects at this point.
- Connect the wire from pin 1 of the attenuator socket (white/blue) to the resistance bridge.
- Replace the 500-ohm meter-zero pot with a 10,000-ohm pot. Re-wire as shown in Fig. 12-8.
- If desired, the existing meter can be replaced (Fig. 12-7), but it's not necessary.
- Install the four-position swr function switch in the 5/16" drilled hole, and wire to the resistance bridge as shown in the schematic of Fig. 12-8.
- A chart for determining vswr is shown in Fig. 12-9. Copy it or cut it out and tape it to your unit. Calibration is simple. After the unit is complete, turn it on. Do not have a load connected to the coax connector. Adjust the incident voltage (INC) for full scale, then switch to the SWR position. The meter should read the same in each position. If it doesn't, trim RX until the meter reads the same. Now, switch to the 1:1 position. The meter should fall to zero, indicating a balanced 50-ohm load into the bridge. Switch to the 2:1 position. The meter should read approximately 30 percent of full scale, indicating about 2:1 vswr. If all this happens, box it up.

Now the beauty of this device takes hold! Because the generator's output is constant, when you vary the frequency dial, the meter will track your vswr curve over the entire amateur band you're testing. The first thing you will notice is that somwhere there is a dip, which is the resonant point of the antenna. The amount of dip indicates the impedance (in vswr). This dip will then give the necessary clue for any adjustments.

There are several possibilities for the generator-bridge. A 500-ohm pot could be substituted for the 50-ohm Ra resistor to give the bridge an impedance range of 0-500 ohms, or a complete LC impedance bridge could be installed for maximum use of the generator-bridge. Old coax can be tested with the generator-bridge by shorting one end of the piece being tested and measuring the vswr. The loss in decibels can then be calculated. If nothing else, the generator-bridge is a very good test rf generator to have in your ham shack. You can adjust beams with gamma matches, verticals (both trapped and monoband), dipoles, inverted vees and even mobile antennas using this generator-bridge.

UHF SWR Indicator

There are many inexpensive standing wave ratio indicators that are listed as good up to two meters and do function well. At frequencies above approximately 150 megahertz the results leave something to be desired.

There are several valid reasons why the swr indicators are inadequate. In order to have good sensitivity at the lower frequencies (80 meters), the coupling loops used for forward and reflected wave samplings are usually 6″ long, and are close-coupled to the center conductor of the transmission line. If the sampling loops are shortened and loosely coupled, then the sensitivity for the lower frequencies is greatly reduced.

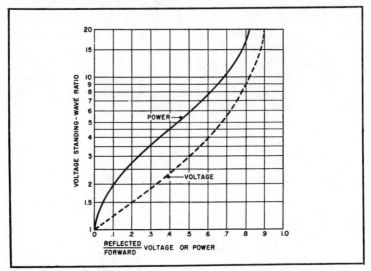

Fig. 12-9. Chart for finding vswr when ratio of reflected to forward voltage is known.

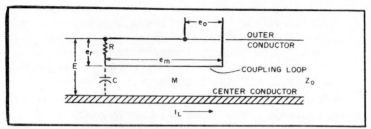

Fig. 12-10. Distributive capacity element.

Another reason for poor operation is that the coupling loops required for low frequency operation represent almost a quarter wave at the three-quarter meter band ($\approx$500 MHz). Distributed capacity between the long coupling loops and the inner coaxial conductor is excessive.

The diode rectifier characteristics become more important at the higher frequencies. The position of the pickup loops with regard to spacing, direction and concentricity must be very carefully aligned if their electrical characteristics are to be as similar as possible. Once aligned or positioned, the loops must remain rigidly affixed if a level of confidence in the indicator is to remain.

If the coupling loop is to be a small percentage of a wavelength (1/20 or about 5 peccent), a length of 1" is about the maximum to be considered. If the indicator is to be used with low power transmitters (½ to 2 watts), the loop should approach the 1" length. For high power transmitters, the loop may be as short as ⅛". This is that portion of the loop that is parallel to the coaxial center conductor.

The coupling loops form a circuit with a resistor such that the mutual coupling is positive (+) in one case, and negative (−) in the other. The same effect could be obtained if a single loop was used for sampling, then rotated 180 degrees and then sampled. Inspection of the equations that follow will show this is taken into account.

That portion of the loop parallel to the center conductor and the resistor form a third component, the distributive capacity element (Fig. 12-10).

The output rf voltage e_0 is made up of e_r and e_m. A voltage divider is formed by distributive element C and R. Then,

$$e_r = \left[\frac{R}{R + X_c}\right] E$$

and if R is much less than X_c,

$$e_r = \frac{RE}{X_c}$$

and since jwC equals $1/W_c$,

$$e_r = REjwC$$

$$e_m = I \left[jw \ (\pm M)\right]$$

by induction.

The sum of e_r and e_m is (factoring out jw)

$$e_o = jw \, (CRE \pm MI)$$

The directivity of the indicator, its ability to discriminate between the forward and reflected wave components, depends upon the relationship CR = M/Z_o, where Z_o is line impedance.

Substituting for CR,

$$e_o = \frac{jw \, (EM \pm MI)}{Z_o} = \frac{jwM \, (E \pm I)}{Z_o}$$

Another relationship must be established before again substituting in the rf output voltage equation for e_o. It is important to note capital letter E is used to designate the line voltage.

The voltage E at *any* point on a transmission line is the sum of the forward and reflected voltages, or

$$E = e_f + e_r$$

It should be remembered that e_r during the next few equations represents the reflected voltage; it should not be confused with e_r, the voltage across resistor R.

Then,

$$I = \frac{e_f}{Z_o} - \frac{e_r}{Z_o}$$

The minus sign is because the reflected wave travels in the opposite direction.

$$I = I_f + I_r$$

where

$$I_r = -\frac{e_r}{Z_o}$$

There are two cases to consider: (a) when the resistor is toward the load; and (b) when the resistor is toward the source. So, substituting for E and I in the e_o equation for each case,

$$e_o = jwM \left(\frac{e_f + e_r}{Z_o} + \frac{e_f - e_r}{Z_o} \right) = \frac{jwM}{Z_o} \, (2e_f)$$

$$e_o = jwM \left(\frac{e_f + e_r}{Z_o} - \frac{e_f - e_r}{Z_o} \right) = \frac{jwM}{Z_o} \, (2 \, e_r)$$

The equations show that the rf voltage from the loop before rectification is directional and proportional to the voltage in the transmission line (due to the forward and reflected wave respectively). Figure 12-11 shows that even though E_s is zero, E_r and E_f are still present.

347

The frequency limits of the indicator are exceeded when resistor R is not very much lower than distributive element C, and when mutual inductance M is not nearly purely reactive.

Construction Details. A length of copper water pipe is used for the wave sampling section of the transmission line. The use of tubing helps to maintain a constant line impedance, keep the inner field distortion to a minimum and radiation leakage down. For the center conductor, a #7 or a #3 AWG copper wire will provide a line impedance of 75 or 50 ohms, respectively. Thin wall model builder's brass tubing of equivalent o.d. was used in constructing the indictor for this project (Figs. 12-12 and 12-13).

To accommodate the sampling loop, drill three ⅛″ diameter holes ¾″ apart in the copper tubing. Form the insulated sampling loop, and insert the leads through the three holes. Single conductor, tinned, #22 AWG, with thin wall insulated sleeving added, maintains its shape better than stranded wire for the sampling loop. With ⅜″ lead protruding out of each hole, temporarily bend each load so as to make the loop captive. This will help

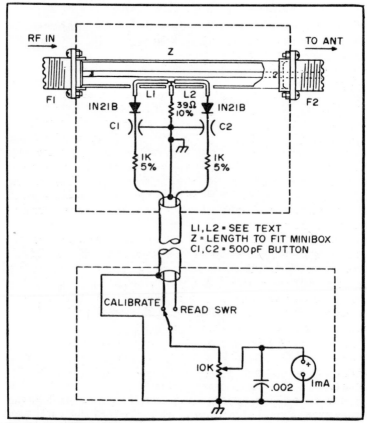

Fig. 12-12. UHF swr indicator.

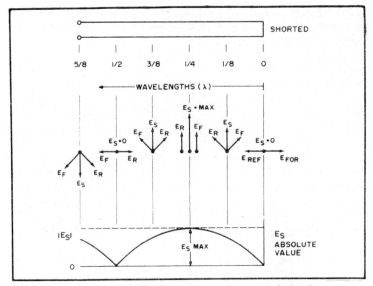

Fig. 12-11. Diagram of voltage standing waves on transmission line.

prevent the loop from slipping out of position during the fastening of the end fittings.

Solder the end of the conductor selected for the center lead to a mica-filled SO-239 coaxial fitting. Other types of fittings such as the N or the BNC may be used, but different mechanical arrangements will be required.

The other SO-239 fitting requires modification if no other holes are to be drilled in the outer copper tubing. The modification consists of removing the center pin of the connector. Depending upon the brand and vintage of the connector, some pins are removable by a simple "C" ring clip. Others have a rolled-in ridge for fastening, etc. After the pin has been removed intact, use a drill to establish clearance so that the pin can be freely inserted from the rear side of the connector.

Temporarily slide the copper tubing and loop subassembly over the center conductor with the fitting previously connected to one end. Determine through inspection the proper length the center conductor should be to permit the recently removed center pin to slide into the fitting with the slightly enlarged hole without protruding. The fitting must also butt against the end of the outer copper line. After the proper length has been determined, solder the pin to the center conductor.

Make two rings about ⅜″ long by cutting the ends off a standard copper elbow or coupler section. With all burrs removed, the rings, line and end fittings are ready for soldering. A 250 watt gun or iron will speed the assembly.

If the assembly is to fit inside a ready-made minibox, the box selected and the length of the line should be compatible. The box will help keep the sampling loop from being disturbed once positioned. A ¾″ hole in each end of the box is required, with one hole slightly elongated, so that when the box ends are sprung back slightly, the line assembly may be inserted. The assembly of the rest of the components is straightforward.

To adjust the loops, the line should be inserted into a transmission line, first in one direction and then the other until the two sets of readings are almost identical (except reversed). Once obtained, place a drop of quick-drying model cement at each of the sampling loop exit leads, without disturbing their placement. Figure 12-14 is a typical meter scale.

If the 1N21Bs and resistors have been matched even with the humble ohmmeter, and the parts layout is reasonably symmetrical, little effort should be required to position the pickup loops.

The line and meter may be mounted together in a larger single box if desired. It may also be desirable to use a set of connectors in the interconnecting cable to insert an extension cable when needed to provide remote readings during backyard antenna matching sessions.

The directivity of this indicator, and its ability to discriminate between forward and reflected wave components from 50 MHz to 500 MHz (and maybe even higher), is excellent. The use of type N fittings is recommended for frequencies above 500 megahertz.

Simple VHF Monitor

This little converter, when used with an inexpensive portable transistor radio, will monitor a wide range of VHF frequencies. It can be built in a few minutes using junk box parts. Even if the parts have to be bought, they will cost only a few dollars. It doesn't provide communications receiver quality, but it is useful when tuning up a transmitter or for general coverage reception. It mounts next to the ferrite loop of the transistor radio to make up a complete receiving system. Reception over a wide range of frequencies is possible by altering the tuning coils. The components specified here cover the range of about 120 to 150 Mhz. This range provides a lot of fun in

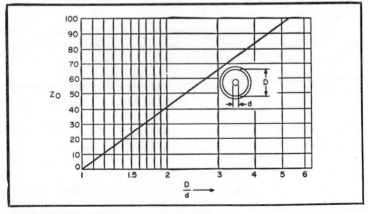

Fig. 12-13. Ratio of diameters to line impedance.

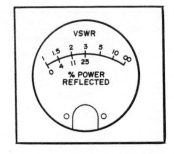

Fig. 12-14. Typical vswr meter scale.

listening to aircraft radio communications, the 2 meter ham band and other services.

The circuit (Fig. 12-15) is that of a regenerative converter. The incoming signal, tuned by the tank coil L-1 and capacitor C-2, is mixed in the transistor with an oscillator frequency controlled by L-3 and C-7. The difference frequency is adjusted to fall in the standard broadcast band.

The converter is built on a small piece of perforated board. Parts layout is not critical. However, all leads should be kept as short as possible and the oscillator and input tank coils should be mounted at right angles to each other. The completed assembly is mounted in a small open backed metal box to minimize the effects of hand capacitance. Tuning is accomplished by drilling a ¼" hole through the box in line with the adjustment screw on the oscillator tuning capacitor. A small piece of ⅛" wooden doweling is then cemented to the tuning screw with a drop of Eastman 910 (or equivalent) cement.

A small dual gang variable capacitor could be used but would be more expensive. Tuning of the input tank circuit is very broad and does not require adjustment after it has been initially set. A hole could be drilled in line with C-2 to permit peaking, using an insulated screwdriver, after the unit is packaged. A BNC connector or RCA type female phonograph connector is mounted at the top of the box for connecting the antenna. Good reception on local signals was obtained with a piece of #14 wire 19" long for the antenna.

Coils L-2 and L-4 can be 100 microhenry VHF rf chokes, or if you don't have any in your junk box you can use TV peaking coils or simply scramble wind about 20" of fine wire on a 100k ¼ watt resistor.

The completed converter is fastened to the back of the radio using vinyl tape or rubber bands. Placement is not critical, but try to locate coil L-3 as close to the ferrite loop as possible. Connect the battery and turn the converter on. Tune the radio until a loud hissing sound is heard. Typically this will occur on at least two spots on the dial. Tune the receiver to the loudest hiss. Turn the converter power off to be sure that you are tuned to the output of the converter. Next tune in a signal with C-7 and peak C-2 for maximum signal strength.

Tuning is best done with C-7, although a certain amount of peaking can be done with the transistor radio tuning. As with most regenerative type devices, a high level signal can readily capture the tuning. Although the

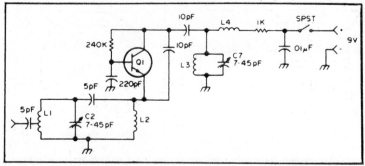

Fig. 12-15. Schematic of low cost VHF monitor. L-1: 4 turns #22 wire on ¼' diameter form. Center tapped. L-2, L-4: See text. L-3: 4 turns #22 wire on ¼' diameter form. Q-1: NPN VHF transistor 2N2222 or equivalent.

converter has limitations, you can get excellent reception from aircraft over 200 miles away, and from ground stations up to 30 miles away.

VHF Noise Snooper

Are you quite sure that nothing can be done about that noise level at your QTH? Just one of the problems with noise is its frequency content. Here is a new approach to noise tracing, using a piece of equipment you most likely already have.

Simply cut or unsolder one end of either of the FM detector diodes in an AM/FM portable, and you have an ultra portable, ultra sensitive noise detector. A schematic is usually unavailable for these Japanese sets, but you can usually spot the diodes sitting side by side between the last i-f cans and the audio transformers. Shorting one diode will probably work as well.

First find a blank spot between stations and start out. You'll probably have to walk to keep out all tire noises. As you move, the noise will go in and out like airplane flutter, slowing down and getting steady when you are very near. Then you can point the end of the whip at the noise source for a null. One noise was found so accurately that the vertical position on the pole was pinpointed for the power company linemen.

Of course there are other problems. When you rid the neighborhood of all those power leaks, you really notice the cars.

RF and Modulation Monitor

Around most ham shacks, several uses can be found for a simple rf and modulation monitor. This one is little more than a glorified crystal detector, but it monitors outgoing rf, aids in the tune-up of transmitters and antennas and lets you hear how your transmitted signal sounds.

Of course, single sideband—if the carrier is properly suppressed—will sound through this monitor like you're talking with a mouthful of wet cement. Amplitude modulated signals should sound clean and natural.

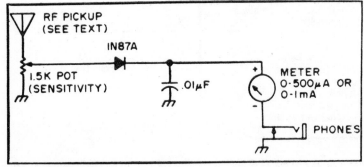

Fig. 12-16. Circuit.

This circuit (Fig. 12-16) can be built in a small phenolic box measuring 3¾" × 2⅝" × 1⅜". A metal minibox also can be used. The rf prove is 5 feet or so of #22 enameled copper wire folded and taped into a *whip* about a foot long and soldered to a phono plug. At low frequencies and low power levels, a longer piece of wire may be necessary for sufficient rf pickup. Almost any germanium diode should work, but a 1N87 gives higher output than a couple of other general purpose diodes.

Build a Simple Field-Strength Meter with Class

The manual for one rig suggests that a "better procedure" for loading it up is to use an swr bridge on forward or a field strength meter to peak plate and loading controls for maximum output power. The field strength meter seemed the easy way out, since a bridge costs $15. The meter would cost next to nothing to build considering the abundance of gear that is probably in your junk box. Take the junk box approach to measure output power.

When you go scrounging parts for this project, remember about the choke. Anything that will choke rf works. So, substitute freely, but don't change from a diode to a resistor.

The design is standard (Fig. 12-17), but the antenna is quite unconventional. Instead of using a piece of 12 gauge bare copper or coat hanger wire, we used 22 gauge tinned copper wire bent into the form of initials.

Construction. Select the amount of wire you feel necessary to complete your initials or whatever. Straighten the wire with a vise and a pair of pliers. Starting at the top of the left initial, make a bend to form the beginning of the

Fig. 12-17. Schematic.

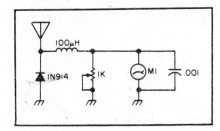

1 diode, 1N914 or just about anything else	
1 50 uA meter	
1 rf choke	
1 pot, 1k or thereabouts	**Table 12-2. Parts List.**
1 .001 μF capacitor	
1 mini box	
1 banana plug	
1 banana jack	
2 shouldered washers	
Plenty of twenty-two gauge wire	

antenna. Try to use one continuous piece of wire to make the antenna. To insure this, draw your idea on a piece of paper first. If you use one piece of wire the antenna will look a lot better, and you will eliminate having to solder on extra pieces of wire. The joints make bulges in the antenna which just don't look as nice on the finished meter. Solder a few spots to hold this thing together. You can also attach a banana plug.

With the antenna done, we built the guts of the meter. The easiest way to go is to mount the pot and use its terminals as tie points. The wiring is the easiest you'll ever do and is almost impossible to mess up.

Conclusion. Taking a lead from modern art, you can make just about anything for an antenna on your FSM. You've got to admit, now that we have an alternative, that bare copper does look a bit uncouth! This meter may not be the best or the smallest, but it does have class! For the parts list, see Table 12-2.

Chapter 13

Bench and Lab Power Supplies

Surprisingly Low-Cost Lab Supply

The μA723, as first introduced by Fairchild, is one of the oldest and most accepted regulator ICs in the industry. The μA723 is second-sourced by almost every linear IC house in the U.S. and still enjoys enormous usage. Thus, this IC has become about as inexpensive as a linear IC can be. The μA723 equivalent circuit is shown in Fig. 13-1.

Because V^{ref} (the internal reference voltage) is about 7 volts, the μA723 is usually configured in one of two basic circuits—one for output voltages below 7 volts and the other for output voltages above 7 volts. These basic circuits are shown in Fig. 13-2. The usual external current-increasing transistors are not shown here, in order to keep the circuits simple.

One of the linear IC houses which makes the equivalent of the μA723, Teledyne, has published an application note in which is shown a novel circuit that allows output voltages above and below 7 volts. That basic circuit is shown in Fig. 13-3. This circuit has in it a correction that was made from the original (which would have prevented the *current limit* from functioning). A complete lab-regulated supply providing 2 to 20 volts DC with current-limiting at 300 mA is shown in Fig. 13-4, with an NPN power transistor to increase current capability. The IC pin connections in Fig. 13-4 are for the dual inline package version of the μA 723 only; the "TO-5 can" packaged 723 would work as well, but pin numbers would have to be changed. The rectifier-filter section utilizes a Triad F91X transformer and a full-wave rectifier with capacitor input. The rectifier diodes are Motorola HEP

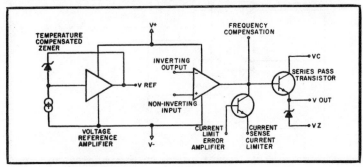

Fig. 13-1. Equivalent circuit.

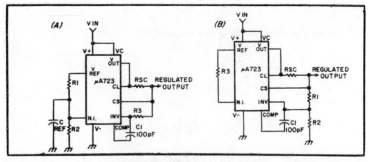

Fig. 13-2. Basic low-voltage regulator (Vout = 2 to 7 volts). Basic high-voltage regulator (Vout = 7 to 37 volts).

R0051s, but almost any two 1 amp silicon rectifier diodes (1N4002s, for instance) could also be used. Q1 is a Motorola HEP S5000, but it has several equivalents—RCA SK3041, Sylvania ECG152 or 2N5191.

All the parts are board mounted, with the exception of the transformer, filter capacitor and 25k voltage-control pot. In off-board mounting of the 25k pot (say on the front panel of one's own power supply cabinet), the lead lengths of the wires to the pot should be minimized.

It is intended that the AC switch, fuse holder, meters (if used) and binding posts also be mounted on the cabinet housing the circuit board. LMB and Bud each have a line of small box cabinets that would be suitable for making a finished bench supply using this circuit. The choice of which particular cabinet will usually be dictated by the meters one has on hand, because the circuit board is only 2″ × 2½″. An LMB W1N box cabinet (10″ × 4″ × 3½″) was used for the metering circuit shown in Fig. 13-5. Only meters that were on hand were used.

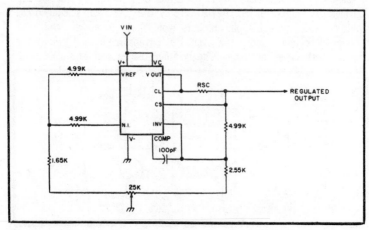

Fig. 13-3. Basic wide output voltage regulator circuit with 723.

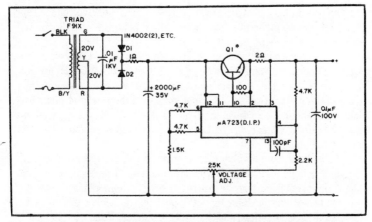

Fig. 13-4. Regulated lab supply. D1, D2 = Motorola HEP R0051; Q1 = Motorola HEP S5000. *Heat sink with washer and silicone grease.

For all its use of inexpensive or available parts, this regulated supply has more than proven itself on the bench. The 2-volt lower limit allows even RTL and low-voltage CMOS logic to be operated from the supply, and the 20-volt maximum output voltage allows for most other logic and linear IC circuits.

Adjustable Bench Supply

How about constructing an adjustable voltage power supply that can have up to 1.5 amperes output with good load voltage regulation and full overload protection at minimal cost? Admittedly, a $5 estimate depends a lot on what parts are available from one's junk box, but for just a few dollars spent on a new IC, one can have the "heart" of a very versatile power supply.

The new IC is the LM317 by National Semiconductor. This IC promises to be as famous as the LM309, which is universally used in power supplies for digital circuitry.

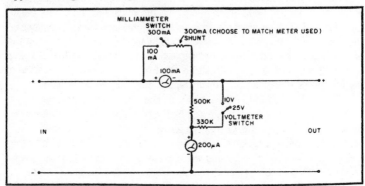

Fig. 13-5. Metering section.

The new LM317 is an adjustable, three-terminal positive voltage regulator. Its simple external connections rather belie the complexity and performance features of the unit. As shown in Fig. 13-6, it has only simple in/out connections and a minimum of three simple external components are required. The output voltage is set by the ratio of two resistors, R1 and R2. By making R2 variable, one can adjust the output voltage to be any value from a few volts less than the DC input voltage to the regulator down to a minimum of about 1.2 volts output. Thus, if the input DC voltage were 40 volts, the output voltage can be continuously varied from about 37 volts down to 1.2 volts.

Although the output voltage is determined only by a resistor setting, the output voltage is regulated at *any* given setting. The regulation will be about 0.1 per cent going from no load to full load (1.5 amperes, assuming the transformer/rectifier used for the DC input voltage handles this current). The LM317 is also overload and thermally protected. If the current limit is exceeded, such as by a short circuit, the LM317 will simply "shut down." If the regulator gets too hot, either because of excessive load current and/or inadequate heat dissipation, it will also protect itself. Although one can destroy the LM317 like any other IC, it is pretty hard to do with any sort of reasonable care.

The manufacturer suggests two additional capacitors (C2 and C3) be used, which may prove useful in some applications. C2 is used to bypass the adjustment terminal to ground to improve ripple rejection. This bypass prevents ripple from being amplified as the output voltage is increased. About 60 dB ripple rejection is achieved without this capacitor, but it can be improved to about 80 dB by adding it. A 10 mF or greater unit can be used, but values over 10 mF do not offer any significant advantage in further ripple improvement. The manufacturer particularly recommends the use of a solid tantalum capacitor type since they have low impedance even at high frequencies. An alternative is the use of the more readily available and inexpensive aluminum electrolytic, but it takes about 25 mF of the latter type to equal 1 mF of the tantalum type for good high frequency bypassing. C3 is added to prevent instability when the output load presents a load capacitance of between 500 and 5000 pF. By using a 1 mF bypass at the output (solid tantalum again or aluminum electrolytic equivalent), any load capacitance in the 500 to 5000 pF range is swamped and stability is ensured. Both the C2 and C3 will not be required for many applications where the LM317 is being used with a specific load circuit. But is the LM317 is used as the heart of a general purpose bench type power supply, they should be included.

Figure 13-7 shows a PC board layout and component placement diagram. This layout has been suggested by the manufacturer, but there is no need to follow it exactly as long as all of the external components are grouped around the regulator with solid short leads. The diagram show the LM317 in a TO-220 plastic case which is designated the LM317T. Most amateurs will probably prefer to buy the LM317 in the familiar TO-3 metal case and, in this case, it is the LM317K. But, when using the unit, note an important difference as compared to the old LM309K. The case on the

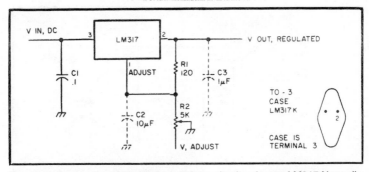

Fig. 13-6. Basic adjustable voltage regulator circuit using an LM317.Normally only three external components are needed, but C2 and C3 may be useful in certain situations as explained in the text.

LM309K was ground so one could simply bolt the thing down on a chassis for heat sinking. The case on the LM317K is the output terminal, so it must be properly insulated from a chassis.

Various power supply ideas and considerations can suggest themselves for the LM317. For instance, R2, instead of being a variable resistor, can be replaced by switchable fixed resistors to obtain some of the commonly used supply voltages such as 6, 9, 12, 15 volts, etc. This idea, plus a continuously variable output voltage position, is featured in the practical realization of a power supply using the LM317 as shown in Fig. 13-8. This supply will deliver fixed output voltages of 6, 9, 12 and 15 volts (depending upon how the trim potentiometers are set), plus a continuously variable output of 1.2

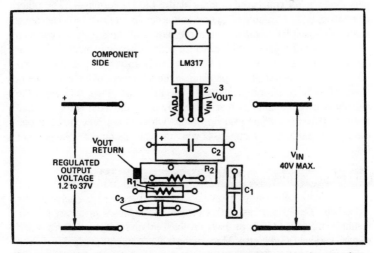

Fig. 13-7. This is a PC board layout for the regulator suggested by the manufacturer. R2 is shown as a multi-torn pot for ease of adjustment. The figure also shows the pin connections for an LM317 if it is obtained in the TO-220 plastic case.

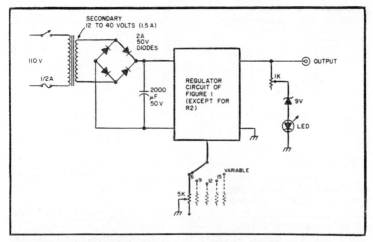

Fig. 13-8. A complete power supply using the LM317. The switch simply selects different 5k Ohm pots which are set for 6, 9, 12, 15, and a variable voltage output. The latter 5k pot is front panel mounted. The function of the LED is described in the text

to about 24 volts. All outputs can deliver at least 1.5 amperes with the components specified. The supply is simple to build in any size metal enclosure suitable for the components used. The only precautions to observe are to firmly heat sink the LM317 to one side of the metal enclosure and to keep the 0.1 mF capacitor going from pin 3 to ground, the 10 mF capacitor going from pin 1 to ground and the 120 ohm resistor going between pins 2 and 1, *all* connected directly at the LM317 terminals. The other components may be mounted wherever it is convenient to do so. The zener diode/resistor/LED combination at the output of the supply serves as a crude but useful voltage output indicator without having to build a regulator voltmeter in the supply. The LED just starts to glow when the output voltage is about 9-10 volts (depending on the tolerances of the components used). The 1k resistor is adjusted so the LED just glows fully when the *maximum* output voltage is reached. So by using the fixed output voltage positions (which are adjusted using a good VOM) and watching the LED, one can obtain a fairly good estimate of what the variable output voltage is set for.

Easy, Quick and Inexpensive Variable DC Supply

It is extremely handy to have a source of variable DC energy in the workshop. The usefulness of this device encompasses many areas. For instance, many of today's projects are built around semiconductors which require various DC voltages. Much of the military surplus available requires either 12v DC or 24v DC sources. You may not want your final conversion to operate on this source, but it can be very helpful to see if the equipment will function on its original source voltage before converting it. Those people active as mobilers certainly could use a husky supply to test that new mobile

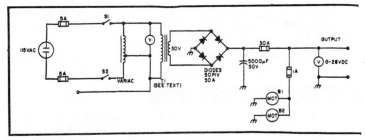

Fig. 13-9. Schematic.

rig. Obviously, anything that is battery operated could be run, tested or repaired with the help of this electronic wonder. It might even help you start your car some cold winter morning.

This is simply a variable autotransformer feeding a husky 30v transformer into a bridge rectifier (Fig. 13-9). The filter is a large capacitor. It would be nice to have a 30 A choke, but this filter has been satisfactory for all of these uses.

Well, now that these two major facts have been established, let's build it.

You may skip this part if you have a transformer or are going to purchase one. The transformer for this unit was a modified TV transformer. It was the heaviest one in the junk box.

First remove the outer shell and the laminations. It may be necessary to use a little force to accomplish this. Never hit the laminations directly with a hammer. Use a mallet or a block of wood with your hammer. Take care to preserve the leads from the windings.

Next, remove the secondary windings. Generally they are the outer windings. When you remove the 6.3v filament winding, count the number of turns. It probably will consist of approximately 12 turns. Usually this type of transformer has a two turns per volt ratio. Retain this information for rewinding the secondary.

Finally, rewind the secondary. Use number 12 copper wire with a thick thermoplastic insulation, or any enamel or formvar insulated wire. As you are winding the secondary, check to be sure that the laminations will fit. You should be able to get about 60 turns on the secondary—hence 30 volts. Using wire with thinner insulation, you can probably get several more turns.

The diodes were surplus units which were mounted on homemade heat sinks. The heat sinks were made from ⅛″ thick aluminum scrap. About 12 square inches were used for each diode. The heat sinks were mounted on 1 inch ceramic pillars above the chassis to promote convection. Several holes were drilled in the chassis below the heat sinks. There was a bottom plate attached and two small 24v DC surplus blowers were mounted on the back of the chassis. The output of the power supply was used to power the blowers. Even at five volts input, the blowers would move some air. The blowers aren't necessary until the supply is run at high current output over a five to 10 minute period of time.

The toggle switch, S2, which removes the variac from the transformer and places the line voltage there instead, is used when there is a load which will greatly surpass the rating of the variac.

This unit can be used for everything from repairing transistors radios to operating dynamotor powered mobile equipment of the 12v, 30a variety.

Super Low Voltage Power Supply

You'll love this project if you have had many disasters while experimenting with unfamiliar ICs and untested circuit designs. These disasters can happen when the meter probe slips or sometimes because of the incorrect design of the circuit. In nearly all cases, however, things burn out because of excessive current drawn for too long a time. The remedy is a power supply that abruptly removes the voltage from the circuit once a preset current is exceeded. This project will be superior to the usual type of current limiting in which the power supply delivers a constant current to the load and less voltage once the current limit is reached. The objection to the latter is that the user may not be aware that the current limit has been reached and that the voltage is no longer regulated, especially if it happens for only a very brief interval. A circuit may not operate correctly with the unregulated voltage during this interval and the user would be hard pressed to discover the reason for the malfunction.

Basic Layout. Three power supplies are constructed as shown in Fig. 13-10. The first is a dual tracking supply with variable output voltage 0 to ±20 volts and current to 100 mA on each output (200 mA total current capacity). Also available is a +12, −6 volt option. Current sensing is done in both the positive and negative legs, and when the current exceeds a preset level, a signal is developed to shut down the output from the voltage regulator. This signal latches so that ouptut voltage can only be restored by pressing a reset switch.

The second supply has variable output from 2.6 to 25 volts and current to 1 ampere. Up to 34 volts is available at reduced current. This supply also has adjustable current sensing and, like the first supply, the output voltage shuts down when the current exceeds a preset level. Voltage is restored by pressing the reset switch.

The third supply provides a fixed 5 volt output at currents to 1 ampere for operating TTL circuits. This supply has output voltage sensing and will shut down if the voltage moves outside a preset range from 4.75 to 5.25 volts.

The first supply provides the power for the sensing circuits used in all three supplies. Also, if any one supply shuts down, the other two will shut down also.

All three supplies use voltage regulators that are short circuitproof, an added safety bonus in the event that the current sensing circuits are manually disabled or in the event of the failure of some component in the current sensing networks.

Current Sensing. The current sensing network in Fig. 13-11 operates as follows: Assume that initially no current is drawn from the supply. With

R2 set to 500Ω, R2 + R3 = 21k and R4 + R5 = 21k. With the wiper of R4 set closest to R3, the voltage at pin 11 of voltage comparator IC1A will be 14 volts, exactly half the voltage across C1. Assuming for the moment that no current flows in R1, the voltage across R6 and R7 will be 28 volts and the voltage at pin 10 of IC1A will be 14 volts also. When current is drawn from the positive leg of the supply, a voltage drop develops across R1 and the voltage at pin 10 of IC1A drops below 14 volts. This drives pin 13 of IC1A positive and the resulting current in R21 charges C3. Q1 fires, sending a pulse through C4 to SCR1. SCR1 turns on, operating relay K1 and forcing Q2 to switch on. Q2 shorts out R27, thus reducing the output of IC3 to nearly zero volts. K1 interrupts the current to IC6 in Fig. 13-12. Q1 also sends a pulse to C14. This pulse turns on SCR2, forcing Q4 to switch on. This action reduces the output of IC5 to zero volts.

When the load is removed from the output of IC3, the power can be restored by opening S2A and S2B (normally closed switches). By moving the wiper of R4 closer to R5, the voltage at pin 11 of IC1A is lowered. It then requires a greater voltage drop across R1 (more current in the load at output of IC3) to lower the voltage at pin 10 of IC1A so that pin 13 will go positive. Thus the setting of the wiper of R4 determines what current will drive pin 13 of IC1A high.

An identical network consisting of R8 to R14 and IC2 senses the current in the negative leg of the supply. The ouptut of IC2 switches between 0 volts and −26 volts approximately. Since IC1B will not operate nomally with any input below −0.3 volts, the voltage from pin 6 of IC2 is

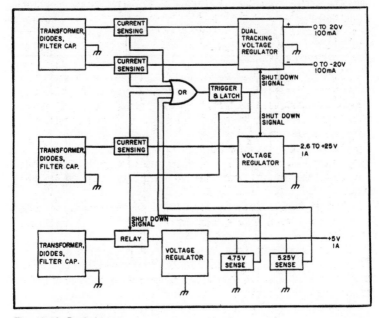

Fig. 13-10. Basic layout.

divided down by R15 and R17 so that the voltage across R17 switches between 0 volts and −0.25 volts. R16 and R18 form another voltage divider which provides −0.15 volts to pin 8 of IC1B. Thus IC1B switches like IC1A in response to an overcurrent in R14. D5 and D6 form an OR gate, hence isolating the outputs of IC1A and IC1B from one another.

In Fig. 13-12, current sensing is done in the same manner as described for the positive leg of Fig. 13-11. Since the maximum current for this supply is 10 times greater than for the first supply, resistance values have been adjusted accordingly. D9 forms another part of the OR gate that feeds R21.

Voltage Sensing. For the 5 volt supply in Fig. 13-13, it is more desirable to have output sensing than current sensing. This is because there are wide variations in the current demanded by TTL circuits when they are switching from state to state. The current limit point would always have to be set rather high, and consequently only gross overcurrents could be sensed. On the other hand, a circuit that senses when the voltage falls below 4.75 volts, the lower operating limit for 7400 series TTL, is quite useful. Suppose, for example, that you are operating near the 1 ampere limit of IC6; a brief current pulse could exceed this limit and the internal circuit of IC6 would then allow the output voltage to drop. Without voltage sensing this could easily go unnoticed and your circuit would malfunction.

In Fig. 13-13, D14 provides a reference voltage. R41 acts as a voltage divider and is set to 5.25 volts. R42 is another voltage divider and is set to 4.75 volts. IC1C and IC1D compare the output of IC6 to these voltages and, if the output moves outside the window from 4.75 to 5.25 volts, pin 1 or pin 2 will go high. This signal goes to R21 of Fig. 13-11 and eventually shuts down all the supplies.

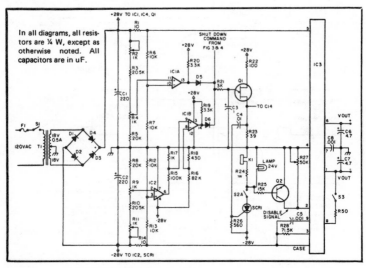

Fig. 13-11. Dual tracking regulated supply.

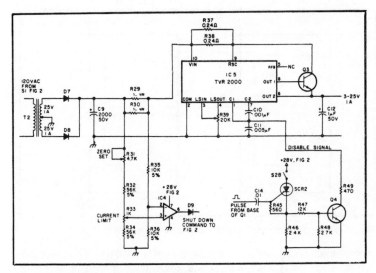

Fig. 13-12 Variable voltage power supply.

Response Time. R21 and C3 determine the response time to the circuit. With R21= 3k and C3 = 1 uF, the circuit responds to an overcurrent, overvoltage or undervoltage that lasts 3 milliseconds or more. K1 adds an additional 7.5 ms to the time required for the 5 volt supply to shut down. By reducing C3 to 0.1 uF, response time can be made as low as 0.3 ms. R21 can be increased to as much as 10 megohms if desired to lengthen the response time, but should not be reduced below 3k.

The Voltage Regulators. The 419TK regulator is readily available. It is internally current limited at about 350 mA when the positive output is

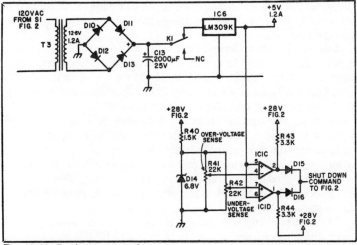

Fig. 13-13. 5 volt power supply.

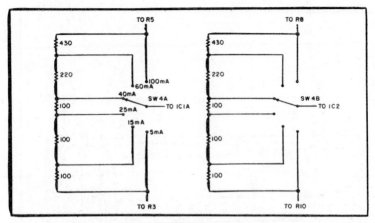

Fig. 13-14. Switch selected resistors replace R4 and R9.

shorted to ground. It also has internal thermal limiting that will reduce the output when it gets too hot. A small heat sink is required when the operating current is 100 mA in each leg of the output. In Fig. 13-11, S3 is normally open. When S3 is closed, R27 can be adjusted to give +12, −6 volts output for the operation of certain types of voltage comparators.

The 309K also has current limiting and thermal limiting. It will provide a little over 1 ampere when mounted on a heat sink with the circuit shown.

The TVR2000 has been available from Poly Paks for a number of years and is quite inexpensive. In Fig. 13-12, a foldback current limiting option is not used. Instead, simple short circuit sensing is used. R37 and R38 set the short circuit current to a value of about 1.2 amperes. The relationship here is R_{sc} 1 out ≈ 0.1 volt, where R37 and R38 in parallel make up R_{sc}. R39 sets the output voltage. Q3 acts as a current booster and is mounted on a heat sink. C10 stabilizes the current limiting circuitry and C11 stabilizes the regulator section of IC5. Different values from those shown may be required to drive high capacitance loads.

Selecting Resistors. Resistors of 1 per cent tolerance are best for R1, R3, R5 to R8, R10 and R12 to R14. This will make the final adjustments simpler and will keep tracking errors in R4 and R9 to a minimum. In Fig. 13-12, 5 per cent resistors will suffice for R32, R34, R35 and R36, providing you choose them such that R32 ≤ R34 and R36 ≥ R25.

Regarding the tracking of R4 and R9—since they form a tandem control, it is important that they both exhibit approximately the same resistance between their wipers and their ends for all rotations of the shaft. Failure to do so will mean that the positive will trip at different currents. If you want very good tracking, replace both R4 and R9 with a series of 5 per cent resistors and use a two pole rotary switch to select the current limit you want as shown in Fig. 13-14.

Construction. All three supplies were constructed on a single 4″ × 5″ printed circuit board as shown in Figs. 13-15 and 13-16. IC3 does not plug directly into the board. The holes in the board have been spaced out to

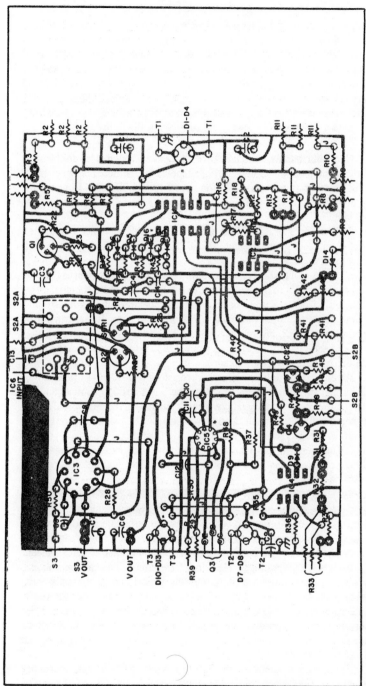

Fig. 13-15. Parts layout.

assure clean etching. Solder a short wire to the outside of each pin of IC3. Insert the wires into the PC board and solder. A piece of aluminum was bolted to IC3 as a heat sink. There are so many connections to the PC board from the external switches, controls, transformers, etc., that it was not possible to arrange for an edge connector on a board of this size. Instead there are about 35 wires soldered at various points around the edge of the board and all are routed to one end of the board can be hinged outward from the chassis if parts on it need to be replaced in the future.

All components fit nicely on a chassis 10″ × 6″ × 2″.

Final Adjustments. Switch S2 to reset. Leave S3 open. This disables the shutdown mechanism. Connect a high impedance voltmeter between pin 7 of IC1D and ground. Adjust R42 for a reading of 4.75 volts. Connect the voltmeter between pin 4 of IC1C and ground. Adjust R41 for a reading of 5.25 volts.

Set the wiper of R33 to the end closest to R32. Connect a voltmeter between pin 6 of IC4 and ground. Adjust R31 so that the reading just goes to zero.

If you are using a dual potentiometer for R4 and R9, proceed as follows: Set the wiper of R4 to the end closest to R3; the wiper of R9 should then be at the end closest to R10. Connect a voltmeter between pin 13 of IC1A and ground. Adjust R2 until the reading just drops to zero. If you run out of adjustment with R2, interchange R6 and R7 and try again.

Connect the voltmeter between pin 14 of IC1B and ground. Adjust R11 until the reading just drops to zero.

If you elect to use the switched resistors in Fig. 13-13 proceed as follows: Set the switch in Fig. 13-13 to the 5 mA position. Connect a load between the positive and negative output terminals of the supply and adjust the output voltage so that the load draws 5 mA. With a voltmeter from pin 13 of IC1A to ground, adjust R2 until the voltage just drops to zero. If you run out of adjustment with R2, interchange R6 and R7 and try again. Connect the voltmeter between pin 14 of IC1B and ground. Adjust R11 until the reading just drops to zero.

Precision +10.000v DC Voltage Reference Standard

This is a simple, easy to build DC voltage reference standard whose temperature coefficient (tempco) can be tailored to the hobbyist's individual requirements. It can be used by itself or as an individual circuit element.

Design Considerations. The circuit described uses an operational amplifier in a non-inverting circuit utilizing a single positive supply (Fig. 13-17). The output of the reference standard supplies the zener reference diode current, providing a stable supply voltage for the zener and decoupling from the positive supply voltage. The positive supply voltage should be regulated by either a three-terminal regulator or zener diode and should be from ±15 to ±18 volts.

Temperature stability is expressed in per cent/°C or ppm (parts per million)/°C. (.0001 per cent/°C is equivalent to 1 ppm/°C.) A 1 ppm/°C

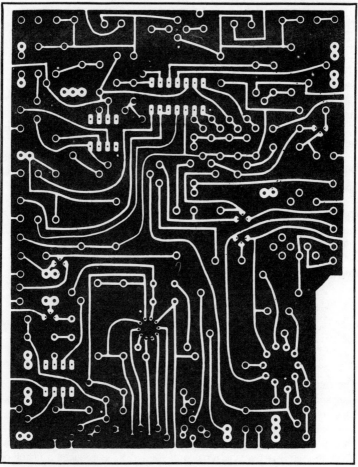

Fig. 13-16. PC board.

change referred to an output of 10.000 volts is 10 uV/°C. Both terms, per cent/°C and ppm/°C, will be referred to in this project.

The zener reference diode, CR1, should not be confused with the more common zener regulator diode. The zener reference diode is intended for use in applications where it is important to maintain stable DC voltages under severe combinations of temperature, shock and vibration. The temperature stability of the zener reference diode is due in part to the combination use of reverse-biased and forward-biased silicon p-n junctions, taking advantage of their opposing tempco characteristics. Application notes and design data sheets are available from the larger suppliers of zener reference diodes (e.g. Motorola, Dickson, Centrolab) and are invaluable as reference material.

The 1N821 through 1N829 family of zener reference diodes is used for CR1. The diodes in this family exhibit tempcos from .01 per cent/°C

(1N821) to .0005 per cent/°C (1N829) at a nominal zener current of 7.5 mA. The nominal zener volatge is 6.2v ±5 per cent. R1, which sets the nominal zener current (7.5 mA in this case), should be a 100 ppm/°C metal film resistor. Other families of zener reference diodes can be used, such as the 1N4565 through 1N4584, with a corresponding change in nominal voltage and current.

As a general rule, the gain resistors used in the reference standard should have tempcos similar to the zener reference diode selected. For example, if the 1N821 is selected, gain resistors R2 and R3 should have tempcos of .01 per cent/°C (100 ppm/°C).

The overall tempco of the reference zener CR1, gain resistors R2 and R3 and to some extent, the zener current resistor R1 (providing it has a 100 ppm/°C tempco) and A1.

The gain required of the circuit is dependent upon the zener voltage. The nominal voltage of the 1N821 is 6.2 V, necessitating a gain of 1.613 for an output of +10.000 volts. The total resistance of R2 and R3 should be 5k to 10k, limiting the current through the gain resistors to from 1 mA to 2 mA.

If R1 is a 100 ppm/°C as suggested, here is an easy way to estimate the overall tempco of the completed reference standard:

- Write down the tempcos of CR1, R2 and R3 in ppm/°C
- Square each one that you have written down
- Add the squares of R2 and R3 together, dividing the answer by seven
- Add the square of CR1 and the answer of step 3
- Find the square root of the answer in step 4.

The last step will be the approximate overall tempco of the reference standard in ppm/°C. This will give you a good idea if the components you have selected fit your tempco requirements. If CR1, R2 and R3 have the same tempco, just multiply the tempco of *one* of them by 1.1 to determine the overall tempco of the reference standard. The reason for the division by seven in step 3 is that the tempco of R2 and R3 is reduced a factor of approximately 2.7 (depending on the gain) by the low gain of the amplifier.

The output of the reference standard can be trimmed to +10.000 volts by paralleling a resistor across R2 and/or R3 as required. Tempco of the gain trim resistors, R4 and R5, is critical, and should not alter the tempco of the gain resistors. For example, if R2 was a 6.04k 25 ppm/°C resistor and a 62k ½ watt carbon resistor was paralleled across it having a 1000 ppm/°C tempco (typical tempco of carbon resistors), the result would be the same as using a 5.5k 100 ppm/°C resistor. The effective tempco of the trim resistor in this case would be 1/10 its real tempco. A 620k ½ watt carbon, in the same example, would have an effective tempco of 1/100 or 10 ppm/°C, and a 6.2 megohm would have an effective tempco of 1/1000 or 1 ppm/°C. Some discretion must be used when selecting the value and tempco of the trim resistors.

Practical Design. The next step is to build a working reference standard. The following components were selected for a reference standard having a calculated tempco of 22 ppm/°C and a measured tempco of 17 ppm/°C after assembly. An LM301A operational amplifier was selected for A1. The LM301A will supply approximately 20 mA and has a typical tempco of 1 ppm/°C. A 1N825 reference zener was selected for CR1 which has a 20 ppm/°C tempco and a nominal voltage of 6.2 volts at 7.5 mA. An RN55 100 ppm/°C 511 ohm metal film resistor was selected for R1. The actual reference zener voltage at 7.5 mA measured 6.286 volts, necessitating a gain of 1.591 for +10.000 volts output. For a gain of 1.591, an RN55E 25 ppm/°C 6.04k metal film resistor was selected for R2 and an RN55E 25 ppm/°C 3.57k metal film selected for R3. A ½ watt 2.2k 5 per cent carbon resistor was selected for R6. After assembly, R3 had to be trimmed for an output of +10.000 volts.

The most difficult task after assembly for the average hobbyist will be finding access to a four or five digit digital voltmeter so that the reference standard output can be trimmed to +10.000 volts. No special equipment is required up to the trimming operation. The accuracy of the reference standard will be as good as the equipment used to trim it.

Conclusion. It has been demonstrated that a precision reference standard can be constructed by utilizing available reference zener diodes and suitable gain resistors. This project should act as a tool to enable some of those interested in acquiring a stable DC source to design their own. Different output voltages can be obtained by either changing the gain of the amplifier or adding a voltage divider across the output of the reference standard (current cannot be supplied in this mode). Layout is not critical, cost is minimal and performance can be determined to some extent before assembly.

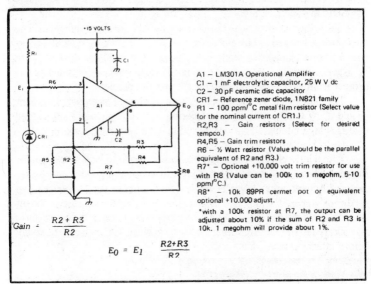

A1 — LM301A Operational Amplifier
C1 — 1 mF electrolytic capacitor, 25 W V dc
C2 — 30 pF ceramic disc capacitor
CR1 — Reference zener diode, 1N821 family
R1 — 100 ppm/°C metal film resistor (Select value for the nominal current of CR1.)
R2,R3 — Gain resistors (Select for desired tempco.)
R4,R5 — Gain trim resistors
R6 — ½ Watt resistor (Value should be the parallel equivalent of R2 and R3.)
R7* — Optional +10.000 volt trim resistor for use with R8 (Value can be 100k to 1 megohm, 5-10 ppm/°C.)
R8* — 10k 89PR cermet pot or equivalent optional +10.000 adjust.

*with a 100k resistor at R7, the output can be adjusted about 10% if the sum of R2 and R3 is 10k. 1 megohm will provide about 1%.

$$Gain = \frac{R2 + R3}{R2}$$

$$E_O = E_I \frac{R2 + R3}{R2}$$

Fig. 13-17. Voltage reference standard schematic.

Versatile Power Distribution System

How many times have you entertained the thought of constructing an AC receptacle that will be portable on your bench and at the same time have some fuse protection for equipment under repair or construction.

Another nagging workbench item is the soldering iron. It would be difficult to determine how many times it sits on the bench of hobbyists cooking for days, or how many times experimenters accidentally burned themselves or damaged material lying on the workbench.

You can now have a portable switched three outlet AC power source with two of the outlets fused directional control light source, soldering iron power source with high/low heat and removable heat shield stand for the soldering iron.

This is another junk box project from start to finish. Selection of components is not critical as long as they can handle 115 volts at 5 to 9 amps. In addition to the basic tools a small punch set and nibbler are needed. If you don't have access for borrowing one, Lafayette Radio usually has them in stock at a modest price.

Figure 13-18 is schematic circuitry of the magic box. A neon indicator light is always on as long as the box is plugged in. The indicator lights for outlets 2 and 3 are wired behind the fuses to indicate when a fuse is blown. If the builder is working with very low current fuses, the indicator light load should be placed in front of the fuse so as not to increase the current drain at the outlet.

Use an aluminum case 3" × 5" × 2". Try to place all the wiring in the lower half to simplify construction. The cradle microswitch support can actually be made from the frame of a broken rocker switch, if you happen to have one in your junk box.

The soldering iron cradle can be constructed from sheet aluminum. The microswitch is operated by a plunger-type rear support. A standoff spacer with a bolt for plunger will work well. The plunger should be designed with ⅛" travel, allowing plenty of movement to insure switch operation and still protect from excess weight being placed on the cradle.

The light source can be an old desk lamp fixture. Placement of the light should allow free access of the soldering iron for right- and left-handed users. A high intensity light unit would make a better light source.

Parts placement is not critical.

With a little effort, your magic box can be made to look store-bought. Remove all dirt and grease from aluminum and use a good primer base. Aerosol lacquer spray paint is easily found in auto supply houses and hardware stores. Painting and labeling should be done before component assembly. Allow a day or two after painting for finish to reach maximum set.

Don't let the simple wiring circuitry mislead you into not checking before attaching to AC line.

Unique Power Supply Tester

This supply tester uses a load bank. The first question asked might be "What is a load bank?" The term is taken from the electrical power industry

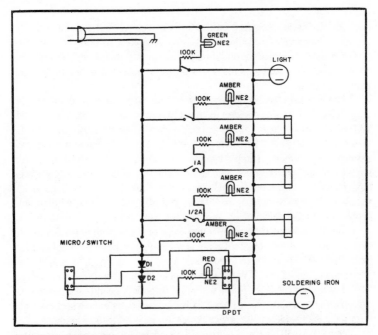

Fig. 13-18. Schematic circuitry.

to explain a device to simulate a load on various power sources in order to check the load performance of those sources. This project covers the building of two simple load banks or dummy loads for small and rather large sized DC power supplies that are finding present usage with experimenters and hobbyists. It can be done economically and with readily available parts. Also, using the basic ideas presented here, load banks for almost any type of low voltage power supply can be built.

In order to properly evaluate the performance of a power supply (whether home built or commercially built), the supply must be tested at various degrees of loading—that is, no load, half load, full load, etc. The major requirement for a suitable load bank is that it be capable of providing these various load conditions. Thus, several load elements are required, or at least one large variable element is required, in order to cover several different load conditions. These load elements would typically be large fixed or variable power resistors. If the power supply is of any size, the load elements have to be of high power rating (10, 25 or even 50 watt). A look at a parts catalog will show that power resistors get expensive as the power rating goes up and the different ohmic values available go down. Also, power resistors present the problem of how to get rid of the heat and how to conveniently mount the resistors.

Instead of using the regular power resistor in the design of these load banks, automotive bulbs and pilot lamps were used. These are readily available at auto parts houses and even department store electrical parts

sections. Bulbs are selected on the basis of voltage and current requirements and typically work out very nicely in regard to both values. The
problem of power dissipation is very minimal and mounting can be very
simple. With the load elements decided, the actual construction of the load
bank can begin.

The first load bank requirement was for a rather large 13.8 volt DC 24
amp DC supply. It was decided to check one third, two thirds, full load and
133 per cent of full load. Voltage and ripple were observed under the various
load conditions. The #1073 automative bulb was selected with a nominal
rating of 12.8 volts @ 1.8 amps. Typical cost is less than 30¢. Since the power
supply voltage was to be a volt higher, the current at that voltage was
approximated to be about 2 amps. This is a fine feature of these bulbs—the
voltage can be increased up to 20 per cent of nominal rating without any
problem in this application.

The bulbs were arranged in four groups or sections of four each. Each
group presents a load of about 8 amps at nominal voltage to simulate four
load conditions. In order to keep costs down and construction simple, the
wires consisting of #14 TW wire were soldered directly to the bulb bases.
The start and end of each section wire were wrapped around a nail which was
driven into the wooden base board. This eliminated expensive sockets and,
after all, a load bank is not a device that is normally on display in the shack.

Since this load bank was to be used for a considerable amount of testing
and since four surplus relays were available, each load section was wired
through relay contacts as shown in the diagram. Any combination of loading
can be quickly selected. This extra feature can be eliminated with just the
use of solder connections or switches with proper current rating. The
switching feature is certainly handy if considerable testing is to be done, but
adds considerably to the parts cost if the junk box is not well stocked.

The second load bank was for a much smaller DC power supply with a
rating of 5 volts @ 1 amp. Following the same design features of the previous
load bank, a #502 bulb was selected as load element in four sections with
two bulbs in each section, giving 30 per cent, 60 per cent, 90 per cent and
120 per cent of full load. Since the current requirements were much less,
about 0.3 amps per section, regular #22 solid hookup wire was directly
soldered to the bulb bases and toggle switches (or even slide switches) used
for controlling the load sections. A small wood base board was again used for
mounting the bulbs and wiring with a small ⅛" pressed board panel (nailed to
the edge) to support the switches. This supply is most useful for checking
logic power supplies used with ICs. Overload or current foldover characteristics can be checked on supplies designed with that feature.

There are many advantages to these simple load banks besides the cost
and ease of construction. The bulbs present a large amount of light under full
load to leave no doubt that the supply is working. Also, there are no burn
marks from power resistors on the workbench or, even worse, the dining
room table. Buying the bulbs by the box reduces cost and automotive type
bulbs are readily available, even at gas stations.

The purists may argue that the bulb is not a constant load resistance due to the filament characteristics. This sudden heavier than normal load when voltage is first applied to the bulb is very short in duration and provides an even stricter load test of the supply being tested.

The load bank ideas presented here can be extended to other sizes and types of load banks by changing the bulb types and the number of bulbs in a section, and even the number of load sections.

These load banks have been used to check voltage and ripple conditions on a variety of home brew and commercial supplies. The cost is certainly cheap if proper shopping around is done on the choice and source of bulbs.

The Smoke Tester Power Supply Checker Outer

This tester allows full power (smoke) testing of large or small power supplies *before* they are put on the rig. It will tell how many volts at how many amps the supply can deliver and what kind of voltage loss to expect under load. It can also help determine some of the attributes of low voltage power transformers, without having to wire them into a power supply.

It consists of two or three transistors, a pot, a couple of meters, a few diodes and some loose hardware. Don't forget the heat sink. If you wish to be formal about it, then add a chassis of some kind.

As the wiper is moved toward the plus, more current flows in the base circuit. The external circuit looking into the collector/emitter sees a lower and lower resistance. The ammeter and voltmeter give an idea of watts cooking. The diode allows less than perfect attention to the polarity of the incoming voltage. The resistor in series with the base limits base current to a safe value.

Although the transistor manual indicates that the 2N3055 is a "15 ampere" transistor, two or three of them cleared like fuses when the collector current was slowly increased above about 11 or 12 amps. The 10 amp fuse was put in the emitter, and collector current excursions were limited after that.

Since power supplies capable of giving more than 20 amps were to be tested, two 2N3055s were put in parallel. The fuses gave enough resistance in the emitters to help the transistors share the current in a more or less equal manner (at least there were no more creamed transistors).

A Darlington or piggyback configuration is used to reduce the power dissipation requirements of the pot. Ten amps in the collector can require as much as ½ amp in the base. Half an amp times several volts equals several watts. The Darlington system reduces the input current by as much as 50 times. That figures out to about 10-20 mA for 10-20 amps in the collectors, and makes the system practical.

Note that if the power supply is delivering 15 amps at 13 volts, then something is going to have to dissipate 195 watts. Be *sure* that the 2N3055s are on a large enough heat sink.

If a full wave bridge rectifier is placed in front of the tester, power transformers can be checked for voltage out versus current. That information would have saved a lot of time in the past.

Of course, there are single transistors around that meet the voltage, current and power ratings of this tester.

If PNP transistors are used, reverse the polarity of the meters and diodes.

Index